T0259036

APPLIED FOOD SCIENCE AND ENGINEERING WITH INDUSTRIAL APPLICATIONS

APPLIED FOOD SCIENCE AND ENGINEERING WITH INDUSTRIAL APPLICATIONS

Edited by
Cristóbal Noé Aguilar, PhD
Elizabeth Carvajal-Millan, PhD

Apple Academic Press Inc.
3333 Mistwell Crescent
Oakville, ON L6L 0A2 Canada

Apple Academic Press Inc.
9 Spinnaker Way
Waretown, NJ 08758 USA

© 2019 by Apple Academic Press, Inc.

First issued in paperback 2021

Exclusive worldwide distribution by CRC Press, a member of Taylor & Francis Group

No claim to original U.S. Government works

ISBN 13: 978-1-77463-393-9 (pbk)
ISBN 13: 978-1-77188-706-9 (hbk)

Library and Archives Canada Cataloguing in Publication

Applied food science and engineering with industrial applications / edited by Cristóbal Noé Aguilar, PhD, Elizabeth Carvajal-Millan, PhD.

Includes bibliographical references and index.
Issued in print and electronic formats.
ISBN 978-1-77188-706-9 (hardcover).--ISBN 978-1-351-04864-4 (PDF)
1. Food industry and trade--Technological innovations. I. Aguilar, Cristóbal Noé, 1970-, editor II. Carvajal-Millan, Elizabeth, editor

TP370.A67 2018	664	C2018-905117-5	C2018-905118-3

Library of Congress Cataloging-in-Publication Data

Names: Aguilar, Cristóbal Noé, editor. | Carvajal-Millan, Elizabeth, editor.
Title: Applied food science and engineering with industrial applications / editors, Cristóbal Noé Aguilar, Elizabeth Carvajal-Millan.
Description: Toronto; New Jersey: Apple Academic Press, 2019. | Includes bibliographical references and index.
Identifiers: LCCN 2018041950 (print) | LCCN 2018042947 (ebook) | ISBN 9781351048644 (ebook) | ISBN 9781771887069 (hardcover : alk. paper)
Subjects: | MESH: Food Analysis | Food Technology | Food Industry
Classification: LCC R855.2 (ebook) | LCC R855.2 (print) | NLM QU 50 | DDC 338.4/76606--dc23
LC record available at https://lccn.loc.gov/2018041950

ABOUT THE EDITORS

Cristóbal Noé Aguilar, PhD

Cristóbal Noé Aguilar, PhD, is Dean of the School of Chemistry at the Universidad Autónoma de Coahuila, México. Dr. Aguilar has published more than 160 papers in indexed journals, more than 40 articles in Mexican journals, and 250 contributions in scientific meetings. He has also published many book chapters, several Mexican books, four editions of international books, and more. Dr. Aguilar is a member level III of S.N.I. (National System of Researchers of Mexico). He has been awarded several prizes and awards, the most important being the National Prize of Research 2010 from the Mexican Academy of Sciences; the "Carlos Casas Campillo 2008" Prize from the Mexican Society of Biotechnology and Bioengineering; National Prize AgroBio—2005; and the Mexican Prize in Food Science and Technology.

Dr. Aguilar is a member of the Mexican Academy of Science, the International Bioprocessing Association, Mexican Academy of Sciences, Mexican Society for Biotechology and Bioengineering, and the Mexican Association for Food Science & Biotechnology. He has developed more than 21 research projects, including six international exchange projects. He has been an advisor for PhD, MSc, and BSc theses.

Dr. Aguilar earned his chemist designation from the School of Chemistry at the Universidad Autónoma de Coahuila, México. He attended the MSc Program in Food Science and Biotechnology that was held at the Universidad Autónoma de Chihuahua, México. His PhD in Fermentation Biotechnology was awarded by the Universidad Autónoma Metropolitana, Mexico.

Elizabeth Carvajal-Millan, PhD

Elizabeth Carvajal-Millan, PhD, is Research Scientist at the Research Center for Food and Development (CIAD) in Hermosillo, Mexico, since 2005. She obtained her PhD in France at Ecole Nationale Supérieure Agronomique à Montpellier (ENSAM), her MSc degree at CIAD, and undergraduate degree at the University of Sonora in Mexico. Her research interests are focused on biopolymers, mainly in the extraction and characterization of high-value-added polysaccharides from coproducts recovered from the food industry and agriculture, especially ferulated arabinoxylans. In particular,

Dr. Carvajal-Millan studies covalent arabinoxylan gels as functional systems for the food and pharmaceutical industries. Globally, Dr. Carvajal-Millan is a pioneer in *in vitro* and *in vivo* studies on covalent arabinoxylan gels as carriers for oral insulin focused on the treatment of type 1 diabetes. She has published 57 refereed papers, 23 book chapters, over 80 conference presentations, and has registered one patent, with two more submitted.

CONTENTS

LIST OF CONTRIBUTORS

Cristóbal Noé Aguilar
Group of Bioprocesses and Bioproducts, Food Research Department, School of Chemistry, Universidad Autonoma de Coahuila, 25280 Saltillo, Coahuila, Mexico. E-mail: cristobal.aguilar@uadec.edu.mx

Yesmin Ara Begum
Department of Food Engineering and Technology, Tezpur University, Napaam 784028, Sonitpur, Assam, India

José Juan Buenrostro-Figueroa
Group of Bioprocesses and Bioproducts, Food Research Department, School of Chemistry, Universidad Autonoma de Coahuila, 25280 Saltillo, Coahuila, Mexico
Research Center in Food and Development AC, 33089 Cd. Delicias, Chihuahua, Mexico

Gloria Castellano
Departamento de Ciencias Experimentales y Matemáticas, Facultad de Veterinaria y Ciencias Experimentales, Universidad Católica de Valencia San Vicente Mártir, Guillem de Castro-94, E-46001 València, Spain

Arup Jyoti Das
Department of Food Engineering and Technology, Tezpur University, Napaam 784028, Sonitpur, Assam, India

Debasis Das
Department of Food Engineering and Technology, Tezpur University, Napaam 784028, Sonitpur, Assam, India

Manas Jyoti Das
Department of Food Engineering and Technology, Tezpur University, Napaam 784028, Sonitpur, Assam, India

Sankar Chandra Deka
Department of Food Engineering and Technology, Tezpur University, Napaam 784028, Sonitpur, Assam, India. E-mail: sankar@tezu.ernet.in

Heliodoro de la Garza-Toledo
Group of Bioprocesses and Bioproducts, Food Research Department, School of Chemistry, Universidad Autonoma de Coahuila, 25280 Saltillo, Coahuila, Mexico

Sajad Ahmad Ganai
Department of Chemistry, National Institute of Technology, Srinagar, Jammu and Kashmir, India. E-mail: sajadali16@gmail.com

Porteen Kannan
Department of Veterinary Public Health and Epidemiology, Madras Veterinary College, Chennai 600007, Tamil Nadu, India

Amjad Mumtaz Khan
Department of Chemistry, AMU, Aligarh, Uttar Pradesh 202001, India

Singamayum Khurshida
Department of Food Engineering and Technology, Tezpur University, Napaam, Sonitpur, Assam 784028, India

G. Leena
Department of Veterinary Public Health and Epidemiology, Veterinary College, Bengaluru 560024, Karnataka, India

Khwairakpam Chanu Salailenbi Mangang
Department of Food Engineering and Technology, Tezpur University, Napaam 784028, Sonitpur, Assam, India

Hari Niwas Mishra
Department of Food Engineering and Technology, Tezpur University, Napaam 784028, Sonitpur, Assam, India

Tatsuro Miyaji
Department of Materials and Life Science Faculty of Science and Technology Shizuoka Institute of Science and Technology, 2200-2 Toyosawa, Fukuroi, Shizuoka 437-8555, Japan

H. V. Mohan
Department of Veterinary Public Health and Epidemiology, Veterinary College, Bengaluru 560024, Karnataka, India. E-mail: mohanhv@gmail.com

Sangita Muchahary
Department of Food Engineering and Technology, Tezpur University, Napaam 784028, Sonitpur, Assam, India

C. Nishanth
Department of Veterinary Public Health and Epidemiology, Veterinary College, Bengaluru 560024, Karnataka, India

L. Arul Pragasan
Department of Environmental Sciences, Bharathiar University, Coimbatore 641046, Tamil Nadu, India. E-mail: arulpragasan@yahoo.co.in

M. Nithya Quintoil
Department of Veterinary Public Health, Rajiv Gandhi Institute of Veterinary Education and Research, Puducherry 605009, India

Raúl Rodrìguez
Group of Bioprocesses and Bioproducts, Food Research Department, School of Chemistry, Universidad Autonoma de Coahuila, 25280 Saltillo, Coahuila, Mexico

S. Wilfred Ruban
Department of Livestock Products Technology, Veterinary College, Hebbal 560024, Bengaluru, Karnataka, India

R. Sarath
Department of Environmental Sciences, Bharathiar University, Coimbatore 641046, Tamil Nadu, India

Dibyakanta Seth
Department of Food Engineering and Technology, Tezpur University, Napaam 784028, Sonitpur, Assam, India

Garima Sharma
Department of Food Engineering and Technology, Tezpur University, Napaam 784028, Sonitpur, Assam, India

Francisco Torrens
Institut Universitari de Ciència Molecular, Universitat de València, Edifici d'Instituts de Paterna, PO Box 22085, E-46071 València, Spain. E-mail: torrens@uv.es

Cristian Torres-León
Group of Bioprocesses and Bioproducts, Food Research Department, School of Chemistry, Universidad Autonoma de Coahuila, 25280 Saltillo, Coahuila, Mexico

LIST OF ABBREVIATIONS

ABTS	2,2'-azinobis-(3-ethylbenzthiazoline-6-sulphonate
AME	*A. myriophylla* bark
ANOVA	analysis of variance
AOA	antioxidant activity
ASB	*Arcobacter*-selective enrichment broth
ASM	*Arcobacter*-selective semisolid medium
AYB	African yam bean
BD	breakdown
BEP	black-eyed pea
BRDF	banana rhizome dietary fiber
CAs	cluster analyses
CF	cellulose fiber
CHO	carbohydrates
CI	Carr's Index
CNP	cellulose nanopaper
CTL	cytotoxic T-lymphocyte
DCF	dry chips flour
DE	dextrose-equivalence
DF	dietary fiber
DMW	demineralized water
DPPH	2,2-diphenyl-1-picrylhydrazyl
DSC	differential scanning calorimetry
dsDNA	double-stranded DNA
EBT	Eriochromr black-T
EC	epicatechin
EDTA	ethylene diamine tetra acetic acid
EGC	epigallocatechin
EGCg	epigallocatechin gallate
EMJH	Ellinghausen–McCullough–Johnson–Harris
EO	essential oil
FAs	fatty acids
FCF	fermented cassava flour

FT-IR	Fourier transform infrared spectroscopy
FV	final viscosity
GA	gallic acid
GAE	gallic acid equivalent
HA	hemicellulose A
HB	hemicellulose B
HBA	4-hydroxybenzoic acid
HPLC	high-performance liquid chromatography
HR	Hausner ratio
IDF	insoluble dietary fiber
IF	insoluble fiber
JM	Johnson–Murano
LAB	lactic acid bacteria
LPS	lipopolysaccharide
NEM	northeast monsoon
NK	natural killer
OHC	oil-holding capacity
PC	pyrocatechol
PCAs	principal component analyses
PCC	Pearson's correlation coefficient
PCD	partial correlation diagram
PCR	polymerase chain reaction
PL	phospholipids
PT	pasting temperature
PV	peak viscosity
PY	pectin yield
RB	rice beer
RVA	Rapid Visco Analyzer
SC	starter cake
SD	standard deviation
SDF	soluble dietary fiber
SEM	scanning electron microscopy
SEM-EDX	scanning electron microscopy-energy dispersive X-ray analysis
SF	soluble fiber
SP	swelling power
SS	summer season
SSA	*Salmonella–Shigella* agar

SWM	southwest monsoon
SYP	sweetened yoghurt powder
TBA	thiobarbituric acid
TEM	transmission electron microscopy
TGA	thermogravimetric analysis
TLC	thin-layer chromatography
TPC	total polyphenol content
UE	ultrasound extraction
WHC	water-holding capacity
WS	winter season
XRD	X-ray diffraction

PREFACE

Of all the basic needs, food is the foremost need of everyone; that is the reason it is regarded as the prime endowment among all. Food inspires creativity and pleasure, and provides much needed peace and fullness. Although the world produces around 50% more food for the entire population, hunger still prevails, as nearly one-third of food produced is wasted. According to the UN, one in every nine persons goes hungry each night. On the other hand, new and more effective techniques in the processing of food is seen as the best solution to overcome the wastage of food, which is much needed to overcome world hunger.

This book, *Applied Food Science and Engineering with Industrial Applications,* is an edited collection with contributions from eminent researchers across the globe that will be of interest to students, professional teachers, researchers, and industry. Our goal was to assimilate the knowledge on applied food science and engineering with industrial applications on different commodities, which is of interest to academia and industry, and to motivate nonprofessionals to learn new techniques of food processing.

This book presents a broad selection of new research in applied food science and engineering with industrial applications and reflects the diversity of recent advances in the field. It provides a broad perspective that will be highly useful to scientists and engineers as well as to graduate students in this dynamic field.

This book is a rich resource on recent research innovations. It presents a practical, unique, and challenging blend of principles and applications for comprehensive learning.

CHAPTER 1

BACTERIOPHAGE: A NOVEL BIOCONTROL AGENT AGAINST FOOD PATHOGENS: MYTH OR REALITY?

S. WILFRED RUBAN[1*], PORTEEN KANNAN[2], and M. NITHYA QUINTOIL[3]

[1]Department of Livestock Products Technology, Veterinary College, Hebbal 560024, Bengaluru, Karnataka, India

[2]Department of Veterinary Public Health and Epidemiology, Madras Veterinary College, Chennai 600007 Tamil Nadu, India

[3]Department of Veterinary Public Health, Rajiv Gandhi Institute of Veterinary Education and Research, Puducherry 605009, India

[*]Corresponding author. E-mail: rubanlpt@gmail.com

ABSTRACT

Bacteriophage, also known as phage, is a type of virus that infects only bacteria cells. These phages are grouped into two broad classes based on their life cycle, namely, the lytic and lysogenic phages. Of these two categories, the major phage of interest is the lytic phage which offers several potential applications in the food industry because of their high host specificity and safety as they are harmless to mammalian cells as well as the normal microbiota of the food. Their application has been well documented as biocontrol agents, bio-preservatives as well as tools for detecting pathogens. This chapter will review in detail the various facts and findings regarding the application of phages as agent against foodborne pathogens in a process known as "biocontrol."

1.1 INTRODUCTION

Foodborne illnesses have been on the rise worldwide and have been a major cause of substantial morbidity and mortality annually, often associated with severe outbreaks and food contamination. According to Centers for Disease Control and Prevention (2014), foodborne illness is known to be a ubiquitous, costly, yet preventable public health concern. The data pertaining to foodborne disease are generally country or pathogen specific only and in developing countries the scenario is still worse due to lack of reporting and documentation of these diseases. The World Health Organization states that food safety remains a continuous challenge to everyone especially in the management of both infectious and noninfectious foodborne hazards (Rocourt et al., 2003). In spite of the currently available technologies and good manufacturing practices in the food industry, the safety of food and its products is being constantly threatened by the factors, such as changes in lifestyle, consumer eating habits, food and agriculture manufacturing processes, and also the increased international trade (Newell et al., 2010).

In the present day context, there has been a continuous increase in occurrence of foodborne outbreaks, majority of which are caused by bacterial pathogens such as *Salmonella*, *Staphylococcus aureus*, *Campylobacter*, *Escherichia coli*, and *Listeria* in different parts of the world despite employment of several modern technologies and interventions in food-production chain. The major intervention strategies employed at present involves physical methods like UV light, high pressure, dry heat, and steam. However, these methods have negative impact on organoleptic characteristics of the products thereby decreasing their acceptability by consumers. Hence, there is a need for development of novel strategies to minimize bacterial load foods and at the same time satisfy the demands of the consumer. Approaches to using natural preservatives, either from plants or microbes, has been the topic of discussion in the recent past and the most often adopted approach to date has been to use biocontrol agents in counteracting the foodborne pathogens, with minimal effect on the organoleptic characteristics of the food.

The major biocontrol agents used in foods include use of bacterial predators, bacteriocins (peptides secreted by one species of bacteria that inhibit another), siderophores (molecules that sequester iron), quorum sensing (where the presence of an autoinducer of one species may influence the growth of another), competitive organisms, bacteriophages, and various microbe and plant-derived antimicrobials. This chapter will provide an insight on various biocontrol strategies employed in control of food pathogen and will discuss in detail the use of bacteriophages as biocontrol agent against food pathogens.

1.2 BACTERIOPHAGES—A POTENTIAL BIOCONTROL AGENT

Bacteriophage applications in the "farm-to-fork" continuum can be categorized in the following categories:

1. Phage therapy—through the prevention or reduction of colonization and diseases in livestock,
2. Phage biosanitation—the decontamination of carcasses and processed products and disinfection of equipment and contact surfaces, and
3. Phage biopreservation—the extension of the shelf life of perishable manufactured foods.

Bacteriophage, also known as phage, is a type of virus that infects only bacteria cells. Phages are very tiny and measure 20–200 nm, which is approximately 100 times smaller than most bacteria. They are obligate intracellular parasites that infect bacteria and reproduce by hijacking their host's biosynthetic pathway. Phages are extremely diversified group and they are known to be the most abundant and self-replicating organisms on Earth (approximately 10^{30}–10^{31} particles compared to 10^7 humans) with the fact that they are 10 times more than their bacterial host (Bergh et al., 1989). Most of the phages are tailed bacteriophage, which accounts for 96% of all phages present on earth, belonging to the order Caudovirales (Ackermann, 2003). According to International Committee on Taxonomy of Viruses, the phages are classified into three families (Niu et al., 2014):

1. the *Myoviridae* (long contractile tail),
2. the *Siphoviridae* (long noncontractile tail), and
3. the *Podoviridae* (short noncontractile tail).

Phages are abundant in the environment and are suggested to have likely been infecting bacteria since they first evolved (Kennedy and Bitton, 1987; Thiel, 2004). Phages were first discovered independently by Frederick Twort (1915) and Felix d'Herelle (1917) during early 20th century and the bactericidal activity of phages was first observed by Hankin in 1896, who noticed that filtered water from the Ganges and Jumna rivers in India had antibacterial properties against *Vibrio cholerae*. Due to its ability to kill bacteria, d'Herelle named the agent "phage" from the Greek word "phagin," meaning "to devour or eat" (d'Herelle, 1917).

1.3 CHARACTERIZATION OF PHAGES

Typically, phages consist of an outer-protein capsid enclosing nucleic acid genomes. The genomes of phages can be either linear or circular molecules of double-stranded DNA (dsDNA), single-stranded DNA, or single-stranded RNA between 5 and 500 kb pairs long (Goodridge and Abedon, 2003). Phage can vary in many different sizes and shapes with the structure varying from complex polyhedral capsid with a tail and tail fibers to a comparatively simple polyhedral capsid (Ackermann, 2009). Most (i.e., 96%) of phages described to have tails and appears to be the oldest known virus group (Ackermann, 2003). Morphology, along with the nature and size of phage genome, are used in the classification of phages.

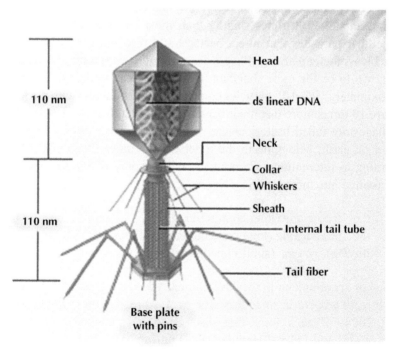

A representative schematic of structure of bacteriophage

1.4 CLASSIFICATION OF PHAGES BY MORPHOLOGY AND GENOME

A. BASED ON GENOME

1. DNA phages

    (a) *Myoviridae*—nonenveloped contractile tail (linear)
    (b) *Siphoviridae*—nonenveloped long noncontractile tail (linear)
    (c) *Podoviridae*—nonenveloped short noncontractile tail (linear)
    (d) *Tectiviridae*—nonenveloped isometric (linear)
    (e) *Corticoviridae*—nonenveloped isometric (circular)
    (f) *Lipothrixviridae*—enveloped rod-shaped (linear)
    (g) *Plasmaviridae*—enveloped pleomorphic (circular)
    (h) *Fuseloviridae*—nonenveloped lemon-shaped (circular)
    (i) *Inoviridae*—nonenveloped filamentous (circular)
    (j) *Microviridae*—nonenveloped isometric (linear)

2. RNA phages

    (a) *Reoviridae—nonenveloped isometric* (segmented)
    (b) *Leviviridae—nonenveloped isometric* (linear)
    (c) *Cystoviridae*—enveloped spherical (segmented)

B. BASED ON MORPHOLOGY

1. Tailed phages

    (a) *Myoviridae*
    (b) *Siphoviridae*
    (c) *Podoviridae*

2. Polyhedral phages

    (a) *Microviridae*
    (b) *Tectiviridae*
    (c) *Corticoviridae*
    (d) *Leviviridae*

3. Filamentous phages

    (a) *Cystoviridae*
    (b) *Lipothrixviridae*

    (c)   *Inoviridae*
    (d)   *Reoviridae*

4. Pleomorphic phages
    (a)   *Fuselloviridae*
    (b)   *Plasmaviridae*

1.5   LIFECYCLE AND MECHANISM OF THE ACTIVITY OF PHAGES

Most phages fall into two groups, virulent or temperate. The characteristic lifecycles, lytic and lysogenic cycles, of phage is illustrated in Figure 1.1. Virulent phages go through a lytic cycle, which can release up to several hundred progeny per cell. The *typical lytic cycle of phages* containing dsDNA can be described as a four-stage process.

1.5.1 ADSORPTION OF THE PHAGE TO SPECIFIC RECEPTORS ON THE BACTERIAL SURFACE

The ability of a particular phage to infect a bacterium is almost limited to a single bacterial species and often to a few strains of that species. That means the interaction between phages and bacteria is highly specific. The main factor contributing to this specificity is the existence of receptors on the bacterial surface for phage adsorption. Replication will only follow if the cell contains specific receptor sites for the phage. On the host cell's surface, many types of accessible molecules may serve as specific phage receptors. Proteins exploited by phages as receptors generally have important roles for the normal functioning of the cell.

The environment is an important factor in adsorption and presence of inorganic salts is essential for initial contact of phages with bacterial cell. Some phages have specific cationic requirements, such as Ca^{2+} and Mg^{2+} (Maloy et al., 1994).

1.5.2 PASSAGE OF THE DNA FROM THE PHAGE THROUGH THE BACTERIAL CELL WALL

The transfer of phage-genetic material into the cytoplasm of the host cell likewise calls for a complex interplay between the receptor and phage structure, a process that requires energy input (Gottesman et al., 1994). Contact of phage with the outer-membrane receptors initiates conformational changes in the phage structure which cause the tail sheath to contract, forcing the hollow inner tube into the cell. These events also change the conformation of proteins in the base plate, tail tube, collar, and possibly the head, allowing release of the genome through the tube into the cell. The injection of the DNA is driven by proton motive force maintained by the bacterial cell.

1.5.3   EXPRESSION OF VIRAL GENES

Following the entry of the viral chromosome, the genes expressed early code for proteins which are involved in phage genome replication and which modify the cellular machinery so that the synthetic capacity of the cell is subverted to the reproduction of the phage. These early gene products are rarely found in the completed phage. The remaining genes that specify "late" functions are then expressed to initiate the synthesis of the various protein subunits of the capsid, which are assembled into intermediate structures.

1.5.4   ASSEMBLY OF THE PHAGE PARTICLES AND RELEASE OF NEWLY SYNTHESIZED PHAGE

The genomes are packaged within capsid and ancillary proteins, and mature viruses appear. Usually, 50–100 phage particles are produced per cell, the number depends on the particular phage and the physiology of the host cell (Maloy et al., 1994). Late in the infection cycle, most phages synthesize enzymes that lyse the host cell. Two enzymes are typically released: an enzyme called *holin*, which disrupts the cytoplasmic membrane, and an enzyme called *lysozyme*, which degrades the cell wall peptidoglycan. The phages are released into the surrounding medium by disruption of the cell membrane and the cell way caused by these enzymes (Gottesman et al., 1994). At the same time, the bacterial intracellular components (e.g., bacterial ATP and adenylate kinase) are released.

The *lysogenic life cycle* is an alternative reproductive pathway to the lytic cycle. The most common lysogenic pathway can be described in the following way. After the adsorption of phage to its target bacterium, the linear phage DNA molecule is injected into a bacterium. Following a brief period of mRNA synthesis, which is needed to synthesize a repressor protein (which inhibits the synthesis of the mRNA species that encode the lytic functions) and a site-specific recombination enzyme, phage mRNA synthesis is turned off by the repressor (Maloy et al., 1994). Then, recombination between the phage DNA molecule and the DNA of the infected bacterium inserts the phage DNA into the bacterial chromosome. This infected bacterium continues to grow and multiply, and the phage genes replicate as part of the bacterial chromosome. A bacterium containing a complete set of phage genes is called a lysogen. Lysogens are resistant to reinfection by a phage of the type that first lysogenized the cell.

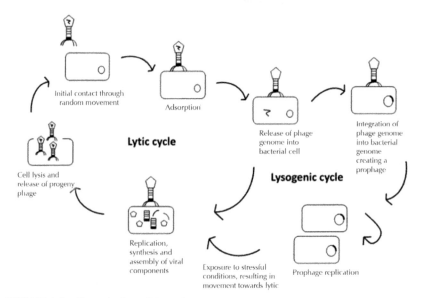

FIGURE 1.1   Bacteriophage life cycle.

1.6   ADVANTAGES OF PHAGES AS BIOCONTROL AGENT

- High specificity to target their host determined by bacterial cell wall receptors, leaving untouched the remaining microbiota, a property that favors phages over other antimicrobials that can cause microbiota collateral damage
- Self-replication and self-limiting, meaning that low or single dosages will multiply as long as there is still a host threshold present, multiplying their overall antimicrobial impact
- As bacteria develop phage defense mechanisms for their survival, phages continuously adapt to these altered host systems
- Low inherent toxicity, since they consist mostly of nucleic acids and proteins

- Phages are relatively cheap and easy to isolate and propagate but may become time consuming when considering the development of a highly virulent, broad-spectrum, and nontransducing phage
- Withstand food processing environmental stresses
- They have proved to have prolonged shelf life
- Phages are readily abundant in foods and have been isolated from a wide variety of raw products (e.g., beef, chicken), processed food (e.g., pies, biscuit dough, and roast turkey), fermented products (e.g., cheese, yoghurt), and seafood (e.g., mussels and oysters, suggesting that phages can be found in the same environments where their bacterial host(s) inhabit, and that phages are daily consumed by humans.

1.7   CRITERIA FOR SELECTION OF PHAGES

For selection of suitable phage for use as biocontrol agents in foods phages should possess the following desirable properties (Gill and Hyman, 2010; Hagens and Loessner, 2010):

1. Phage should be strictly lytic
2. Phage should have a broad host range
3. Lack of transduction of bacterial DNA
4. Oral feeding studies of phages should show no adverse effects
5. Phage preparation should be stable over storage and application
6. Phage should be suitable for commercial production

1.8   ISOLATION AND HOST RANGE CHARACTERIZATION OF BACTERIOPHAGES

The methods that are used routinely to isolate, purify, and characterize bacteriophages from environments have been described in detail by Middelboe et al. (2010).

1.8.1   SPOTTING ON TARGET HOST CELLS

The lysis of the bacterial cells by the phages could be visualized by plaque formation in soft agar. This principle can be used for the detection and subsequent isolation of specific lytic phages in environmental samples.

1.8.2   BACTERIOPHAGE ISOLATION USING ENRICHMENT CULTURES

A generally more efficient way of isolating lytic phages from environments is by the use of enrichment cultures. In this approach, the prefiltered water sample that is to be screened for phages against a given target bacterium is enriched with a bacterial growth medium and amended with that target bacteria.

1.9   USE OF BACTERIOPHAGES IN THE BIOCONTROL

1.9.1   PASSIVE TREATMENT

In this approach, bacteriophages are added in sufficient quantities to overwhelm all target organisms by primary infection or by lysis from without. Although much higher numbers of the bacteriophages are required, they should be able to eliminate even sparse populations of susceptible bacteria. One other advantage of this approach is that, since much of the effect is a result of lysis from without, natural resistance due to restriction enzymes present in host bacteria will not be an issue.

1.9.2   ACTIVE TREATMENT

A relatively small dose of bacteriophages may be required for efficacious elimination of the undesirable bacteria, since most are killed by secondary infections due to replication and transmission from neighboring organisms. This is dependent on the bacteriophages being able to spread between susceptible bacterial hosts, which may be hindered by the surrounding material being viscous or by the presence of outnumbering inert bacteria.

Three scenarios have been proposed for the use of bacteriophages in biocontrol:

- Control of pathogenic bacteria in foods
- Prevention of bacterial food spoilage
- Reduction of antibiotic resistance by suppressing resistance gene expression by using bacteriophages to deliver antisense DNA. This is purely in the experimental phase.

1.9.3 PREHARVEST PHAGE APPLICATIONS

Berchieri et al. (1991) used phages isolated from human sewage to reduce experimental *Salmonella typhimurium* colonization of the chicken intestine by approximately 1 $\log_{10}$ cfu and significantly reduced mortality compared with untreated animals. However, the phages only persisted whilst *Salmonella* spp. could also be recovered.

Fiorentin et al. (2005) used bacteriophages isolated from free-range chickens to reduce *Salmonella enteritidis* PT4 colonization of broiler chicks. Day old broilers were challenged with *S. enteritidis* and treated with 10^{11} PFU of a cocktail of three bacteriophages 7 days afterward. A 3.5 $\log_{10}$ reduction in cecal carriage was recorded 5 days after phage treatment (compared with the control). An appreciable reduction in *Salmonella* colonization in the PT group remained for up to 25 days after phage treatment.

Atterbury et al. (2007) demonstrated that the cecal colonization of both *S. enteritidis* and *S. typhimurium* in broiler chickens could be reduced by up to 4.2 $\log_{10}$ cycles through the oral administration of high-titer (10^{11} PFU) phage suspensions. Pathogenic *E. coli* can cause significant morbidity and mortality in poultry flocks.

Barrow et al. (1998) used phages to prevent *E. coli* septicemia in chickens when the mortality rate in untreated birds was almost 100%. Delaying PT until the onset of clinical symptoms greatly reduced the severity of infection.

The effect of PT on the survival of chickens challenged with *E. coli* strain 3–1 was compared with antibiotic chemotherapy (chloromycetin, 1 mg per 10 g body weight). Phage therapy was found to reduce the incidence of diarrhea in the chickens to 26%, compared with 51.6% for the control group. Survival in the PT group was approximately 10-fold higher than the control group and sixfold higher than the antibiotic-treated group. Noteworthy, the birds in the PT group had higher bodyweights than those in the antibiotic-treated group.

Campylobacter numbers have been reduced in the ceca of broiler chickens by 1–2 $\log_{10}$ cfu following repeated doses of phages (Wagenaar et al., 2005).

Loc Carrillo et al. (2005) also evaluated the ability of phages to reduce the intestinal colonization of campylobacters. Significant reductions of up to 5 $\log_{10}$ cfu g^{-1} were recorded for some phage–host combinations in several parts of the intestinal tract. The authors suggested that such reductions could be highly beneficial as a preslaughter treatment to reduce *Campylobacter* contamination on the carcass.

Callaway et al. (2008) used a cocktail of phages to significantly reduce numbers of *E. coli* O157:H7 in the intestinal tract of sheep. In contrast to some other PT studies, the highest phage multiplicity of infection (MOI) used (100 and 10) were less efficacious than a MOI of 1.

1.9.4 POSTHARVEST APPLICATION OF PHAGES

Phages have not only been used to reduce the burden of zoonotic pathogens in the intestinal tracts of animals but have also been applied directly to contact surfaces and carcasses. Phages have been applied in a variety of food matrices. However, most studies have concentrated on chicken and beef. Phages have been isolated from the gastrointestinal tracts of animals and are present in a wide variety of foods. As such, it is likely that phages are consumed regularly by humans without ill effects (**Kennedy and Bitton**, 1987).

Atterbury et al. (2003) reported that the application of phages onto the surface of chicken skin artificially contaminated with *Campylobacter jejuni* led to a reduction of $1-1.3$ $\log_{10}$ cfu within 24 h. Combining phage treatment with freezing the skin sections at $-20°C$ was more effective than either treatment used independently.

Chicken carcasses artificially contaminated with *Salmonella* were used as a model for surface disinfection using phages. They found that a cocktail of phages could reduce *Salmonella* recovery by >1000-fold compared with untreated controls.

The preservative effect of *Pseudomonas* phages in raw chilled beef has been thoroughly examined in experimentally inoculated meat, showing a significant extension of the retail shelf life of phage-treated beef (**Greer**, 1988).

Bigwood et al. (2008) investigated the use of phages against *S. Typhimurium* and *C. jejuni* in cooked and raw meat at different temperatures. The greatest reduction in pathogen numbers was obtained when both the population density of target bacteria and MOI were high. Incubation temperature also appeared to be important, with greater reductions in pathogen numbers occurring at the higher temperatures used in the study ($\sim24°C$). The reduction in pathogen numbers following phage treatment could be maintained for up to 8 days when the meat samples were incubated at $5°C$. This was despite no recorded increase in bacteriophage numbers after 24 h.

1.9.4.1   LIMITATIONS OF PHAGE

Despite the various advantages the phages offer as biocontrol tool, there are few limitations that need to be addressed to improve the efficiency of application in food system.

1. Development of bacterial resistance or cross-resistance mechanisms over a period of time. This limitation can be overcome by use of cocktail of phages or by use of immobilization technologies to prevent residual effect.
2. Conversion of lytic phages to lysogenic phages due to occurrence of mutational changes (Greer, 2005).
3. Higher microbial load in food system may result in increase in nonspecific binding sites, resulting in lower efficiency of phage therapy (Garcia et al., 2008).

1.10   CONCLUSION

Consumer expectations of pathogen-free food with an acceptable shelf life without the use of synthetic chemicals present major challenges for food manufacturers, given the currently available technologies. The biocontrol approaches reviewed here offer some particular advantages by being more targeted than conventional treatments, both leaving the good bugs alone and reducing wastage, and are perceived as more "natural" and "green." While no one biocontrol approach currently offers a complete solution, collectively they create a toolbox of options employing various mechanisms of attack which could be applied individually or in combination.

Application of phages in biocontrol of pathogens will be a step forward; however, complete phage genome analysis is required to ensure safety and effectiveness of use of phages. Finally, food industry acceptance and consumer preference are critical hurdles to be overcome for their commercial application.

In addition, combination of activities might be a future direction in which virus-derived tools could play an important role. In particular, one can imagine designing a virus targeting, for example, a Gram-negative host and carrying genes that are toxic for Gram-positive species (such as endolysins) or fungi. In this regard, it is an interesting case of the so-called protein antibiotics, the gene products of some small phages that do not produce endolysins and are capable of inhibiting the synthesis of the cell wall.

Applied Food Science and Engineering with Industrial Applications

Coupling these different properties may offer the possibility to come up with new highly efficient broad antimicrobial strategies.

"The enemy of our enemy is our friend; utilizing the natural killers (bacteriophages) against bacteria will improve food safety."

KEYWORDS

- bacteriophage
- lytic phage
- lysogenic phage
- bio-preservatives
- biocontrol agents

REFERENCES

Ackermann, H. W. Phage Classification and Characterization. In *Bacteriphages: Methods and Protocols*; Clokie, M. R. J., Kropinski, A. M., Eds.; Humana Press: Totowa, NJ, 2009; pp 127–140.

Ackermann, H. W. Bacteriophage Observations and Evolution. *Res. Microbiol.* **2003**, *154* (4), 245–251.

Atterbury, R. J.; Connerton, P. L.; Dodd, C. E.; Rees, C. E.; Connerton, I. F. Application of Host-Specific Bacteriophages to the Surface of Chicken Skin Leads to a Reduction in Recovery of *Campylobacter jejuni*. *Appl. Environ. Microbiol.* **2003**, *69*, 6302–6306.

Atterbury, R. J.; Van Bergen, M. A.; Ortiz, F.; Lovell, M. A.; Harris, J. A.; De Boer, A.; et al. Bacteriophage Therapy to Reduce *Salmonella* Colonization of Broiler Chickens. *Appl. Environ. Microbiol.* **2007**, *73*, 4543–4549.

Barrow, P.; Lovell, M.; Berchieri, A. Jr. Use of Lytic Bacteriophage for Control of Experimental *Escherichia coli* Septicemia and Meningitis in Chickens and Calves. *Clin. Diagn. Lab. Immunol.* **1998**, *5*, 294–298.

Berchieri, A. Jr.; Lovell, M. A.; Barrow, P. A. The Activity in the Chicken Alimentary Tract of Bacteriophages Lytic for *Salmonella typhimurium*. *Res. Microbiol.* **1991**, *142*, 541–549.

Bergh, O.; Borsheim, K. Y.; Brabak, G.; Heldal, M. High Abundance of Viruses Found in Aquatic Environments. *Nature* **1989**, *340*, 467–468.

Bigwood, T.; Hudson, J. A.; Billington, C.; Carey-Smith, G. V.; Hememann, J. A. Phage Inactivation of Food-Borne Pathogens on Cooked and Raw Meat. *Food Microbiol.* **2008**, *25*, 400–406.

Callaway, T. R.; Edrington, T. S.; Brabban, A. D.; Anderson, R. C.; Rossman, M. L.; Engler, M. J.; et al. Bacteriophage Isolated from Feedlot Cattle Can Reduce *Escherichia coli* O157:H7 Populations in Ruminant Gastrointestinal Tracts. *Foodborne Pathog. Dis.* **2008**, *5*, 183–191.

Centers for Disease Control and Prevention. *Estimating Foodborne Illness: An Overview*, 2014.

d'Herelle, F. C. R. Sur un Microbe Invisible Antagoniste des Bacilles dysentériques. *Acad. Sci. Ser.* **1917**, *165*, 373.

Fiorentin, L.; Vieira, N. D.; Barioni, W. Oral Treatment with Bacteriophages Reduces the Concentration of *Salmonella Enteritidis* PT4 in Caecal Contents of Broilers. *Avian Pathol.* **2005**, *34*, 258–263.

Garcia, P., et al. Bacteriophages and Their Application in Food Safety. *Lett. Appl. Microbiol.* **2008**, *47* (6), 479–485.

Gill, J. J.; Hyman, P. Phage Choice, Isolation, and Preparation for Phage Therapy. *Curr. Pharm. Biotechnol.* **2010**, *11*, 2–14.

Goodridge, L.; Abedon, S. T. Bacteriophage Biocontrol and Bioprocessing: Application of Phage Therapy to Industry. *Feat. Art.* **2003**, *53* (6), 254–262.

Gottesman, M.; Oppenheim, A. Lysogeny and Prophage. In *Encyclopedia of Virology*; Webster, R. G., Graof, A., Eds.; Academic Press: New York, 1994; pp 814–823.

Greer, G. G. Effects of Phage Concentration, Bacterial Density and Temperature on Phage Control of Beef Spoilage. *J. Food Sci.* **1988**, *53*, 1226–1227.

Greer, G. G. Bacteriophage Control of Foodborne Bacteria. *J. Food Protect.* **2005**, *68* (5), 1102–1111.

Hagens, S, Loessner, M. J. Bacteriophage for Biocontrol of Foodborne Pathogens: Calculations and Considerations. *Curr. Pharm. Biotechnol.* **2010**, *11*, 58–68.

Kennedy, J. E.; Bitton, G. Bacteriophages in Foods, In *Phage Ecology*; Goyal, S. M., Gerba, C. P., Bitton, G., Eds.; John Wiley and Sons: New York, 1987; pp 289–316.

Loc Carrillo, C.; Atterbury, R. J.; el-Shibiny, A.; Connerton, P. L.; Dillon, E.; Scott, A.; Connerton, I. F. Bacteriophage Therapy to Reduce *Campylobacter jejuni* Colonization of Broiler Chickens. *Appl. Environ. Microbiol.* **2005**, *71*, 6554–6563.

Maloy, S. R.; Cronan, J. E.; Freifelder, D. *Microbial Genetics*, 2nd ed.; Jones and Bartlett Publishers: London, 1994; pp 81–86.

Middelboe, M.; et al. *Manual of Aquatic Viral Ecology*; American Society of Limnology and Oceanography, Inc., 2010; Chapter 13, pp 118–133.

Newell, D. G.; Koopmans, M.; Verhoef, L.; Duizer, E.; Aidara-Kane, A.; Sprong, H.; Opsteegh, M.; Langelaar, M.; Threfall, J.; Scheutz, F.; van der Gieessen, J.; Kruse, H. Food-borne Diseases—The Challenges of 20 Years Ago Still Persist While New Ones Continue to Emerge. *Int. J. Food Microbiol.* **2010**, *139* (Suppl. 1), S3–S15.

Niu, Y. D.; McAllister, T. A.; Nash, J. H.; Kropinski, A. M.; Stanford, K. Four *Escherichia coli* O157:H7 Phages: A New Bacteriophage Genus and Taxonomic Classification of T1-Like Phages. *PLoS One* **2014**, *9*, e100426.

Rocourt, J.; Moy, G.; Vierk, K.; Schlundt, J. *The Present State of Foodborne Disease in OECD Countries*; Food Safety Department, WHO: Geneva, 2003.

Thiel, K. Old Dogma, New Tricks—21st Century Phage Therapy. *Nat. Biotechnol.* **2004**, *2* (1), 31–36.

Wagenaar, J.; Van Bergen, M. A.; Mueller, M. A.; Wassenaar, T.; Carlton, R. Phage Therapy Reduces *Campy lobacterjejuni* Colonization in Broilers. *Vet. Microbiol.* **2005**, *109*, 275–283.

CHAPTER 2

ARCOBACTER SP.: AN EMERGING FOODBORNE ZOONOTIC PATHOGEN

H. V. MOHAN[1*], C. NISHANTH[1], G. LEENA[1], S. WILFRED RUBAN[2], and PORTEEN KANNAN[3]

[1]*Department of Veterinary Public Health and Epidemiology, Veterinary College, Bengaluru 560024, Karnataka, India*

[2]*Department of Livestock Products Technology, Veterinary College, Hebbal 560024, Bengaluru, Karnataka, India*

[3]*Department of Veterinary Public Health and Epidemiology, Madras Veterinary College, Chennai 600007, Tamil Nadu, India*

Corresponding author. E-mail: mohanhv@gmail.com

ABSTRACT

Foodborne diseases (both infections and intoxications) are responsible for considerable morbidity and mortality in both industrialized and developing countries; so, foodborne pathogens are a major issue in food safety worldwide. Foodborne zoonotic pathogens are of great importance with regard to consumer health and protection. The globalization of the food supply has presented new challenges for food safety and has contributed to the national and international public health problem of foodborne diseases. Foodborne zoonotic pathogens are of great importance with regard to consumer health and protection. Introduction of new technologies in food industry and changes in the susceptibility of population to disease have all highlighted the problem of foodborne emerging pathogens. During the past two decades, refinements in the isolation techniques in veterinary science led to the discovery of *Arcobacter* spp. as animal and human pathogen. *Arcobacter* sp. are emerging as an important foodborne zoonotic pathogens, causing infections in humans and animals. There are 25 species of *Arcobacter* genus that are being reported. *Arcobacter* sp. have been reported from chickens, domestic animals (cattle, sheep, pigs, horses, and dogs), meat (poultry, pork, and beef), milk, and vegetables in different countries. *Arcobacter* sp. are implicated as causative agents of diarrhea, mastitis, and abortion in animals and gastroenteritis, diarrhea, bacteremia, endocarditis, and peritonitis in humans.

2.1 INTRODUCTION

Arcobacters are potential emerging food and waterborne pathogens. *Arcobacter* is part of the *Epsilonproteobacteria* (rRNA superfamily VI of the *Proteobacteria*) and belongs to the family *Campylobacteriaceae*. The family includes the genera *Campylobacter*, *Sulfospirillum*, and *Arcobacter.* Their sizes range from 0.2 to 0.9 μm in width and 1–3 μm in length. *Arcobacter* was very first isolated and described from aborted bovine fetal tissues (Ellis et al., 1977) and later from porcine fetuses (Ellis et al., 1977, 1978; Aydin et al., 2007). The bacteria are Gram negative, curved to S-shaped rods. They are motile and have one unsheathed single polar flagellum at one or both ends of the cells. Arcobacters are able to grow under aerobic and anaerobic conditions over a wide temperature range (15–42°C). Optimal growth occurs under micro-aerobic conditions (3–10% O_2) and the bacteria do not require hydrogen. It has been revealed that these organisms can be distinguished from *Campylobacter* by their ability to grow between 15 and 30°C temperature aerobically and needs micro-aerophilic condition for primary isolation (Ferreira et al., 2016). Arcobacters have respiratory type of metabolism, are nonsaccharolytic, oxidase, and catalase positive. Like *Campylobacter*, *Arcobacter* species have been indicated as foodborne microorganisms since they have been detected in foods of animal origin, especially in products from chicken all over the world (Son et al., 2006).

Till date, 25 *Arcobacter* sp. have been reported with a significant genetic diversity among and within the species (Ramees et al., 2017) (Table 2.1).

TABLE 2.1 List of *Arcobacter* spp. and Their Origin.

Sl. No.	*Arcobacter* sp.	First time isolated from
1	A. nitrofigilis	Roots of *Spartina alterniflora*
2	A. cryaerophilus	Bovine abortus fetus

TABLE 2.1 *(Continued)*

Sl. No.	*Arcobacter* sp.	First time isolated from
3	*A. butzleri*	Human feces
4	*A. skirrowii*	Sheep feces
5	*Candidatus Arcobacter sulfidicus*	Coastal seawater
6	*A. cibarius*	Chicken meat
7	*A. halophilus*	Hypersaline lagoon
8	*A. mytili* sp. nov.	Mussels
9	*A. thereius* sp. nov.	Porcine abortion
10	*A. marinus* sp. nov.	Seawater, seaweed, and a starfish
11	*A. trophiarium*	Pig feces
12	*A. defluvii* sp. nov.	Sewage samples
13	*A. molluscorum* sp. nov.	Shellfish
14	*A. ellisii* sp. nov.	Shellfish
15	*A. venerupis* sp. nov.	Shellfish
16	*A. bivalviorum* sp. nov.	Shellfish
17	*A. cloacae* sp. nov.	Sewage
18	*A. suis* sp. nov.	Pork meat
19	*A. anaerophilus* sp. nov.	Estuarine sediment
20	*A. ebronensis* sp. nov.	Mussels
21	*A. aquimarinus* sp. nov.	Sea water
22	*A. lanthieri* sp. nov.	Cattle and pig manure
23	*A. pacificus*	Seawater
24	*A. acticola* sp. nov.	Seawater
25	*Arcobacter lekithochrous* sp. nov.	Great scallop (*Pecten maximus*) larvae and tank seawater

Adapted from Ramees et al. (2017).

2.2 VIRULENCE FACTORS

There are scanty reports on the virulence factors in arcobacters. By previous studies, it is understood that adhesion, invasion of the pathogen, secretion of toxin, and pro-inflammatory cytokine (IL-8) plays a major role in establishing infection in the host (Collado and Figueras, 2011).

2.2.1 EPIDEMIOLOGY

Arcobacters have been reported worldwide from chickens, domestic animals (cattle, pigs, sheep, dogs, horses), reptiles, meat (poultry, pork, lamb, beef, goat, rabbit), vegetables, and from humans in different countries like Belgium, Denmark, Brazil, United States of America, Australia, Italy, Netherlands, Japan, Spain, Malaysia, Korea, and India (Patyal et al., 2011; Ramees et al., 2014a, 2014b, 2014c; Mohan et al., 2014). *Arcobacter* is a potential food and waterborne pathogen, and thus pose serious public health concerns (Ferreira et al., 2016). Among the 25 recognized species till date, *A. butzleri*, *A. cryaerophilus*, and *A. skirrowii* are the species of veterinary importance. Isolation of *Arcobacter* from feces of healthy pigs and from dogs has been reported. In dogs, *A. cryaerophilus* was more commonly recovered (Houf et al., 2008). Various isolation rates have been reported from fecal samples of different species. From India, Patyal et al. (2011) recorded a prevalence rate of 21% in pigs, 15% in poultry droppings, and 3% in human diarrheal samples. Mohan et al. (2014) reported isolation rate of 8% in poultry feces, 10% from bovine fecal samples, 12% from pigs, and 2% in human diarrheal samples and Ramees et al. (2014a) recorded an isolation rate of 2% from human diarrheal samples. From Belgium, Van Driessche et al. (2003) recorded *Arcobacter* isolation of 44% (pigs), 39% (bovine), 16% (ovine), and 15% (equine) in fecal samples collected at slaughterhouses. *A. thereius* has also been isolated from liver and kidneys of pigs aborted spontaneously (Houf et al., 2009). Feco-oral route of transmission play an important role in establishing infection in cattle (Scarano et al., 2014). Poultry act as potential reservoirs for arcobacters and these organisms have been recovered from live poultry skin, droppings, and meat (Kabeya et al., 2004; Lehner et al., 2005). Collado et al. (2014) reported the prevalence of *Arcobacter* sp. from different marine species like clams (88%), followed by razor clams (65%), mussels (33%), clams (24%), scallops (18%), and oysters (15%). A moderate rate of prevalence has been reported from sea foods (21%) from India (Patyal et al., 2011).

2.2.2 TRANSMISSION IN HUMANS

Arcobacters are mainly transmitted through contaminated food (chicken, pork meat, and vegetables) and water sources which may be contaminated through sewage. Several reports regarding the presence of *Arcobacter* sp. in water have been documented and hence consumption of contaminated water acts as an efficient source of infection (Gonzalez et al., 2010; Collado et al., 2011). Chicken meat, pork, and sea foods which are consumed uncooked or partially cooked can also lead to infection as these are the major sources of arcobacters (Collado et al., 2009).

2.2.3 FOODBORNE TRANSMISSION

Chicken meat has been reported with highest prevalence for *Arcobacter* sp. followed by pork and beef (Shah et al., 2011). Different prevalence rates have been recorded in chicken meat samples in India ranging from 12% to 33% (Patyal et al., 2011; Mohan et al., 2014; Ramees et al., 2014b). Highest incidence of *A. butzleri* was reported with 83% from chicken meat followed by beef (20%) and pork (15%) in Poland (Zacharow et al., 2015). In Germany, Lehmann et al. (2015) recorded prevalence of *Arcobacter* sp. from fish meat (34%), poultry meat (27%), and minced meat (2%). Similarly, De Smet et al. (2010) reported the presence of *Arcobacter* from pre- and postchilled bovine carcasses indicating the need for hygienic practices to break the transmission cycle.

Shah et al. (2012) from Selangor, Malaysia, recorded 5.8% prevalence of *Arcobacter* spp. from cow milk samples, wherein *A. butzleri* was the predominant species (60%), followed by *A. cryaerophilus* (40%). Person-to-person transmission of *A. butzleri* was reported in an Italian school with symptoms of recurrent abdominal cramps (Vandamme et al., 1992a; Gonzalez and Ferrus, 2011).

2.2.4 OCCURRENCE IN WATER

Apart from poultry and other meat, *Arcobacter* spp. (mostly *A. butzleri*) was isolated from a drinking water reservoir (Jacob et al., 1993), river or surface water, ground water, and sewage. Thus, consumption of contaminated water is a likely source of exposure to Arcobacters. *A. butzleri* can easily attach to water distribution pipe surfaces (stainless steel, copper, and plastic), which makes regrowth in the water distribution system. This can become a significant problem in drinking water and food-processing plants with respect to public health. Contaminated water is considered as an important source of *Arcobacter* infection to human beings (Collado et al., 2010).

2.2.5 DISEASE IN ANIMALS CAUSED BY ARCOBACTERS

Most of the infections caused by *Arcobacter* sp. are asymptomatic in nature and only few exhibit clinical disease. *Arcobacter* sp. in animals causes abortions, diarrhea, and mastitis (Logan et al., 1982; Neill et al., 1985). In cattle, pigs, and horses, *A. butzleri* causes enteritis and diarrhea; in sheep and cattle, *A. skirrowii* causes diarrhea and hemorrhagic colitis (Ho et al., 2006, Vandamme et al., 1992b). *A. cryophillus* is the predominant species isolated from aborted animals, but it has also been isolated from preputial washings of healthy bovine (Gill, 1983) and from vaginal swabs of cows (Kabeya et al., 2003). *A. butzleri* and *A. skirrowii* have been less frequently reported from aborted animals (Collado and Figueras, 2011). However, pathology and pathogenesis of *Arcobacter* sp. infection in animals needs to be elucidated properly by using different animal models involving both in vitro and in vivo studies.

2.2.6 HUMAN INFECTIONS WITH ARCOBACTERS

A. cryaerophilus was the first *Arcobacter* sp. to be identified from humans in the year 1988 (Tee et al., 1988; Collado et al., 2011). *Arcobacter* has been isolated from feces and blood samples of humans and symptoms may range from diarrhea to septicemia (Fisher et al., 2014). Gastrointestinal manifestations are the common signs in humans and may be exhibited as watery diarrhea in case of *A. butzleri* infection, while bloody diarrhea is usually noticed in *Campylobacter jejuni* (Collado et al., 2011). *Arcobacter* spp. has also been attributed to one of the bacterial agents of traveler's diarrhea along with *E. coli*, *Shigella*, *Salmonella*, and *Campylobacter* (McGrego and Wright, 2015).

The enteritis caused by *Arcobacter* is an acute diarrhea lasting for about 3–15 days, sometimes becoming persistent or recurrent for more than two weeks or even as long as two months (Vandenberg et al., 2004). The condition is often accompanied by abdominal pain and nausea, fever, chills, vomiting and weakness (Vandamme et al., 1992a). *A. butzleri* is considered as the fourth most common *Campylobacter*-like organism next to *C. jejuni*, *C. coli* and *C. upsaliensis* isolated from the stool samples of human patients in France and Belgium (Vandenberg et al., 2004).

2.3 ANTIBIOTIC SENSITIVITY AND EMERGING DRUG RESISTANCE

Resistance pattern against antibiotics like cephalothin, vancomycin, and novobiocin has been reported by several workers, whereas in most cases, it was found sensitive against azithromycin, gentamicin, and nalidixic acid (Mohan et al., 2014). Antimicrobial susceptibilities of 71 *Arcobacter* isolates were tested against 14 drugs using the disk diffusion method, all the *Arcobacter* isolates tested were found to be resistant to one or more antimicrobial agents. Resistance to cephalothin and vancomycin (96%) was the most common finding, followed by resistance to methicillin, azithromycin, and ampicillin. All the *Arcobacter* isolates were susceptible to streptomycin, kanamycin, gentamicin, and tetracycline (Rahimi, 2014). Van den Abeele et al. (2014) reported the antimicrobial susceptibility of *A. butzleri* and *A. cryaerophilus* strains isolated from Belgian patients. Majority of the *Arcobacter* strains were susceptible to gentamicin (99%) and tetracycline (89%). However, erythromycin (78%), ciprofloxacin (72%), and doxycycline (76%) showed moderate activity against *Arcobacter* spp. Only 9% of the strains were susceptible to ampicillin. Majority of the *A. butzleri* strains were susceptible to ciprofloxacin (87%), whereas half of the *A. cryaerophilus* isolates (51%) showed high-level resistance. These results suggest that *A. cryaerophilus* showed an acquired resistance to fluoroquinolones. Macrolides are not drug of choice for *Arcobacter* infections. Tetracyclines can be suggested for treatment of *Arcobacter* sp. induced gastrointestinal infections. *Arcobacter* isolates were resistant to vancomycin, rifampicin, ceftriaxone, trimethoprim, and cephalothin, whereas isolates were highly susceptible to oxytetracycline, tetracycline, ciprofloxacin, erythromycin, kanamycin, amikacin, enrofloxacin, and gentamicin. Tetracycline and aminoglycosides can be used for treatment of human *Arcobacter* infections.

2.3.1 DIAGNOSIS

Identification of *Arcobacter* sp. by different biochemical tests is difficult as these organisms are metabolically inert (Collado and Figueras, 2011); thus, cultural isolation of these organisms remains gold standard. Current advances in the area of molecular diagnostics have provided various molecular tools for rapid and specific detection and differentiation of arcobacters at genus as well as species level with higher sensitivity and specificity.

2.3.2   MEDIA FOR ISOLATION OF ARCOBACTERS

Arcobacters were first recovered from bovine and porcine fetal tissues cultured in semisolid Ellinghausen–McCullough–Johnson–Harris (EMJH) medium containing 5-fluorouracil, which was originally designed for the isolation of *Leptospira* sp. (Ellis et al., 1977). De Boer et al. (1996) developed an *Arcobacter*-selective enrichment broth (ASB) and an *Arcobacter*-selective semisolid medium (ASM) for the recovery of *Arcobacter* from food. Atabay and Corry (1998) described the *Arcobacter*-enrichment broth which incorporated with CAT selective supplement for the recovery of *A. butzleri*, *A. cryaerophilus*, and *A. skirrowii*.

Johnson and Murano (1999) developed Johnson–Murano (JM) broth and agar plates, and Houf et al. (2001) developed an *Arcobacter*-specific isolation method by using *Arcobacter* medium (Oxoid) with addition of five antimicrobials (cefoperazone, trimethoprim, amphotericin, novobiocin, and 5-fluorouracil). De Boer et al. (1996) developed an ASB and ASM for the recovery of *Arcobacter* sp. from foods of animal origin.

2.3.3   DETECTION OF ARCOBACTER spp. BY MOLECULAR METHODS

Cultural identification is very difficult between *Arcobacter* sp. and *Campylobacter* sp. because of their phenotypic similarity. Hence, molecular methods are more useful (Collado and Figueras, 2011). Polymerase chain reaction (PCR) combined with cultural methods has revealed good results in detection of *Arcobacter* sp. from seawater in Italy (Maugeri et al., 2005). There are many different molecular detection methods like PCR, multiplex PCR and real-time PCR, restriction fragment length polymorphism and fluorescence in situ hybridization are developed for rapid diagnosis of *Arcobacter* sp. (Collado and Figueras, 2011; Patyal et al., 2011; Mohan et al., 2014; Ramees et al., 2017).

2.4   GENOTYPING AND DETERMINING GENETIC DIVERSITY OF ARCOBACTER spp.

In epidemiological surveillance, genotyping plays a crucial role to know the risk factors and routes of transmission. Typing methods includes, the use of different PCR-based methods like enterobacterial repetitive intergenic consensus-polymerase chain reaction (ERIC-PCR) which is most commonly used to trace the different outbreaks and to characterize different isolates and to study genetic diversity among them (Collado et al., 2010). ERIC-PCR and random amplification of polymorphic DNA-polymerase chain reaction (RAPD-PCR) are both valuable techniques for characterizing *A. butzleri*, *A. cryaerophilus*, and *A. skirrowii* isolates. Both methods have satisfactory typeability and discriminatory power. The fingerprints generated with ERIC-PCR were more reproducible and complex than the fingerprints generated with RAPD-PCR (Houf et al., 2002).

2.5 PREVENTION AND CONTROL

A definitive link between *Arcobacter* spp. and human disease has not been established yet, but public health concerns have been raised. This is due to the fact that there are several opportunities that exist for human exposure to *Arcobacter*. A major focus is currently on raw meat products. Furthermore, unpasteurized dairy products may also pose a risk of *Arcobacter* transmission to the human population. Therefore, while the role of *Arcobacter* in human disease awaits further evaluation, a precautionary approach is advisable. Measures aimed at reduction or eradication of *Arcobacter* from the human food chain should be encouraged. The fecal carriage of foodborne pathogens among livestock animals at slaughter is strongly correlated with the hazard of carcasses contamination at the slaughter line. To reduce the risk represented by zoonotic agents to the consumer health, it is essential to reduce contamination of carcasses during the slaughtering processes. Therefore, the maintenance of slaughter hygiene is consequently of central importance in meat production. It can be measured in daily practice by "in-process-controls" and regular microbiological monitoring of carcasses as a verification system according to the hazard analysis critical control point principles. Moreover, preventive measures, such as implementation of codes of good-manufacturing practices, increased care during hiding and evisceration should be encouraged. Since *Arcobacter* spp. are nonspore formers, standard cooking regimes (heat treatment to 70°C) will kill the organism. In the era of emerging and rising antibiotic resistance against several commonly used antibiotics, there is a demanding need for exploring alternative and novel therapeutic options for countering infection with arcobacters in a better way, as like being explored to tackle other foodborne pathogens.

KEYWORDS

- **foodborne diseases**
- **intoxications**
- **food safety**
- **consumer health and protection**
- **pathogens**

REFERENCES

Atabay, H. I.; Corry, J. E. L. Evaluation of a New *Arcobacter* Enrichment Medium and Comparison with New Media Developed for Enrichment for *Campylobacter* spp. *Int. J. Food Microbiol.* **1998,** *41,* 53–58.

Aydin, F.; Gumussoy, K. S.; Atabay, H. I.; Ica, T.; Abay, S. Prevalence and Distribution of *Arcobacter* Species in Various Sources in Turkey and Molecular Analysis of Isolated Strains by ERIC-PCR. *J. Appl. Microbiol.* **2007,** *103,* 27–35.

Collado, L, Figueras, M. J. Taxonomy, Epidemiology and Clinical Relevance of the Genus *Arcobacter. Clin. Microbiol. Rev.* **2011,** *24,* 174–192.

Collado, L.; Guarro, J.; Figueras, M. Prevalence of *Arcobacter* in Meat and Shellfish. *J. Food Protect.* **2009,** *72,* 1102–1110.

Collado, L.; Kasimir, G.; Perez, U.; Bosch, A.; Pinto, R.; Saucedo, G.; Huguet, J. M.; Figueras, M. J. Occurrence and Diversity of *Arcobacter* spp. along the Llobregat River Catchment, at Sewage Effluents and in a Drinking Water Treatment Plant. *Water Res.* **2010,** *44,* 3696–3702.

Collado, L.; Jara, R.; Vasquez, N.; Telsaint, C. Antimicrobial Resistance and Virulence Genes of *Arcobacter* Isolates Recovered from Edible Bivalve Mollusks. *Food Control.* **2014,** *46,* 508–512.

De Boer, E.; Tilburg, J. J. H. C.; Woodward, D. L.; Loir, H.; Johnson, W. M. A Selective Medium for the Isolation of *Arcobacter* from Meats. *Lett. Appl. Microbiol.* **1996,** *23,* 64–66.

De Smet, S.; De Zutter, L.; Van Hende, J.; Houf, K. *Arcobacter* Contamination on Pre- and Post-Chilled Bovine Carcasses and in Minced Beef at Retail. *J. Appl. Microbiol.* **2010,** *108,* 299–305.

Ellis, W. A.; Neill, S. D.; O' Brien, J. J.; Ferguson. H. W.; Hanna, J. Isolation of *Spirillum/Vibrio*-Like Organisms from Bovine Fetuses. *Vet. Record.* **1977,** *100,* 451–452.

Ellis, W. A.; Neill, S. D.; O' Brien, J. J.; Hannna, J. Isolation of Spirullum-like Organisms from Pig Fetuses. *Vet. Record.* **1978,** *102,* 106.

Ferreira, S.; Queiroz, J. A.; Oleastro, M.; Domingues, F. C. Insights in the Pathogenesis and Resistance of *Arcobacter*: A Review. *Crit. Rev. Microbiol.* **2016,** *42,* 364–383.

Fisher, J. C.; Levican, A.; Figueras, M. J.; McLellan, S. L. Population Dynamics and Ecology of *Arcobacter* in Sewage. *Front. Microbiol.* **2014,** *5,* 1–9.

Gill, K. P. Aerotolerant *Campylobacter* Strain Isolated from a Bovine Preputial Sheath Washing. *Vet. Rec.* **1983,** *112,* 459.

Gonzalez, A. M.; Ferrus, M. A. Study of *Arcobacter* spp. Contamination in Fresh Lettuces Detected by Different Cultural and Molecular Methods. *Int. J. Food Microbiol.* **2011,** *145,* 311–314.

Gonzalez, A.; Suski, J.; Ferrus, M. A. Rapid and Accurate Detection of *Arcobacter* Contamination in Commercial Chicken Products and Waste Water Samples by Real-Time Polymerase Chain Reaction. *Foodborne Pathog. Dis.* **2010**, *7*, 327–338.

Ho, H. T.; Lipman, L. J. A.; Van der Graaf-van Bloois, L.; Van Bergen, M.; Gaastra, W. Potential Routes of Acquisition of *Arcobacter* Species by Piglets. *Vet Microbiol.* **2006**, *114*, 122–133.

Houf, K.; Devriese, L. A.; De Zutter, L.; Van Hoof, J.; Vandamme, P. Susceptibility of *Arcobacter butzleri*, *Arcobacter cryaerophilus*, and *Arcobacter skirrowii* to Antimicrobial Agents Used in Selective Media. *J. Clin. Microbiol.* **2001**, *39* (4), 1654–1656.

Houf, K.; De Zutter, L.; Van Hoof, J.; Vandamme, P. Assessment of the Genetic Diversity among Arcobacters Isolated from Poultry Products by Using Two PCR-based Typing Methods. *Appl. Environ. Microbiol.* **2002**, *68*, 2172–2178.

Houf, K.; Sarah, D. S.; Julie, B.; Sylvie, D. Dogs as Carriers of the Emerging Pathogen: *Arcobacter*. *J. Vet. Microbiol.* **2008**, *130*, 208–213.

Houf, K.; On, S. L.; Coenye, T.; Debruyne, L.; De Smet, S.; Vandamme, P. *Arcobacter thereius* sp. nov.; Isolated from Pigs and Ducks. *Int. J. Syst. Evol. Microbiol.* **2009**, *59*, 2599–2604.

Jacob, J.; Loir, H.; Feuerpfeil, I. Isolation of *Arcobacter butzleri* from Drinking Water Reservoir in Eastern Germany. *Int. J. Hyg. Environ. Med.* **1993**, *193*, 557–562.

Johnson, L. G.; Murano, E. A. Comparison of Three Protocols for the Isolation of *Arcobacter* from Poultry. *J. Food Prot.* **1999**, *62*, 610–614.

Kabeya, H.; Maruyama, S.; Morita, Y.; Kubo, M.; Yamamoto, K.; Arai, S.; Izumi, T.; Kobayashi, Y.; Katsube, Y.; Mikami, T. Distribution of *Arcobacter* Species among Livestock in Japan. *Vet. Microbiol.* **2003**, *93*, 153–158.

Kabeya, H.; Maruyama, S.; Morita, Y.; Ohsuga, T.; Ozawa, S.; Kobayashi, Y.; Abe, M.; Katsube, Y.; Mikami, T. Prevalence of *Arcobacter* Species in Retail Meats and Antimicrobial Susceptibility of the Isolates in Japan. *Int. J. Food Microbiol.* **2004**, *90*, 303–308.

Lehmann, D.; Alter, T.; Lehmann, L.; Uherkova, S.; Seidler, T.; Golz, G. Prevalence, Virulence Gene Distribution and Genetic Diversity of *Arcobacter* in Food Samples in Germany. *Berl. Munch. Tierarztl. Wochenschr.* **2015**, *128*, 163–168.

Lehner, A.; Tasara, T.; Stephan, R. Relevant Aspects of *Arcobacter* spp. as Potential Food-Borne Pathogens. *Int. J. Food Microbiol.* **2005**, *102*, 127–135.

Logan, E. F.; Neill, S. D.; Mackie, D. P. Mastitis in Dairy Cows Associated with an Aerotolerant *Campylobacter*. *Vet Res.* **1982**, *110*, 229–230.

Maugeri, T. L.; Irrera, G. P.; Lentini, V.; Carbone, M.; Fera, M. T.; Gugliandolo, C. Detection and Enumeration of *Arcobacter* spp. in the Coastal Environment of the Straits of Messina (Italy). *New Microbiol.* **2005**, *28*, 177–182.

McGrego, A. C.; Wright, S. G. Gastrointestinal Symptoms in Travellers. *Clin Med.* **2015**, *15*, 93–95.

Mohan, H. V.; Rathore, R. S.; Dhama, K.; Ramees, T. P.; Patyal, A.; Bagalkot, P. S.; Wani, M. Y.; Bhilegaonkar, K. N.; Kumar, A. Prevalence of *Arcobacter* spp. in Humans, Animals and Foods of Animal Origin in India Based on Cultural Isolation, Antibiogram, PCR and Multiplex PCR Detection. *Asian J. Anim. Vet. Adv.* **2014**, *9*, 452–466.

Neill, S. D.; Cambell, J. N.; O'brien, J. J.; Weatherup, S. T. C.; Ellis, W. A. Taxonomic Position of *Campylobacter cryaerophila* sp. Nov. *Int. J. Syst. Bacteriol.* **1985**, *35*, 342–356.

Patyal, A.; Rathore, R. S.; Mohan, H. V.; Dhama, K.; Kumar, A. Prevalence of *Arcobacter* spp. in Humans, Animals and Foods of Animal Origin Including Sea Food from India. *Transbound Emerg. Dis.* **2011**, *58*, 402–410.

Rahimi, E. Prevalence and Antimicrobial Resistance of *Arcobacter* Species Isolated from Poultry Meat in Iran. *Br. Poult. Sci.* **2014**, *55*, 174–180.

Ramees, T. P.; Rathore, R. S.; Bagalkot, P. S.; Mohan, H. V.; Kumar, A.; Dhama, K. Multiplex PCR Detection of *Arcobacter butzleri* and *Arcobacter cryaerophilus* in Skin of Poultry. *J. Pure Appl. Microbiol.* **2014a**, *8*, 1755–1758.

Ramees, T. P.; Rathore, R. S.; Bagalkot, P. S.; Mohan, H. V.; Kumar, A.; Dhama, K. Detection of *Arcobacter butzleri* and *Arcobacter cryaerophilus* in Clinical Samples of Humans and Foods of Animal Origin by Cultural and Multiplex PCR Based Methods. *Asian J. Anim. Vet. Adv.* **2014b**, *9*, 243–252.

Ramees, T. P.; Rathore, R. S.; Bagalkot, P. S.; Ravi Kumar, G. V. P. P. S.; Mohan, H. V.; Anoopraj, R.; Kumar, A.; Dhama, K. Realtime PCR Detection of *Arcobacter butzleri* and *Arcobacter cryaerophilus* in Chicken Meat Samples. *J. Pure Appl. Microbiol.* **2014c**, *8*, 3165–3169.

Ramees, P. T.; Dhama, K.; Karthik, K.; Rathore, R. S.; Kumar, A.; Saminathan, M.; Tiwari, R.; Malik, Y. S.; Singh, R. K. *Arcobacter*: An Emerging Food-Borne Zoonotic Pathogen, Its Public Health Concerns and Advances in Diagnosis and Control—A Comprehensive Review. *Vet. Q.* **2017**, *37* (1), 136–161.

Scarano, C.; Giacometti, F.; Manfreda, G.; Lucchi, A.; Pes, E.; Spanu, C.; De Santis, E. P.; Serraino, A. *Arcobacter butzleri* in Sheep Ricotta Cheese at Retail and Related Sources of Contamination in an Industrial Dairy Plant. *Appl. Environ. Microbiol.* **2014**, *80*, 7036–7041.

Shah, A. H.; Saleha, A. A.; Zunita, Z.; Murugaiyah, M. *Arcobacter*—An Emerging Threat to Animals and Animal Origin Food Products? *Trend Food Sci. Technol.* **2011**, *22*, 225–236.

Shah, A. H.; Saleha, A. A.; Murugaiyah, M.; Zunita, Z.; Memon, A. A. Prevalence and Distribution of *Arcobacter* spp. in Raw Milk and Retail Raw Beef. *J. Food Prot.* **2012**, *75*, 1474–1478.

Son, I.; Englen, M. D.; Berrang, M. E.; Fedorka-Cray, P. J.; Harrison, M. A. Genetic Diversity of *Arcobacter* and *Campylobacter* on Broiler Carcasses during Processing. *J. Food Prot.* **2006**, *69*, 1028–1033.

Tee, W.; Baird, R.; Dyall-Smith, M.; Dwyer, B. *Campylobacter cryaerophila* Isolated from a Human. *J. Clin. Microbiol.* **1988**, *26*, 2469–2473.

Van den Abeele, A. M.; Vogelaers, D.; Van Hende, J.; Houf, K. Prevalence of *Arcobacter* Species among Humans, Belgium, 2008–2013. *Emerg. Infect. Dis.* **2014**, *20*, 1731–1734.

Van Driessche, E.; Houf, K.; Van Hoof, J.; De Zutter, L.; Vandamme, P. Isolation of *Arcobacter* Species from Animal Feces. *FEMS Microbiol.* **2003**, *229*, 243–248.

Vandamme, P.; Pugina, P.; Benzi, G.; Van Etterijck, R.; Vlaes, L.; Kersters, K.; Butzler, J. P.; Lior, H.; Lauwers, S. Outbreak of Recurrent Abdominal Cramps Associated with *Arcobacter butzleri* in an Italian School. *J. Clin. Microbiol.* **1992a**, *30*, 2335–2337.

Vandamme, P.; Vanvanneyt, M.; Pot, B.; Mels, L.; et al. Polyphasic Taxonomic Study of the Emended Genus *Arcobacter* with *Arcobacter butzleri* Comb. Nov. and *Arcobacter skirrowii* sp. Nov., and Aerotolerant Bacterium Isolated from Veterinary Specimens. *Int. J. Syst. Bacteriol.* **1992b**, *42*, 344–356.

Vandenberg, O.; Dediste, A.; Houf, K.; Ibekwem, S.; Souayah, H.; Cadranel, S.; Douat, N. G.; Butzler, Z. J. P.; Vandamme, P. *Arcobacter* Species in Humans. *Emerg. Infect. Dis.* **2004**, *10*, 1863–1867.

Zacharow, I.; Bystron, J.; Waecka-Zacharska, E.; Podkowik, M.; Bania, J. Prevalence and Antimicrobial Resistance of *Arcobacter butzleri* and *Arcobacter cryaerophilus* Isolates from Retail Meat in Lower Silesia Region, Poland. *Pol. J. Vet. Sci.* **2015**, *18*, 63–69.

CHAPTER 3

NANOCELLULOSE-BASED PAPER FROM BANANA PEDUNCLE USING HIGH-INTENSITY ULTRASONICATION

DEBASIS DAS, MANAS JYOTI DAS, SANGITA MUCHAHARY, and SANKAR CHANDRA DEKA*

Department of Food Engineering and Technology, Tezpur University, Napaam 784028, Sonitpur, Assam, India

*Corresponding author. E-mail: sankar@tezu.ernet.in

ABSTRACT

Banana is the second largest produced fruit after citrus, and usually after harvesting of banana, the peduncle is discarded as a waste product. Results revealed that the peduncle is very rich in cellulose and its fiber contains approximately 60% cellulose. Reports are available on making packaging material based on cellulose, but there is no such report on cellulose from banana peduncles. The objective of the present work was to make a comparison of nanocellulose paper from peduncle fiber with nanocellulose paper from culinary banana peel and cellulose paper. Extraction of cellulose from peduncle fiber was done and tailored into nanocellulose by ultrasonic treatment and confirmed by transmission electron microscope and dynamic light scattering. The characteristic test of cellulose nanopaper (CNP) evinced increased thermal stability and was evidenced by thermogravimetric analysis. The percent crystallinity also increased in CNP and was analyzed for tensile strength, Fourier transforms infrared spectroscopy, and light transmission. The results have the credentials to support that CNP was better than the cellulose paper and comparable with the CNP from banana peel.

3.1   INTRODUCTION

The waste material utilization is one of the emerging research areas in recent times, and with the increase in population, the production of agricultural crops are also increasing. In a recent survey it was found that around 150 million tons of agrowaste is produced in India alone.

Banana is one of the largest cultivated agrocrops in India. The main objective of banana cultivation is for their fruits; the utilization of leaves and other parts of banana plants in India is very rare and after single harvest of the fruit bunch, the banana plant is discarded and is usually used as animal feed or disposed of. The peduncle which is a part of the fruit bunch is usually disposed of, which is a very rich source of fiber. It has been estimated that about 2.5 metric tons (overall in India) of banana peduncle are discarded and allowed to decompose in the plantation. Very few researches have been undertaken for the valorization of the banana peduncle. It is a rich source of cellulose and is a kind of softwood and comprises almost 8% of the weight of banana bunch harvest.

Cellulose is an organic polymer which is found abundant in plants sources and is present mainly on the cell wall of the plants. Cellulose along with hemicellulose and lignin impacts the structural integrity to the branches. It is made up of 3000 or more D-glucose unit linked by β-1–4-glycosidic bond. Cellulose has a unique molecular structure which justifies its properties like hydrophilicity, chirality, biodegradability, and high functionality. The potential utilization of cellulose is being studied for many years in the field of paper production, food packaging, polymer, etc. However, the tensile strength of the paper and its barrier properties in the cellulose paper is very low compared to the other commercially available packaging materials, namely, low-density polyethylene, high-density polyethylene (HDPE), etc.

Modification of cellulose paper is done in various ways to increase its usefulness as packaging material and one of such attempts is tailoring the size of the cellulose fiber (CF) length. Cellulose being a branched polymer is made up of a group of nanofiber. Several studies to individualize the nanofiber are performed. Different researchers used different techniques to tailor nanofibers; physical technique was used to disintegrate through homogenization (Ruiz et al., 2000), including cryo-crushing (Wang et al., 2013), chemical transformation of cellulose to nanocellulose was done by 2,2,6,6-tetramethylpiperidine-1-oxyl radical-mediated oxidation (Saito and Isogai, 2006), and enzymatic process followed, by high shear refining, cryo-crushing, and dispersion in water by a disintegrator (Janardhnan and Sain,

2007). Refining to increase the accessibility of the cell wall to the subsequent monocomponent by endoglucanase treatment, followed by second refining stage and high-pressure homogenizing was conducted (López-Rubio et al., 2007; Pääkkö et al., 2007). In this study, both chemical and ultrasonication treatments were employed so as to form nanocellulose.

Ultrasonication is an emerging technique and has been used by various researchers (Chen et al., 2011; Khawas et al., 2016) in different raw materials. During ultrasonication, the ultrasonic energy induces cavitation which increases its size with respect to time till it reaches its maximum size and burst, and bursting energy causes the cellulose structure to disintegrate. The resulted energy due to formation and collapse of cavitation in sonochemistry process is approximately found to be 10–100 kJ/mol (Tischer et al., 2010). This energy was employed in this study by using an ultrasonic probe and used to disintegrate the longer fiber of cellulose into nanocellulose.

Studies have been conducted on nanofibers of cellulose as a reinforcing material in nanocomposites. The cellulose-based nanocomposites are being considered to have the potential application in the field of packaging and other research. The nanopaper produced (Henriksson et al., 2007; Yoo and Hsieh, 2010) has far more superior properties, namely, high light transparency with good mechanical properties (Nogi et al., 2009), light and strong (Sehaqui et al., 2010), and high degree of crystallinity (Alemdar and Sain, 2008) compared to cellulose paper. There are no works on nanocellulose production from banana peduncle for its utilization in the form of nanocellulose as a packaging material. Since it is an unexploited source of cellulose, the present study aimed for extracting of nanocellulose by combined chemical and ultrasonic treatments for utilizing as a reinforced polymeric paper for packaging purpose.

3.2 MATERIALS AND METHODS

3.2.1 RAW MATERIALS AND CHEMICALS

The raw material shown in Figure 3.1 was collected from Tezpur University campus, Napaam, Assam (India), and the samples were washed in running water followed by washing with double distilled water until no dust or any foreign materials adhered to the surface of the peduncles. All the chemicals used in this study were of high purity analytical grade and were purchased from Sigma Aldrich, USA.

FIGURE 3.1   Banana peduncle (after harvest).

3.2.2   PROXIMATE ANALYSIS OF BANANA PEDUNCLE

Moisture content, total ash, total fat content, protein, and crude fiber of banana peduncle were determined by Association of Official Analytical Chemists method.

3.2.3   PREPARATION OF PEDUNCLE POWDER

The cleaned peduncles were cut into small pieces (approx. 1 cm^2) to increase the surface area for moisture evaporation. The pieces were immediately dipped in 1% potassium metabisulfite for 10 h to avoid the negative effect of oxidation and enzymatic browning and dried in an air circulatory tray dryer at 55°C for 48 h. The dried peduncle was passed through a pulverizer (Fritsch Pulverisette 14, Germany) and then through a lab grain mill to obtain a fine powder of 0.25 mm and stored in air-tight container for further analysis.

3.2.4 ISOLATION OF CELLULOSE FIBERS AND CELLULOSE NANOFIBERS

The culinary banana peduncle flour revealed a fiber content of 47.5% (dry-weight basis) out of which cellulose, hemicellulose, and lignin content were 54.4%, 33.8%, and 11.8%, respectively. The cellulose was extracted by method described by Khawas et al. (2016). The dried powdered sample (40 g) was taken and mixed with 20% (w/v) NaOH solution and 0.1% anthra-quinone in a ratio of 1:20 (powder to solution ratio) and treated in a digester (KelPlus, Pelican Equipment, India) at 170°C for 1.5 h. This first alkali treatment partially dissolved the pectin, hemicellulose, and lignin. The partially digested soluble peduncle powder was washed using double-distilled water to remove the unwanted pectin, lignin, and hemicelluloses. The remaining insoluble residues of the powder were given bleaching treatment in 2% sodium chlorite and adjusted to pH 5 (with 10% (v/v) acetic acid) to remove the lignin for 1 h at 70°C. The insoluble remains were washed to further remove the lignin present and then neutralized. The previous bleaching treatment was repeated again for further removal of excess lignin and phenolic compounds which resulted in the brightening of the peduncle pulp. The insoluble pulp remained after the second bleaching treatment was given a second alkali treatment for removal of residues, namely, hemicelluloses, in 5% KOH (w/v) at 25°C overnight. The cellulose obtained after giving the above treatment was hydrolyzed in 2% H_2SO_4 for 2 h at 80°C to dissolve most of the amorphous region of cellulose and exposed the crystalline region.

The obtained crystalline nanocellulose after acid hydrolyzation was further tailored into size reduction by ultrasonic treatment by a probe-type ultrasonicator (UW 2070, Bandelinsonoplus, Germany) having 1.5 cm (diameter) cylindrical titanium alloy probe tip at frequency of 25 kHz. Chemically purified nanocellulose solution (0.5% approx. conc.) was ultrasonicated at 800 W for 30 min in ice-water bath which resulted in the required nanocellulose size.

3.2.5 PROPERTIES FOR NANOCELLULOSE

3.2.5.1 TRANSMISSION ELECTRON MICROSCOPY

The samples were analyzed using transmission electron microscopy (TEM) to obtain the length and diameter of the nanocellulose formed after chemical and ultrasonic treatments of the cellulose. The aspect ratio, length, and breadth of the nanofibers were calculated using this technique. The sample was dispersed in water and readings were taken at 2000 nm scale and 14,000× using 120,000 V (TECNAI G2 20 S-TWIN, FEI Company, USA).

3.2.5.2   PARTICLE SIZE ANALYSIS AND ZETA POTENTIAL

The particle size distribution and zeta potential were measured in water as a solvent by particle-size analyzer and zeta-potential analyzer, provided with dynamic light scattering (DLS) (MirotacNanotrac Wave MN401, USA).

3.2.6   PREPARATION OF CELLULOSE NANOPAPER

Cellulose nanopaper (CNP) was made using similar process used for culinary banana peel by Khawas et al. (2016). Aqueous solution of 5 mL (conc. 0.5% mass) nanocellulose was filtered in a vacuum filter and it was hot pressed; after that, the wet paper formed was dried for 5 min in a vacuum oven at 93°C.

The prepared CNP was activated by serially dipping in distilled water, acetone, and dimethyl acetamide (DMAc) for 2 h at 25°C. The CNP was dipped again in DMAc/LiCl (8% solution) for 1 h (LiCl and DMA was dried in an oven to avoid negative dissociation of CNP). The resulted nanopaper was of size 5 cm diameter and of 25 μm thickness.

3.2.7   CHARACTERIZATION OF CELLULOSE PAPER AND NANOCELLULOSE PAPER

3.2.7.1   SCANNING ELECTRON MICROSCOPY

The surface morphology of the cellulose and CNP was observed. The papers were deposited onto a copper holder which had a conductive carbon paint and were coated with gold under vacuum before observation (JSM 6390LV, JEOL, Japan).

3.2.7.2   FOURIER TRANSFORM INFRARED SPECTROSCOPY

The infrared spectra of the cellulose paper and CNP were analyzed in a spectrophotometer [Nicolet Instruments 410 Fourier transform infrared spectroscopy (FT-IR)] equipped with KBr and DTGS detector (Thermo Scientific Instrument, USA). The infrared spectra were recorded in between 400 and 4000 cm^{-1}.

3.2.7.3   X-RAY DIFFRACTION PATTERN

The X-ray diffraction pattern of nanopaper was evaluated using instrument from MiniflexRigaku Corporation, Japan. Samples loaded at Cu $K\alpha$ radiation of wavelength 0.1546 nm and a curved graphite crystal monochromator at a scanning rate of 0.2°/min. Crystalline structures was studied between 10 and 80°C with scanning mode at 2θ/θ, scattering slit of 4.2°, and receiving slit of 0.3°. The crystallinity index (I_c) was calculated using the following equation:

$$I_c = \frac{I_{002} - I_{am}}{I_{002}} \times 100 \tag{3.1}$$

where I_{002} is the diffraction intensity closes to 2θ = 22° represents crystalline material and I_{am} is the diffraction intensity close to 2θ = 18° referring amorphous material in CNP.

3.2.7.4   LIGHT TRANSMISSION AND PAPER TRANSPARENCY

Nanopaper light transmission and paper transparency was measured at 650 nm using a UV–vis spectrometer (Eppendorf BioSpectrometer Fluorescence Hamburg, Germany) (Aulin et al., 2016).

3.2.7.5   MECHANICAL TEST

Mechanical properties of the CNP were determined in reference to its tensile strength. Universal testing machine (Zwich Z010, USA) was used to determine the tensile strength by applying the test standard DIN EN ISO 527-1 for paper and boards loaded with 100 N load cell. The CNP was cut into rectangular pieces of 5 mm width and 5 cm length and the thickness of the paper was 25 μm. The tensile stress–strain curves were recorded at the strain rate of 10% per minute and a crosshead speed of 5 mm/min.

3.2.7.6 WATER VAPOR PERMEABILITY

The water vapor permeability tests of the films were done by using a modified method of ASTM standard (E-96). Water containing test cups (1.5 cm below the film) with silica gel were taken and weight gain was used to determine the water vapor transmission rate (WVTR) and was calculated using the following equation:

$$\text{WVTR} = \frac{g \times 24}{t \times a} \tag{3.2}$$

where g is the weight gain in time; t is the time in h; and a is the exposed area.

3.2.7.7 THERMOGRAVIMETRIC ANALYSIS

The thermogravimetric analysis (TGA) of cellulose and CNP was analyzed (TGA-50 SHIMADZU) which determines the thermal stability. Temperature at 25–600°C with constant heating rate of 10°C/min under nitrogen atmosphere was maintained.

3.3 RESULTS AND DISCUSSIONS

3.3.1 PROXIMATE ANALYSIS

The peduncle has 90% moisture content, 0.86% ash content, 1.3% protein content, 0.99% fat content, 2.1% carbohydrate, and 4.75% crude fiber content.

3.3.2 PROPERTIES OF NANOCELLULOSE

3.3.2.1 TRANSMISSION ELECTRON MICROSCOPE

TEM images (Fig. 3.2) of cellulose nanofibers (CNFs) after 30 min of high-intensity ultrasonication at 800 W confirmed the formation of the nanofiber. It was observed that the chemical treatment dissolved much of the amorphous regions and only needle-like CNFs were present. Isolated nanocellulose from banana peduncle confirmed that the combined chemical and ultrasonic treatments gave better result compared to the enzymatic (Tibolla et al., 2014). The TEM image analysis showed the diameter (D), length (L), and aspect ratio; the length of the CNF was found to be around 453 nm, the diameter around 40 nm, and the aspect ratio were calculated to be 11.37. The credible reason for reduction in size of the cellulose was ultrasonication. The ultrasonication treatment effectively facilitated the disintegration and individualization of the CF. The removal of lignin and hemicellulose by chemical treatment also facilitated the individualization on cellulose because lignin has plasticization effect. The combined chemical and ultrasonication is very cost-effective as compared to the other works done on the nanocellulose and this method has more reproducibility.

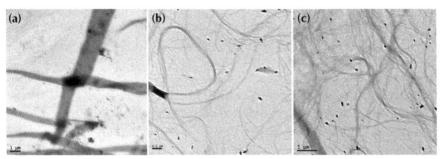

FIGURE 3.2   TEM micrographs: (a) cellulose fiber (scale 1 μm), (b) nanocellulose fiber (scale 0.5 μm), and (c) nanocellulose fiber (scale 1 μm).

3.3.2.2 PARTICLE SIZE AND ZETA POTENTIAL

The particle-size analysis clearly showed (Fig. 3.3) that the average mean particle size after chemical and ultrasonication treatments of cellulose decreased to form the nanocellulose. Therefore, ultrasonication applied to the cellulose converted into nanocellulose; this was very much in line with various authors (Frone et al., 2011).

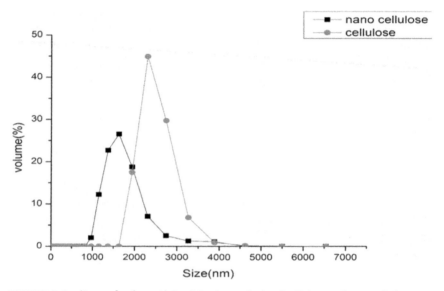

FIGURE 3.3 **(See color insert.)** Particle size analysis of cellulose and nanocellulose.

Zeta potential is the measurement of charge of the surface of a colloidal suspension and describes the electric potential of the suspension in a solid/liquid layer of a material in an aqueous solution. It indicates the degree of repulsion of like charges. The zeta potential that was measured with the help of DLS revealed −5.6 mV for the untreated cellulose and −10.4 mV for the nanocellulose treated with ultrasonication; zeta potential less than −15 mV has stability (Pelissari et al., 2014).

3.3.3 CHARACTERIZATION OF CELLULOSE PAPER AND CELLULOSE NANOPAPER

3.3.3.1 SCANNING ELECTRON MICROSCOPY

The scanning electron micrographs (Fig. 3.4) illustrated the structural morphology of cellulose and nanocellulose paper. The morphology depicted that the bleaching treatment employed to the sample was successful to remove the protein, minerals, lignin, and hemicellulose. To further shorten the chain of cellulose, chemical and ultrasonication treatments were employed and were successful. Ultrasonication caused the longer fiber of the cellulose to shorten due to the cavitation effect of ultrasonication. We did not observe any nanocellulose tubes in untreated cellulose paper (Fig. 3.4a), but after tailoring by chemical and ultrasonic treatments, the nanocellulose was observed in the CNP (Fig. 3.4b). The scanning electron microscopy (SEM) images of CNP showed more densely packed and fragmented nano-cellulose structure; the untreated cellulose was more loosely packed, and no nanobundles can be observed due to larger size of cellulose (Fig. 3.4b).

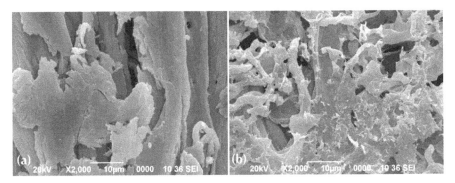

FIGURE 3.4   SEM micrographs: (a) cellulose paper and (b) cellulose nanopaper.

3.3.3.2   FOURIER TRANSFORM INFRARED SPECTROSCOPY

FT-IR (Fig. 3.5) was useful to identify the functional groups present in the sample. In both cellulose and nanocellulose paper, a broad peak was observed at 3448 cm^{-1} which represented the OH stretching and resulted from the vibration of intermolecular hydrogen-bonded hydroxyl group. Around 817 and 863 cm^{-1} (Li et al., 2012), a small spectral band peak was observed for cellulose and nanocellulose, respectively, which represented the COC, CCO, and CCH deformation and stretching; at 622 cm^{-1} (Fan et al., 2012), a peak was noted and represented C–OH out-of-plane bending. A peak was observed at 1065 and 1000 cm^{-1} for cellulose and nanocellulose, respectively, and represented C–C, C–OH, C–H ring and side group vibra-tions (Fan et al., 2012). In the case of nanocellulose at 1378 cm^{-1}, a peak was observed which represented in plane CH bending. In both the samples at 1654 cm^{-1}, a peak was observed which represented OH bending of absorbed water and the peak at 2900 cm^{-1} represented C–H symmetrical stretching (Fan et al., 2012). The infrared profile of the nanocellulose was comparatively same as that of cellulose; thus, it can be confirmed that no chemical change occurred during nanocellulose.

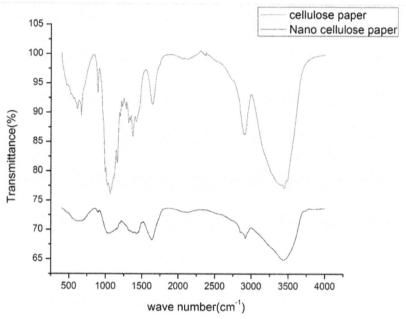

FIGURE 3.5   **(See color insert.)** FT-IR curve of cellulose paper and cellulose nanopaper.

3.3.3.3   X-RAY DIFFRACTION

The diffraction spectra (Fig. 3.6) evaluated the effect of combined chemical and ultrasonic treatments on the crystallinity and the crystal-type structure of the extracted cellulose and the nanocellulose. Both the samples exhibited typical spectra of cellulose I structure with a peak at around $2\theta = 16°$, 22°, and 35° corresponding to the 1 1 0, 0 0 2, and 0 0 4 planes (Fig. 3.6); therefore, the chemical treatment employed for the isolation of the cellulose from banana peduncle was quite successful. The crystalline structure of the cellulose was developed due to the van der Waals force between the closest cellulose molecules. The crystalline index (I_c) calculated for the cellulose and nanocellulose were found to be 40.20% and 64.93%, respectively, and was comparatively less than culinary banana peel (68.45%) studied by Khawas et al. (2016) for the ultrasonic-treated cellulose. The comparative decrease of crystalline index (I_c) in peduncle might be due to longer fiber present in it than peel. The increased percent of crystallinity might be attributed to the chemical and ultrasonication of cellulose that dissolved much of the amorphous cellulose and increased the percentage of crystalline cellulose (Ranganna, 2004).

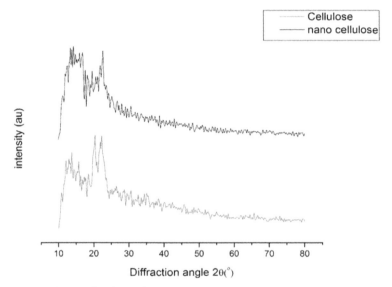

FIGURE 3.6 **(See color insert.)** X-ray diffraction curve of cellulose paper and cellulose nanopaper.

3.3.3.4 LIGHT TRANSMISSION AND PAPER TRANSPARENCY

The paper transparency and light transmittance in CNP was $7.5 \pm 0.114\%$ and 66.83%, respectively. In case of untreated cellulose paper, light transmission and paper transparency was found to be 30.28% and $20.55 \pm 0.43\%$, respectively. Few works reported 62% (Li et al., 2012) light transmittance in 650 nm in CNP.

3.3.3.5 MECHANICAL PROPERTIES

Nanopaper derived from banana peduncle exhibited good tensile properties in addition to the morphology. The stress–strain curve (Fig. 3.7) revealed that the tensile strength exhibited by the CNP was excellent. The tensile strength determines the maximum force required to rupture the CNP. The achieved tensile strength of the CNP was observed as 97.1 MPa and was in range (around 84–223 MPa) with work done by Sehaqui et al. (2010). Compared to the tensile strength of HDPE (37 MPa; 25.4 μm thickness) which is mostly used in food industry, the strength of the developed nanopaper can be considered much better.

The Young modulus of the nanopaper, which is stress by strain within elastic limit, was found to be 10.5 GPa for banana peduncle CNP, whereas other researchers reported different Young modulus for CNP from different sources (Li et al., 2012; Nogi et al., 2009). Banana peel CNP Young modulus was found to be 326.70 MPa. It has been reported that higher tensile strength may be attributed to the plasticization of the residual lignin present, although lignin content was too small, and higher densities of CNP also contributed to the strength of the nanaopaper (Fan et al., 2012).

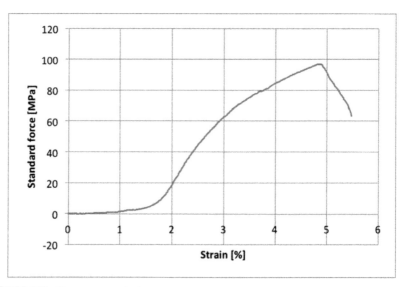

FIGURE 3.7   Stress versus strain curve.

3.3.3.6 WATER VAPOR TRANSMISSION RATE

The WVTR is used to predict the transmission of water vapor through a package and was conducted in 63% RH at 25°C and calculated (Wang et al., 2013). The WVTR for CNP was found to be 151.5 ± 2.0 g/m² day and 1317 ± 2.2 g/m² day for control. These can be compared with the HDPE which has WVTR around 150–200 g/m² day (25.4 μm thickness) and paperboard (740 g/m² day). Few works (Abe et al., 2009) have reported that the WVTR of the nanocellulose paper was 70.1 g/m² day in 50% RH at 23°C, and it increased to more than 2000 g/m² day at 90% RH; therefore, the WVTR of the prepared CNP was comparable with other available nanopaper (Abe et al., 2009).

3.3.3.7 THERMOGRAVIMETRIC ANALYSIS

Initial weight loss was seen (Fig. 3.8) during 24–127°C for CNP; for cellulose, it started at around 23.94°C and gradually decreased. In one of our earlier experiments of banana peel CNP, it was found between 21°C and 27°C (Khawas et al., 2016). The initial weight loss was recorded because of the bound water present in the samples which got evaporated. The thermal depolymerization of hemicellulose attributed to the second steep weight loss for the CNP started at around 230–347°C; for the untreated cellulose paper, it started at around 110–220°C; and for peel CNP, it was around 339–336°C (Khawas et al., 2016). The third weight loss was due to the pyrolysis of cellulose and fiber residues. Therefore, the thermal stability of CNP increased with the removal of lignin, hemicellulose, and amorphous cellulose and is comparable to CNP of culinary banana peel.

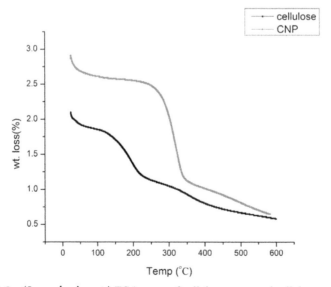

FIGURE 3.8 **(See color insert.)** TGA curve of cellulose paper and cellulose nanopaper.

3.4   CONCLUSION

Banana peduncle cellulose was treated and tailored into nanocellulose which was further molded into CNP and characterized. The paper was found to have a good tensile strength and was comparatively higher than HDPE. The CNP had a very good water vapor barrier property and was around same with HDPE. The TGA showed that the nanocellulose present was highly crystalline, and much of the amorphous part was dissolved in the isolation process and the prepared nanocellulose paper was good in light transparency. The SEM results showed the morphological structure of cellulose and the CNP and nanobundle was observed in CNP. Therefore, banana peduncle is a potential source for preparation of nanopackaging material.

ACKNOWLEDGMENT

The financial help received from DRDO, Ministry of Defence, Govt. of India, is duly acknowledged.

KEYWORDS

- peduncle
- nanocellulose
- CNP
- TEM
- TGA

REFERENCES

Abe, K.; Nakatsubo, F.; Yano, H. High-Strength Nanocomposite Based on Fibrillated Chemi-Thermomechanical Pulp. *Compos. Sci. Technol.* **2009,** *69* (14), 2434–2437.

Alemdar, A.; Sain, M. Biocomposites from Wheat Straw Nanofibers: Morphology, Thermal and Mechanical Properties. *Compos. Sci. Technol.* **2008,** *68* (2), 557–565.

Aulin, C.; Salazar-Alvarez, G.; Lindström, T. High Strength, Flexible and Transparent Nanofibrillated Cellulose–Nanoclaybiohybrid Films with Tunable Oxygen and Water Vapor Permeability. *Nanoscale* **2012,** *4* (20), 6622–6628.

Chen, W.; Yu, H.; Liu, Y.; Chen, P.; Zhang, M.; Hai, Y. Individualization of Cellulose Nanofibers from Wood Using High-Intensity Ultrasonication Combined with Chemical Pretreatments. *Carbohydr. Polym.* **2011,** *83* (4), 1804–1811.

Fan, M.; Dai, D.; Huang, B. Fourier Transform Infrared Spectroscopy for Natural Fibres. In *Fourier Transform—Materials Analysis*; Salih, S. M., Ed.; 2012. DOI: 10.5772/35482. ISBN 978-953-51-0594-7.

Frone, A. N.; Panaitescu, D. M.; Donescu, D.; Spataru, C. I.; Radovici, C.; Trusca, R.; Somoghi, R. Preparation and Characterization of PVA Composites with Cellulose Nanofibers Obtained by Ultrasonication. *BioResources* **2011,** *6* (1), 487–512.

Henriksson, M.; Henriksson, G.; Berglund, L. A.; Lindström, T. An Environmentally Friendly Method for Enzyme-Assisted Preparation of Microfibrillated Cellulose (MFC) Nanofibers. *Eur. Polym. J.* **2007,** *43* (8), 3434–3441.

Janardhnan, S.; Sain, M. M. Isolation of Cellulose Microfibrils—An Enzymatic Approach. *BioResources* **2007,** *1* (2), 176–188.

Khawas, P.; Das, A. J.; Deka, S. C. Production of Renewable Cellulose Nanopaper from Culinary Banana (Musa ABB) Peel and Its Characterization. *Ind. Crops Prod.* **2016,** *86,* 102–112.

Li, J.; Wei, X.; Wang, Q.; Chen, J.; Chang, G.; Kong, L.; Liu, Y. Homogeneous Isolation of Nanocellulose from Sugarcane Bagasse by High Pressure Homogenization. *Carbohydr. Polym.* **2012,** *90* (4), 1609–1613.

López-Rubio, A.; Lagaron, J. M.; Ankerfors, M.; Lindström, T.; Nordqvist, D.; Mattozzi, A.; Hedenqvist, M. S. Enhanced Film Forming and Film Properties of Amylopectin Using Micro-Fibrillated Cellulose. *Carbohydr. Polym.* **2007,** *68* (4), 718–727.

Nogi, M.; Iwamoto, S.; Nakagaito, A. N.; Yano, H. Optically Transparent Nanofiber Paper. *Adv. Mater.* **2009,** *21* (16), 1595–1598.

Pääkkö, M.; Ankerfors, M.; Kosonen, H.; Nykänen, A.; Ahola, S.; Österberg, M.; Lindström, T. Enzymatic Hydrolysis Combined with Mechanical Shearing and High-Pressure Homogenization for Nanoscale Cellulose Fibrils and Strong Gels. *Biomacromolecules* **2007,** *8* (6), 1934–1941.

Pelissari, F. M.; do Amaral Sobral, P. J.; Menegalli, F. C. Isolation and Characterization of Cellulose Nanofibers from Banana Peels. *Cellulose* **2014,** *21* (1), 417–432.

Ranganna, S. *Handbook of Analysis and Quality Control for Fruits and Vegetables*, 2nd ed.; Tata McGraw Hill: New Delhi, India, 2004; pp 452–453.

Ruiz, M. M.; Cavaille, J. Y.; Dufresne, A.; Gerard, J. F.; Graillat, C. Processing and Characterization of New Thermoset Nanocomposites Based on Cellulose Whiskers. *Compos. Interfaces* **2000,** *7* (2), 117–131.

Saito, T.; Isogai, A. Introduction of Aldehyde Groups on Surfaces of Native Cellulose Fibers by TEMPO-Mediated Oxidation. *Colloids Surf. A: Physicochem. Eng. Aspects* **2006,** *289* (1), 219–225.

Sehaqui, H.; Liu, A.; Zhou, Q.; Berglund, L. A. Fast Preparation Procedure for Large, Flat Cellulose and Cellulose/Inorganic Nanopaper Structures. *Biomacromolecules* **2010,** *11* (9), 2195–2198.

Tibolla, H.; Pelissari, F. M.; Menegalli, F. C. Cellulose Nanofibers Produced from Banana Peel by Chemical and Enzymatic Treatment. *LWT—Food Sci. Technol.* **2014,** *59* (2), 1311–1318.

Tischer, P. C. F.; Sierakowski, M. R.; Westfahl, Jr., H.; Tischer, C. A. Nanostructural Reorganization of Bacterial Cellulose by Ultrasonic Treatment. *Biomacromolecules* **2010,** *11* (5), 1217–1224.

Wang, H.; Li, D.; Zhang, R. Preparation of Ultralong Cellulose Nanofibers and Optically Transparent Nanopapers Derived from Waste Corrugated Paper Pulp. *BioResources* **2013,** *8* (1), 1374–1384.

Yoo, S.; Hsieh, J. S. Enzyme-Assisted Preparation of Fibrillated Cellulose Fibers and Its Effect on Physical and Mechanical Properties of Paper Sheet Composites. *Ind. Eng. Chem. Res.* **2010,** *49* (5), 2161–2168.

CHAPTER 4

BANANA FLOWER: A POTENTIAL SOURCE OF FUNCTIONAL INGREDIENTS AND ITS HEALTH BENEFICIAL EFFECTS

YESMIN ARA BEGUM and SANKAR CHANDRA DEKA[*]

Department of Food Engineering and Technology, Tezpur University, Napaam 784028, Sonitpur, Assam, India

[*]*Corresponding author. E-mail: sankar@tezu.ernet.in*

ABSTRACT

Agricultural by-products are gaining more attention for valorization through processing which can be very useful to human health. Flowers of all bananas including culinary banana is a by-product of postharvest cultivation and is a rich source of dietary fiber. This nutritional flower is consumed as cooked vegetables. Besides, consumption as vegetable, it is used to treat against several diseases because of its high medicinal value. Extraction of dietary fiber from banana flower and incorporation into a wide range of products contribute to the development of value-added foods or functional foods that meet the demand of dietary fiber supplement. It is also a good source of antioxidants and number of phytochemicals, which are desirable for human health. The phenolic rich extract of banana flower protects against oxidative stress-related diseases which were evidenced from the in vitro and in vivo studies. The colored bracts that are otherwise considered as agricultural residue are good source of anthocyanin. These bracts can be used as a natural source of food colorant which is currently in high demand. Hence, banana flower could be a promising source for formulating functional food with potential health benefits.

4.1   INTRODUCTION

Banana plant is known as the largest herbaceous flowering plant in the world. It belongs to the family of Musaceae. It is also known as "Apple of the Paradises." It includes banana and plantains (Evans, 2002). Most edible bananas originated from two species, namely, *Musa acuminata* and *Musa balbisiana*. The cultivars are either hybrids among subspecies of *M. acuminata* or between *M. acuminata* (genome type A) and *M. balbisiana* (genome type B). These hybrids are diploid, triploid, or tetraploid. Hybrid diploid are classified as AA, BB, or AB. Hybrid triploids are classified as AAA, AAB, or ABB. Tetraploid may be AAAA, AAAB, AABB, or ABBB (Nelson et al., 2006). There are several hundreds of cultivars of banana spread throughout the world. Two of the common cultivars are *Musa paradisiaca* and *Musa sapientum*. *M. paradisiaca* includes the plantain varieties which has persistent bracts and male flowers with large fruits and eaten after cooking. All other types which had small fruits held compact in a bunch with softening of the pulp when ripen belongs to *M. sapientum* (Imam and Akter, 2011; Sarma et al., 2006). Kachkal is an important group of plantain and belongs to *Musa* ABB variety. It is widely available in Assam and entire North Eastern region. It has got high commercial value and high demand on market due to its culinary purpose. But the flower of this variety of banana is generally disposed as residue after harvesting of banana.

Banana flowers (Fig. 4.1) are large, dark purple-red blossom. This color of bracts is an important taxonomic tool for classification of wild bananas into different species and subspecies (Simmonds, 1962). It grows from the end of a bunch of bananas and also called as banana inflorescence, banana blossom, and banana male bud. The flower cluster is actually an elongated, plump, purple-to-green bud, and sometimes called the "bell" or "heart." As the elongation of bud continues, it exposes semicircular layers of female flowers, then neutral flowers, and finally small, generally nonfunctional (with no viable pollen) male flowers. Each group of flowers is arranged in radial on the stem known as peduncle and it is covered by bracts. About 12–20 flowers are produced for every cluster. Collectively, the flowering parts and fruit are referred to as the bunch. Individual clusters of fruits are known as hands, and individual fruits are known as fingers (Nelson et al., 2006).

FIGURE 4.1 Culinary banana (*Musa* ABB) flower.

4.2 NUTRITIONAL AND MEDICINAL IMPORTANCE OF BANANA FLOWER

4.2.1 NUTRITIONAL IMPORTANCE

Agricultural waste constitutes largely underexploited residues, and in recent years, waste valorization practices have attracted significant amount of attention by the researchers (Luque and Clark, 2013). Banana flower is considered as an agricultural by-product that is often consumed as vegetable in many Asian countries such Malaysia, Indonesia, Sri Lanka, Philippines, and other South-East Asia countries. Banana flower become available upon harvesting the banana because it is usually harvested together. Banana flowers are treated in several cuisines as vegetables. The bracts are used as cattle feed. Banana flower is a nutritional flower and consumed as food additive in Asian countries such as Sri Lanka, Indonesia, and Thailand (Sheng et al., 2010). In Sri Lanka, it is cooked in variety of dishes like curry, boils, or deep-fried (Wickramarachchi and Ranamukhaarachchi, 2005). Moreover, it is used to make various products, namely, dehydrated vegetable, pickle, and canned food (Wickramarachchi and Ranamukhaarachchi, 2005; Yunchalad et al., 1995). Study on the nutritional composition of banana flower (cvs. Baxijiao (AAA) and Paradisiaca (AAB)) indicated that banana flower contain high dietary fiber (DF) and is a good source of minerals, namely, magnesium, iron, and copper. Furthermore, it contains high-quality protein because of its well balance of essential amino acid (Sheng et al., 2010).

4.2.2   MEDICINAL IMPORTANCE

All parts of the banana plant have medicinal applications. Traditionally, banana flower is used to treat dysentery, ulcers, and bronchitis (Kumar et al., 2012). Consumption of banana flower also helps to treat diarrhea, stomach cramps, painful menses, and menopausal bleeding (Leonard, 2006). It is also used for the management of various diseases including diabetes (Pellai and Aashan, 1955). Chinese people use this flower to treat heart pain, asthma, and endocrine problem like diabetes. A report by Mustaffa and Sathiamoorthy (2002) indicated that juice from male bud (banana flower) provides remedy for stomach problem, while Pasupuleti and Anderson (2008) stated that banana flower are used in Indian folk medicine for the treatment of diabetes mellitus. Banana flower can be used as food supplements for diabetics as it was found to ameliorate diabetes and inhibit formation of AGEs (advanced glycation end-products) in streptozotocin-induced diabetic rats (Bhaskar et al., 2011). Banana flower extract treatment also resulted in decrease of free radical formation in the tissues in addition to an antihyperglycemic action (Pari and Umamaheswari, 2000). The anticancer activity of banana flower (*M. paradisiaca*) extract has been evaluated on the cervical cancer cell line (HeLa) and a very good cytotoxic and antiproliferative effects has been reported (Timsina et al., 2014).

4.3   BANANA FLOWER AS A SOURCE OF DIETARY FIBER

4.3.1   DIETARY FIBER

DF is an analogous carbohydrate that is resistant to digestion or absorption in the human small intestine and complete or partial fermentation occur in the large intestine which includes polysaccharides, oligosaccharides, lignin, and associated plant substances (AACC, 2001). Based on solubility in water, DF is of two types: insoluble DF (IDF) which include mainly cellulose, hemi-celluloses, and lignin; soluble DF (SDF) consist of pectins, β-glucans, arabi-noxylans, galactomannans, and other polysaccharides. Both IDF and SDF serve different technological functions. IDF stabilize food systems, improve product density, and minimize shrinkage and also can be used as agents for enhancing food texture and appeal, whereas SDF is used as thickening and gelling agents, foam and emulsion stabilizers, film-forming, and fat-mimetic agents, and also have a potential use in encapsulation (Lecumberri et al., 2007; Gelroth and Ranhotra, 2001). Study on in vitro starch digestibility in a wheat flour gel model substituted with various blends of SDF and IDF demonstrated that SDF is more efficient in reducing in vitro starch digest-ibility and total DF is critical in lowering pGI (predicted glycemic index) (Bae et al., 2016).

4.3.2   DIETARY FIBER AND HEALTH BENEFITS

DF has demonstrated its benefits in health and disease prevention in medical nutrition therapy. Consumption of DF has several benefits such as lowering blood cholesterol levels, normalizing blood glucose and insulin levels, promoting normal laxation, avoid constipation, prevent diverticulosis and diverticulitis, lower the risk of colon cancer and breast cancer, and prevent obesity as well as others diseases (Wickramarachchi and Ranamukhaarachchi, 2005).

The composition of fiber, its organizational structure, physicochemical and surface properties, and associated bioactive (oligosaccharides and bound phenolics) compounds are mainly responsible for the physiological benefits of DF (Macagnan et al., 2016; Chau and Huang, 2004; Elleuch et al., 2011; Tungland and Meyer, 2002). Besides, polyphenols associated with DF is also responsible for beneficial effects of it, especially in the prevention and management of chronic and degenerative diseases (Arun et al., 2017). These phenolic compounds associated with the plant polysaccharides constitute approximately 50% of total dietary antioxidants (Pérez-Jiménez and Saura-Calixto, 2015; Pérez-Jiménez et al., 2015; Saura-Calixto, 2011).

4.3.3   DIETARY FIBER IN BANANA FLOWER

By-product of fruits and vegetables has received high importance in recent years as a newer source of DF and it varies according to the plant species, its genotype, and even within the cultivar of the same species (Rodica et al., 2010). Banana flower is a rich source of DF along with high antioxidant activity. Study showed banana flower (*Musa* sp. var. *elakki bale*) contains 65.6% DF. Further, compositional analysis of DF exhibit higher amounts of cellulosic fraction, water-soluble polysaccharide, and had significant amounts of pectic polysaccharides, hemicellulose A (HA), and hemicellulose B (HB) polysaccharides. Additionally, the study evidenced the presence of bound phenolics with HA and HB fractions of banana flower DF. The HA fraction of banana flower showed presence of phenolic acids such as gallic acid, catechol, protocatechuic acid, gentisic acid, vanillic acid, caffeic acid, syringic acid, *p*-coumaric acid, and ferulic acid, whereas HB fraction showed gallic acid, catechol, protocatechuic acid, gentisic acid, vanillic acid, syringic acid, and epicatechin as major phenolic acids (Bhaskar et al., 2012).

Banana flower (*M. paradisiaca*) was reported as good source of SDF and IDF, namely, 12.45% and 53.31%, respectively. DF from *M. paradisiaca* demonstrated potential glucose and cholesterol absorption capacity. Thus, the study validated that banana flower could be used as neutraceuticals for diabetic patients (Arun et al., 2017).

Culinary banana (*Musa* ABB) contains 3.97% pectin. A study was conducted on synthesis of hydroxyapatite nanoparticles using pectin extracted from culinary banana bract. Hydroxyapatite is the bioactive ceramic material and is the main inorganic component of bone and human hard tissue. It has been widely used in various biomedical fields such as dental implants, alveolar bridge augmentation, orthopedics, maxillofacial surgery, scaffold materials, and drug-delivery agents (Suchanek and Yoshimura, 1998). The results evident that pectin of culinary banana bract is very important biomaterial which could be used as chelating agent for the synthesis of HA nanoparticle. Thus, it was a green approach to use the agricultural waste for biomedical applications (Begum and Deka, 2017b).

4.4 PHENOLICS AND HEALTH BENEFITS

Phenolic phytochemicals are the largest category of phytochemicals with wide distribution in plant kingdom. The three most important groups of dietary phenolics are flavonoids, phenolic acids, and polyphenols (Walton et al., 2003). Oxidative stress is one of the important factors in the pathogenesis of a disease (Bhaskar et al., 2012). Phenolics are potent antioxidant and protect against free-radical-mediated disease processes (Dai et al., 2010). Polyphenols provide a significant protection against development of several chronic diseases such as cardiovascular diseases (CVDs), cancer, diabetes, infections, aging, asthma, etc. (Pandey and Rizvi, 2009).

4.4.1 PHENOLICS IN BANANA FLOWER

In recent years, the use of by-product as a natural source of antioxidant has been gaining increasing attention among the scientific community. Previous studies on banana flower suggest that it is a rich source of phenolics with health beneficial effects. Furthermore, banana flower extract could ameliorate diabetes and associated CVDs which was demonstrated by in vitro and in vivo studies on streptozotocin-induced diabetic rats (Arun et al., 2017; Bhaskar et al., 2011) (Table 4.1).

TABLE 4.1 Phenolics Present in Different Varieties of Banana Flower.

Banana flower	Compound present	Reference
Musa sapientum	Lecocyanidin, cyanidin, malvidin, peonidin, hesperetin, naringenin, hesperetin triacetate, hesperetin dihydrochalcone, naringenin flavanone, naringenin pelargidanon	Ganugapati et al. (2012)
Musa sp var. *elakki bale*	Gallic acid, catechol, protocatechuic acid, gentisic acid, vanillic acid, caffeic acid, syringic acid, epicatechin, *p*-coumaric acid, ferulic acid	Bhaskar et al. (2012)
Musa paradisiaca	Gallic acid, catechol, chlorogenic acid, caffeic acid, syringic acid, *p*-coumaric acid, ferulic acid, ellagic acid, myercetin, cinnamic acid, quercetin, kaempferol, apigenin	Arun et al. (2017)

Banana flower extracts (cvs. Baxijiao (AAA) and Paradisiaca (AAB)) are good source of antioxidant including high total phenol and flavonoid content. The extracts showed high antioxidant efficiency with lower EC_{50} for different assays, namely, 2,2-diphenyl-1-picrylhydrazyl (DPPH) radical scavenging activity, reducing power, 2,2′-azinobis-(3-ethylbenzthiazoline-6-sulphonate (ABTS) radical scavenging activities, and inhibition of lipid peroxidation in egg lecithin through the formation of thiobarbituric acid-reactive substances (Sheng et al., 2011). *Musa paradicicus* belongs to India and it is consumed as vegetable. In vitro antioxidant study on different cultivars (Kathali, Bichi, Shingapuri, Kacha, Champa, and Kalabou) of *Musa paradicicus* flower extract suggests that banana flower can be used as neutraceuticals for prevention of oxidative stress (China et al., 2011). *Musa acuminate* flower extract is a good source of antioxidant and phytochemical screening confirmed presence of phytochemicals such as glycosides, tannins, saponnins, phenols, steroids, and flavonoids. Moreover, it showed good antimicrobial activity against *Staphylococcus aureus*, *Proteus mirabilis*, *Bacillus subtilis*, *Aspergillus niger*, *Candida albicans*, *Micrococcus* sp., and *Salmonella* sp. Besides, nontoxicity of the flower extract was proved on *Artemia salina* by Brine shrimp toxicity test (Sumathy et al., 2011). A study by Joseph et al. (2014) revealed that ethanol extract of *M. paradisiaca* (AAB Nendran variety) showed high antioxidant activity with high content of polyphenol and flavonoids.

Buds and bracts of inflorescence (*M. paradisiaca* cv. Mysore) extracts could be potential source of antioxidant and exhibited wide spectrum of inhibition against foodborne pathogenic bacteria such as *S. aureus*, *Bacillus cereus*, *Listeria monocytogenes*, and *Vibrio parahaemolyticus* (Padam et al., 2012). *M. acuminata* "Nendran" (AAB) is an important cooking banana in Kerala and Tamil Nadu. Fourier transform infrared spectroscopy and UV–vis spectroscopic studies of the bract revealed the presence of alkaloids, flavonoids, terpenoids, coumarins, phenols, tannins, glycosides, steroids, and saponins in different solvents (Gunavathy et al., 2014). *Musa cavendishii* extract showed high content of phenolics and flavonoids which is related to high antioxidant capacity. The highest level of phenolics, flavonoids, and antioxidant activity was obtained by conventional agitation method and optimum condition was 60°C temperature, 50% ethanol concentration, 30 min time, and stirring extraction without the use of ultrasound (Schmidt et al., 2015).

Significant variances were observed in the study of functional components of different parts of cooked and raw banana bud: the outer and inner bract and the male flower. Cooking can modify the functional components of banana bud. Certain functional components such as the antioxidative activity, tannin, phenols, flavonoid, and saponin of the inner banana bract decreased after cooking. But, the anthocyanidin content of the samples increased after cooking. The alkaloid content of the inner banana bract was lost after cooking. Additionally, it was found that the raw outer and inner bract contained high amount of saponin while cooked outer bract, cooked flower, and raw flower contained high amount of flavonoid (Decena et al., 2010). Docking analysis of the banana flower flavonoids indicated that few of the flavonoids may be potential activators of insulin receptor tyrosine kinase and hence banana flower is useful in the treatment of diabetes mellitus (Ganugapati et al., 2012). Banana flower and pseudostem (*Musa* sp. var. *elakki bale*) were found to possess antidiabetic properties and inhibit the formation of AGEs in streptozotocin-induced diabetic rats (Bhaskar et al., 2011). *M. paradisiaca* inflorescence (PI) contain bioactive phytochemicals. Methanolic extract of *M. paradisiaca* exhibited significant ABTS free radical scavenging activity with IC_{50} value 57.2 ± 1.15 g/mL and help in reducing oxidative stress by reducing reactive oxygen species production in L6 myoblasts. Moreover, it demonstrated potential inhibition of carbohydrate-digestive enzymes (α-amylase and α-glucosidase), better antiglycation property, and enhanced glucose uptake in L6 myoblasts (Arun et al., 2017).

4.5 BANANA BRACT AS A SOURCE OF ANTHOCYANIN

4.5.1 ANTHOCYANIN AND HEALTH BENEFITS

Anthocyanins are flavonoid phenolic compounds which are responsible for attractive colors of flowers, fruits, and berries (Longo and Vasapollo, 2004). In recent years, it has been gaining increased attention as a natural colorant in food system. Besides, it possesses some positive therapeutic effect which is again related to its antioxidant properties (Tamura and Yamaganci, 1994; Wang et al., 1999; Wang and Jiao, 2000).

Anthocyanins have several beneficial effects on human health (Weisel, 2006). The most important role of anthocyanins is the antioxidant activity. Anthocyanins possess biological, pharmacological, anti-inflammatory, antioxidant, and chemoprotective properties (De Pascual-Teresa and Sanchez-Ballesta, 2007). Moreover, diet rich in anthocyanin can prevent hyperlipidemia (Xia et al., 2007) and CVDs (Garcia-Alonso et al., 2004). In addition, anthocyanins can lower blood glucose levels by protecting β cells, improving insulin resistance, increasing insulin secretion, improve liver function, and inhibit carbohydrate hydrolyzing enzymes (Gowd et al., 2017). Thus, anthocyanin is a promising candidate to be used as neutraceutical in food industry.

4.5.2 EXTRACTION OF ANTHOCYANIN

4.5.2.1 USE OF ULTRASONIC EXTRACTION

Isolation, identification, and use of anthocyanins are primarily based on the extraction methods used (Lapornik et al., 2005). Conventional extraction methods (solvent extraction or thermal extraction) has some disadvantages such as long extraction hours, low extraction efficiency, degradation of anthocyanin that results in decrease of the antioxidant activity of the extracts (Camel, 2001; Lapornik et al., 2005). Ultrasound extraction (UE) is based on the concept of production of cavitation phenomena. It causes molecular movement of solvent and sample and breakdown of sample micelle or matrix to the intracellular hydrophobic compounds. Thus, UE provides several advantages over the conventional one, that is, improved extraction efficiency, low solvent usage, high level of automation, and reduced extraction time (Wang and Curtis, 2006).

4.5.3 ANTHOCYANIN AND ITS ENCAPSULATION BY SPRAY DRYING

The major problem associated with the storage of anthocyanins is that these compounds are highly unstable. Thus, the stabilization of anthocyanins is an important work. Among the existing stabilization methods of anthocyanins, encapsulation is an interesting mean as it is reliable and cost effective. Encapsulation of anthocyanin improves its stability along with the bioavailability. Spray drying is the one of the oldest encapsulation methods and it is commonly used encapsulation technique in the food industry. This technique provides several advantages such as low operating cost, high quality of capsules in good yield, rapid solubility of the capsules, small size, high stability capsules, and continuous operation (Fang and Bhandari, 2010; Madene, 2006). Maltodextrin is widely used as wall material in spray drying as it satisfies demand, cost effective (Zuidam and Shimoni, 2010), and best thermal defender that preserves the integrity of the anthocyanins during their encapsulation (Ersus and Yurdagel, 2007; Robert et al., 2010).

4.5.4 ANTHOCYANIN IN BANANA BRACT

Different cultivars of banana has different bract color such as red, purple, violet, green, or yellow and the composition of anthocyanin is responsible for these variation. This bract color can be used as a parameter for classification of wild bananas into different species and subspecies (Simmonds, 1962). Anthocyanin identified in different variety of banana bracts is mentioned in Table 4.2.

TABLE 4.2   Anthocyanin Present in Different Variety of Banana Bracts.

Anthocyanin	Banana bracts	References
Delphinidin-3-rutinoside, cyanidin-3-rutinoside, petunidin-3-rutinoside, peonidin-3-rutinoside, and malvidin-3-rutinoside	*M. itinerans*, *M. acuminata* subsp. *siamea* 1, *M. acuminata* subsp. *siamea* 2, *M. acuminata* subsp. *malaccensis* 1, *M. acuminata* subsp. *malaccensis* 2	Kitdamrongsont et al. (2008)
Malvidin-3-rutinoside	*M. acuminata* subsp. *truncate*	
Cyanidin-3-rutinoside and pelargonidin-3-rutinoside	*M. coccinea*	
ND	*M. acuminata* yellow bract and *E. glaucum*	
Delphinidin-3-rutinoside and cyanidin-3-rutinoside	*M. balbisiana*, *M. velutina*, *M. laterita*, and *E. superbum*	
Delphinidin-3-rutinoside, cyanidin-3-rutinoside, petunidin-3-rutinoside, peonidin-3-rutinoside, and malvidin-3-rutinoside	*M. paradisiaca*	Pazmino-Duran et al. (2001)
Cyanidin-3-*o*-glucoside and peonidin-3-*o*-glucoside	*M.* ABB	Begum and Deka (2017a)

ND, Not detected.

FIGURE 4.2   (a) Culinary banana (*Musa* ABB) flower and (b) culinary banana (*Musa* ABB) bracts.

Banana bracts are abundant residue of banana production (Fig. 4.2). During banana harvesting, 300 kg of colored bracts per hectare are disposed as residues (Preethi and Balakrishnamurthy, 2011). *M. paradisiaca* contained 32 mg anthocyanin/ 100 g bracts and the color characteristics (CIE *L*hc*) showed attractive hue of the natural pigments (*L** = 86.8; *h* = 44.2; and *c* = 12.7). Hence, it could be an excellent source of anthocyanins (Pazmino-Duran et al., 2001).

Study by Roobha et al. (2011) suggests *M. acuminate* bract as a potential source of anthocyanin with anthocyanin content of 14 mg/100 g. The study revealed that anthocyanins from *M. acuminata* bract exhibited free radical scavenging activity against DPPH radical, superoxide anions, hydroxyl radical, metal chelating, and hydrogen peroxide radical in a dose-dependent manner. *M. acuminata* bract can be used as neutraceutical which protect against degenerative conditions including CVDs, inflammatory conditions, and neurodegenerative diseases such as Alzheimer's disease, mutations, and cancer. *M. acuminate* bract showed strong antiproliferative activity against MCF-7 cell lines in dose-dependent manner with 12.24% inhibition at 1000 µg/mL concentration. These results support chemopreventive property of bract anthocyanin for human breast cancer.

Among different cultivars of banana bract, namely, Grand Naine, Ney Poovan, Poovan, Karpuravalli, Red Banana, and Virupakshi were used to extract and estimate the anthocyanin content, assess their antimicrobial properties and their suitability as food colorant. The results revealed that Red Banana recorded the highest anthocyanin, phenolic and flavonoid contents (89.73 mg/100 g bracts, 238.93 mg pyrocatachol per 100 mL and 333.37 mg quercetin/100 mL, respectively). The anthocyanin extract from banana bracts exhibited antimicrobial activity against bacteria and fungi. Furthermore, the anthocyanin extract was spray dried and added as food colorant in amla squash which showed the highest stability under refrigerated condition (8°C) for 28 days (Preethi and Balakrishnamurthy, 2011).

Culinary banana flower (*Musa* ABB), widely found in North Eastern region, Assam, also showed as potential source of anthocyanin. Extraction of anthocyanin from this variety was optimized to increase the yield of anthocyanin and optimize condition: 15:0.5 solvent to solute ratio, 53.97 mL/100 mL ethanol concentration, and 49.4°C temperature and anthocyanin content obtained in this condition was 56.98 mg/100 g. Anthocyanin extract was encapsulated using spray drying to improve its stability. The encapsulated anthocyanin showed suitable hygroscopicity, solubility, and good encapsulation efficiency with smooth spherical morphology (2–10 µm).

Encapsulated anthocyanin showed 7 days stability at 30°C and 75% relative humidity (Begum and Deka, 2017a).

4.6 BANANA FLOWER AS A FUNCTIONAL INGREDIENTS IN FOOD SYSTEM

The nutritional and phytochemical study evident that banana flower could positively affect human health and can be used in food industry as an additive. Banana flower has tremendous nutritional value along with health beneficial effects. Banana flower is considered as vegetable and cooked in a variety of dishes in Asian countries like curry, deep-fried, cutlet, and more (Leonard, 2006). But the consumption of this flower is limited due to cumbersome cooking process. Therefore, use of banana flower in making some ready-to-eat product enhances the utilization of banana flower in food product. This value addition introduces banana flower as new source of DF with improve nutritional and health benefits.

Dehydrated banana flower was incorporated in biscuit preparation and the study confirms addition of banana flower powder to biscuits products increases DF intake and decreases of the caloric density of baked goods (Elaveniya et al., 2014). Banana flower is a rich source of DF which accounts for 65.6% (Bhaskar et al., 2011) and more than 50% of functional foods in the market comprises of DF (Giuntini and Menezes, 2011). Extraction of DF and incorporation in food system will meet the demand of DF supplements. The inclusion of potato peel as a source of DF in cake formulation, improves nutritional, technological, and stability of cake (Jeddou et al., 2017). Study by Oh et al. (2014) on in vitro starch digestion of cake prepared with different ratio of IDF and SDF showed that cake prepared with 3 g DF per servings with equal proportion of IDF and SDF reported acceptable quality attributes with lower pGI.

As mentioned in Section 4.5.4, banana bract is an excellent source of anthocyanin. Moreover, its wide availability and nontoxic nature encourage its food applications such as jelly dessert, milk dessert, soft ice cream, hard ice cream, and yogurt (Shi et al., 2003). Banana bract anthocyanin addition in amla squash revealed acceptable organoleptic qualities with 28 days stability under refrigerated condition (Preethi and Balakrishnamurthy, 2011).

Culinary banana (*Musa* ABB) bract was also reported as good source of anthocyanin and its encapsulation by spray drying showed good solubility, encapsulation efficiency, hygroscopicty, and surface morphology (Begum and Deka, 2017a). The encapsulated anthocyanin could be used in food model as natural food colorant to replace the synthetic one.

4.7 CONCLUSION

Banana flower has potential health beneficial effect because of its high DF and phenolic content. Several in vitro and in vivo studies have the credential to evidence the impact of protective effect of banana flower against severe diseases by multiple mechanisms. There is an increasing need for developing methodology to extract DF with improved quality and phenolics from banana flower. Moreover, incorporation of these functional ingredients in food system will help in value addition of the agrowaste and will also reduce the waste disposal problem to certain extent.

KEYWORDS

- **by-products**
- **banana flower**
- **dietary fiber**
- **phytochemical**
- **anthocyanin**
- **functional food**

REFERENCES

AACC. The Definition of Dietary Fiber. *Cer. Foods World* **2001**, *46* (3), 112–126.

Arun, K. B.; Thomas, S.; Reshmith, T. R.; Akhil, G. C.; Nisha, P. Dietary Fibre and Phenolic-Rich Extracts from *Musa paradisiaca* Inflorescence Ameliorates Type 2 Diabetes and Associated Cardiovascular Risks. *J. Funct. Foods* **2017**, *31*, 198–207.

Bae, I. Y.; Jun, Y.; Lee, S.; Lee, H. G. Characterization of Apple Dietary Fibres Influencing the In Vitro Starch Digestibility of Wheat Flour Gel. *LWT—Food Sci. Technol.* **2016**, *65*, 158–163.

Begum, Y. A.; Deka, S. C. Stability of Spray Dried Microencapsulated Anthocyanins Extracted from Culinary Banana Bract. *Int. J. Food Prop.* **2017a**, *20* (12), 3135–3148. DOI:10.1080/10942912.2016.1277739.

Begum, Y. A.; Deka, S. C. Green Synthesis of Pectin Mediated Hydroxyapatite Nanoparticles from Culinary Banana Bract and Its Characterization. *Acta Alimentar. Int. J. Food Sci.* **2017b**, *46* (4), 428–438. DOI:10.1556/066.2017.46.4.5.

Bhaskar J. J.; Shobha, M. S.; Sambaiah, K.; Salimath. P. V. Beneficial Effects of Banana (*Musa* sp. var. *elakki bale*) Flower and Pseudostem on Hyperglycemia and Advanced Glycation End-Products (AGEs) in Streptozotocin-Induced Diabetic Rats. *J. Physiol. Biochem.* **2011,** *67*, 415–425.

Bhaskar, J. J.; Paramahans, S. V.; Nandin, C. D. Banana (*Musa* sp. var. *elakki bale*) Flower and Pseudostem: Dietary Fiber and Associated Antioxidant Capacity. *J. Agric. Food Chem.* **2012,** *60* (1), 427–432.

Camel, V. Recent Extraction Techniques for Solid Matrices Supercritical Fluid Extraction Pressurized Fluid Extraction and Microwave-Assisted Extraction: Their Potential and Pitfalls. *Analyst* **2001,** *126*, 1182–1193.

Chau, C. F.; Huang, Y. L. Characterization of Passion Fruit Seed Fibres—A Potential Fibre Source. *Food Chem.* **2004,** *85* (2), 189–194.

China, R. Dutta, S. Sen, S.; Chakrabarti, R.; Bhowmik, D.; Ghosh, S.; Dhar, P. In Vitro Antioxidant Activity of Different Cultivars of Banana Flower (*Musa paradicicus* L.) Extracts Available in India. *J. Food Sci.* **2011,** *76*, 1292–1299.

Dai, J.; Mumper, R. Plant Phenolics: Extraction, Analysis and their Antioxidant and Anticancer Properties. *Molecules* **2010,** *15*, 7313–7352.

De Pascual-Teresa, S.; Sanchez-Ballesta, M. T. Anthocyanins: From Plant to Health. *Phytochem. Rev.* **2007,** *7*, 281–299.

Decena, J. D. Functional Components of Cooked and Raw Banana (*Musa* sp. cv. *saba*) Bud. Scientific Report Submitted in Partial Fulfillment of the Requirements in HNF. *Food Nutr. Res.* **2010,** *152*.

Elaveniya, E.; Jayamuthunagai, J. Functional, Physicochemical and Anti-Oxidant Properties of Dehydrated Banana Blossom Powder and Its Incorporation in Biscuits. *Int. J. ChemTech Res.* **2014,** *6* (9), 4446–4454.

Elleuch, M.; Bedigian, D.; Roiseux, O.; Besbes, S.; Blecker, C.; Attia, H. Dietary Fibre and Fibre-Rich By-Products of Food Processing: Characterization, Technological Functionality and Commercial Applications: A Review. *Food Chem.* **2011,** *124* (2), 411–421.

Ersus, S.; Yurdagel, U. Microencapsulation of Anthocyanin Pigments of Black Carrot (*Daucus carota* L.) by Spray Drier. *J. Food Eng.* **2007,** *80* (3), 805–812.

Evans, W. C. *Trease and Evans Pharmacognosy*, 16th ed.; Saunders: New York, 2002.

Fang, Z.; Bhandari, B. Encapsulation of Polyphenols—A Review. *Trends Food Sci. Technol.* **2010,** *21* (10), 510–523.

Ganugapati, J.; Baldwa, A.; Lalani, S. Molecular Docking Studies of Banana Flower Flavonoids as Insulin Receptor Tyrosine Kinase Activators as a Cure for Diabetes Mellitus. *Bioinformation* **2012,** *8* (5), 216–220.

Garcia-Alonso, M.; Rimbach, G.; Rivas-Gonzalo, J. C. Antioxidant and Cellular Activities of Anthocyanins and their Corresponding Vitamins A—Studies in Platelets, Monocytes, and Human Endothelial Cells. *J. Agric. Food Chem.* **2004,** *52*, 3378–3384.

Gelroth, J.; Ranhotra, G. R. Food Uses of Fiber. In *Handbook of Dietary Fiber*; Cho, S. S., Dreher, M. L., Eds.; Marcel Dekker: New York, 2001.

Giuntini, E. B.; Menezes, E. W. Fibra alimentar. *Série de Publicações ILSI Brasil—Funções Plenamente Reconhecidas de Nutrientes*; ILSI: São Paulo, 2011; Vol. 18, 23 pp.

Gowd, V.; Jia, Z.; Chen, W. Anthocyanins as Promising Molecules and Dietary Bioactive Components against Diabetes—A Review of Recent Advances. *Trends Food Sci. Technol.* **2017,** *68*, 1–13.

Gunavathy, N.; Padmavathy, S.; Murugavel, S. C. Phytochemical Evaluation of *Musa acuminata* Bract Using Screening, FTIR and UV–Vis Spectroscopic Analysis. *J. Int. Acad. Res. Multidiscipl.* **2014**, *2*, 212–221.

Imam, M. Z.; Akter, S. *Musa paradisiaca* L. and *Musa sapientum* L.: A Phytochemical and Pharmacological Review. *J. Appl. Pharm. Sci.* **2011**, *1* (5), 14–20.

Jeddou, K. B.; Bouaziz, F.; Zouari-Ellouzi, S.; Chaari, F.; Ellouz-Chaabouni, S.; Ellouz-Ghorbel, R.; Nouri-Ellouz, O. Improvement of Texture and Sensory Properties of Cakes by Addition of Potato Peel Powder with High Level of Dietary Fiber and Protein. *Food Chem.* **2017**, *217*, 668–677.

Joseph, J.; Paul, D.; Kavitha, M. P.; Dineshkumar, B.; Menon, J. S.; Bhat, A. R.; Krishnakumar, K. Preliminary Phytochemical Screening and In Vitro Antioxidant Activity of Banana Flower (*Musa paradisiaca* AAB Nendran Variety). *J. Pharm. Res.* **2014**, *8* (2), 144–147.

Kitdamrongsont, K.; Pothavorn, P.; Swangpol, S.; Wongniam, S.; Atawongsa, K.; Svasti, J.; Somana, J. Anthocyanin Composition of Wild Bananas in Thailand. *J. Agric. Food Chem.* **2008**, *56*, 10853–10857.

Kumar, K. P. S.; Bhowmik, D.; Duraivel, S.; Umadevi, M. Traditional and Medicinal Uses of Banana. *J. Pharmacogn. Phytochem.* **2012**, *1* (3), 51–63.

Lapornik, B.; Prosek, M.; Wondra, A. G. Comparison of Extracts Prepared from Plant By-Products Using Different Solvents and Extraction Time. *J. Food Eng.* **2005**, *71*, 214–222.

Lecumberri, E.; Mateos, R.; Izquierdo-Pulido, M.; Ruperez, P.; Goya, L.; Bravo, L. Dietary Fibre Composition, Antioxidant Capacity and Physico-Chemical Properties of a Fiber-Rich Product from Cocoa (*Theobroma cacao* L.). *Food Chem.* **2007**, *104*, 948–954.

Leonard, D. B. *Medicine at Your Feet: Healing Plants of the Hawaiian Kingdom Bidens spp. (Kïnehi).* Roast Duck Produdktion: Kapaa-Princeville, HI, 2006; pp 1–15.

Longo, L.; Vasapollo, G. Extraction and Identification of Anthocyanins from *Smilax aspera* L. Berries. *Food Chem.* **2006**, *94*, 226–231.

Luque, R.; Clark, J. H. Valorization of Food Residues: Waste to Wealth Using Green Chemical Technologies. *Sustain. Chem. Process.* **2013**, *1*, 10.

Macagnan, F. T.; da Silva, L. P.; Hecktheuer, L. Dietary Fibre: The Scientific Search for an Ideal Definition and Methodology of Analysis, and Its Physiological Importance as a Carrier of Bioactive Compounds. *Food Res. Int.* **2016**, *85*, 144–154.

Madene, A. Flavour Encapsulation and Controlled Release—A Review. *Int. J. Food Sci. Technol.* **2006**, *41* (1), 1–21.

Mustaffa, M. M.; Sathyamoorthy, S. Status of Banana Industry in India, Report Publishing in Advancing Banana and Plantain R&D IN Asia and Pacific. In *Proc. 1st BAPNET Steering Committee Meeting* 2002; pp 81–92.

Nelson, S. C.; Ploetz, R. C.; Kepler, A. K. *Musa* Species (Bananas and Plantains). In *Species Profiles for Pacific Island Agroforestry*; Elevitch, C. R., Ed.; Permanent Agriculture Resources (PAR): Hōlualoa, HI, 2006.

Oh, I. K.; Bae, I. Y.; Lee, H. G. In vitro Starch Digestion and Cake Quality: Impact of the Ratio of Soluble and Insoluble Dietary Fibre. *Int. J. Biol. Macromol.* **2014**, *63*, 98–103.

Padam, B. S.; Tin, H. S.; Chye, F. Y.; Abdullah, M. I. Antibacterial and Antioxidative Activities of the Various Solvent Extract of Banana (*Musa paradisiaca* cv. Mysore) Inflorescences. *Int. J. Biol. Sci.* **2012**, *12* (2), 62–73.

Pandey, K. B.; Rizvi, S. I. Plant Polyphenols as Dietary Antioxidants in Human Health and Disease. *Oxidat. Med. Cell. Longev.* **2009**, *2* (5), 270–278.

Pari, L.; Umamaheswari, J. Antihyperglycaemic Activity of *Musa sapientum* Flowers: Effect on Lipid Peroxidation in Alloxan Diabetic Rats. *Phytother. Res.* **2000**, *14*, 136–138.

Pasupuleti, V. K.; Anderson, J. W. *Nutraceuticals, Glycemic Health and Type 2 Diabetes*; Wiley-Blackwell: London, UK, 2008, 512 pp.

Pazmino-Duran, E. A.; Giutsi, M. M.; Wrolstad, R. E.; Gloria, M. B. A. Anthocyanin from Banana Bract (*Musa × paradisiaca*) as Potential Food Colorant. *Food Chem.* **2001**, *73*, 327–332.

Pellai, N.; Aashan, T. N. *Ayurveda Prakashika*; Reddiar and Son: Quilon, India, 1955; pp 97–115.

Pérez-Jiménez, J.; Saura-Calixto, F. Macromolecular Antioxidants or Nonextractable Polyphenols in Fruit and Vegetables: Intake in Four European Countries. *Food Res. Int.* **2015**, *74*, 315–323.

Pérez-Jiménez, J.; ElenaDíaz-Rubio, M.; Saura Calixto, F. Contribution of Macromolecular Antioxidants to Dietary Antioxidant Capacity: A Study in the Spanish Mediterranean Diet. *Plant Foods Hum. Nutr.* **2015**, *70*, 365–370.

Preethi, P.; Balakrishnamurthy, G. Assessment of Banana Cultivars for Pigment Extraction from Bracts, Its Suitability and Stability as Food Colorant. *Int. J. Process. Postharv. Technol.* **2011**, *2* (2), 98–101.

Robert, P.; Tamara, G.; Nalda, R.; Elena, S.; Jorge, C.; Carmen, S. Encapsulation of Polyphenols and Anthocyanins from Pomegranate (*Punica granatum*) by Spray Drying. *Int. J. Food Sci. Technol.* **2010**, *45*, 1386–1394.

Rodica, C.; Adrian, C.; Calin, J. Biochemical Aspects of Non-Starch Polysaccharides. *J. Anim. Sci. Biotechnol.* **2010**, *43*, 368–375.

Roobha, J. J.; Saravanakumar, M.; Aravindhan, K. M.; Suganya, P. D. In Vitro Evaluation of Anticancer Property of Anthocyanin Extract from *Musa acuminate* Bract. *Res. Pharm.* **2011**, *1* (4), 17–21.

Sarma, R.; Prasad, S.; Mohan Bhimkal, N. K. Description and Uses of a Seeded Edible Banana of North Eastern India. *Infomusa* **2006**, *4* (1), 8–9.

Saura-Calixto, F. Dietary Fiber as a Carrier of Dietary Antioxidants: An Essential Physiological Function. *J. Agric. Food Chem.* **2011**, *59* (1), 43–49.

Schmidt, M. M.; Prestes, R. C, Kubota, E. H.; Scapin, G.; Mazutti, M. A. Evaluation of Antioxidant Activity of Extracts of Banana Inflorescences (*Musa cavendishii*). *J. Food* **2015**, *13* (4), 498–505.

Sheng, Z. W.; Ma, W. H.; Jin, Z.; Bi, Y.; Sun, Z.; Dou, H.; Gao, J.; Li, J.; Han, L. Investigation of Dietary Fibre, Protein, Vitamin E and Other Nutritional Compounds of Banana Flowers of Two Cultivars Grown in China. *Afr. J. Biotechnol.* **2010**, *9* (25), 3888–3895.

Sheng, Z. W.; Ma, W. H.; Gao, J. H.; Jin, Z. Antioxidant Properties of Banana Flower of Two Cultivars in China Using 2,2-Diphenyl-1-Picrylhydrazyl (DPPH) Reducing Power, 2,2'-Azinobis-(3-Ethylbenzthiazoline-6-Sulphonate) (ABTS) and Inhibition of Lipid Peroxidation Assays. *Afr. J. Biotechnol.* **2011**, *10* (21), 4470–4477.

Shi, J.; Yu, J.; Pohorly, J.; Young, J. C.; Bryan, M.; Wu, Y. Optimization of the Extraction of Polyphenols from Grape Seed Meal by Aqueous Ethanol Solution. *J. Food Agric. Environ.* **2003**, *1* (2), 42–47.

Simmonds, N. W. Anthocyanins in Wild Bananas. In *The Evolution of Banana*; Longmans: London, 1962; pp. 23–25.

Suchanek, W.; Yoshimura, M. Processing and Properties of Hydroxyapatite Based Biomaterials for Use as Hard Tissue Replacement Implants. *J. Mater. Res.* **1998**, *13*, 94–117.

Sumathy, V.; Lachumy, S. J.; Zakaria, Z.; Sasidharan, S. *In Vitro* Bioactivity and Phytochemical Screening of *Musa acuminata* Flower. *Pharmacologyonline* **2011**, *2*, 118–127.

Tamura, H.; Yamaganci, A. Antioxidative Activity of Monoacylated Anthocyanins Isolated from Muscat Bailey a Grape. *J. Agric. Food Chem.* **1994**, *42*, 1612–1615.

Timsina, B.; Nadumane, V. K. Anti-Cancer Potential of Banana Flower Extract: An In Vitro Study. *Banglad. J. Pharmacol.* **2014**, *9*, 628–635.

Tungland, B. C.; Meyer, D. Polysaccharides (Dietary Fibre): Their Physiology and Role in Human Health and Food. *Comprehens. Rev. Food Sci. Food Saf.—Inst. Food Technol.* **2002**, *1* (3), 73–92.

Walton, N. J.; Mayer, M. J.; Narbad, A. Molecules of Interest: Vanillin. *Phytochemistry* **2003**, *63*, 505–515.

Wang, H.; Nair, M. G.; Strasburg, G. M.; Chang, Y. C.; Booren, A. M.; Gray, J. I.; Dewitt, D. L. Antioxidant and Anti-Inflammatory Activities of Anthocyanins and Their Aglycon, Cyanidin, from Tart Cherries. *J. Nat. Prod.* **1999**, *62*, 294–296.

Wang, L; Curtis, L. W. Recent Advances in Extraction of Nutraceuticals from Plants. *Trends Food Sci. Technol.* **2006**, *17*, 300–312.

Wang, S.; Jiao, H. Scavenging Capacity of Berry Crops on Superoxide Radicals, Hydrogen Peroxide, Hydroxyl Radicals, and Singlet Oxygen. *J. Agric. Food Chem.* **2000**, *75*, 5677–5684.

Weisel, T. Untersuchungen, zur, antioxidativen Wirkung von flavonoid-/polyphenolreichen Mischfrchtsäftenbei Probanden. Ph.D. Thesis; Technische Universität: Kaiserslautern, Germany, 2006.

Wickramarachchi, K. S.; Ranamukhaarachchi, S. L. Preservation of Fibre-Rich Banana Blossom as a Dehydrated Vegetable. *Sci. Asia* **2005**, *31*, 265–271.

Xia, M.; Ling, W.; Zhu, H. Anthocyanin Prevents CD40-Activated Pro-Inflammatory Signaling in Endothelial Cells by Regulating Cholesterol Distribution. *Arterioscl. Thromb. Vasc. Biol.* **2007**, *27*, 519–524.

Yunchalad, M.; Thaveesook, K.; Hiraga, C.; Stonsaovapak, S.; Teangpook, C.; Jatujiranont, N. Processing of Canned Banana Flower and Heart of Pseudostem. *Kasetsart J. Nat. Sci.* **1995**, *29*, 55–63.

Zuidam, N.; Shimoni, E. Overview of Microencapsulates for Use in Food Products or Processes and Methods to Make Them. In *Encapsulation Technologies for Active Food Ingredients and Food Processing*; Zuidam, N. J., Nedovic, V., Eds.; Springer: New York, 2010; pp 3–29.

CHAPTER 5

CHARACTERIZATION OF RICE BEER AND ITS INGREDIENTS OF NORTHEAST INDIA

ARUP JYOTI DAS[1], TATSURO MIYAJI[2], and
SANKAR CHANDRA DEKA[1*]

[1]*Department of Food Engineering and Technology, Tezpur University, Napaam, Sonitpur, Assam 784028, India*

[2]*Department of Materials and Life Science Faculty of Science and Technology Shizuoka Institute of Science and Technology, 2200-2 Toyosawa, Fukuroi, Shizuoka 437-8555, Japan*

*Corresponding author. E-mail: sankar@tezu.ernet.in

ABSTRACT

Some of the rural areas where rice beer is predominantly prepared were visited and the process of preparation was observed and documented. The methodologies followed by the Karbi, Deori, Ahom, Mising, Bodo, and Dimasa communities of Assam were studied. Also, the methodologies practiced by the Angami tribe of Nagaland, Khasi tribe of Meghalaya, and Adi-Galo tribe of Arunachal Pradesh, which are neighboring states, were studied for comparison. The plant species used for starter cake preparation were collected from the places visited and their taxonomical identification was carried out. The starter cakes and rice beer prepared by the various tribes were also collected.

The rice beer samples were studied for the content of organic acids, carbohydrates, and amino acids by high-performance liquid chromatography (HPLC). The aromatic compounds were detected by GC–MS method. The analysis evinced a wide variation in content of the major organic acids. Lactic acid was found in high concentration in all the samples, while the other organic acids were present in variable amounts. Among the carbohydrates, glucose was predominant and some other monosaccharaides were also detected. Most of the essential amino acids were found to be present, and among them, aspartic acid was the most abundant. All the samples contained the volatile or semivolatile aromatic compounds with phenyl ethyl alcohol being the most abundant compound. The overall study revealed that this form of drink has important nutritional values and dietary requirements.

A comparative study of rice beer prepared using the various starter cakes was done based on their physicochemical, biochemical, and microbiological properties. Significant variations in the density, hardness, and color of the starter cakes were found. The moisture was low in all the cakes and carbohydrate was the major component. The pH, titrable acidity, alcohol content, sugars, and starch did not vary much among the prepared rice beer. Polyphenols were present in the final product in various concentrations and the rice beers also evinced considerably high antioxidant activity. Yeasts and lactic acid bacteria were dominant in all the samples and spoilage microbes were absent. The study revealed significant variations among the starter cakes and the final product.

5.1 INTRODUCTION

All of the tribes in Assam prepare their indigenous alcoholic beverages at home using round to flattened solid ball-like mixed dough inocula or starter (Jeyaram et al., 2008; Tamang et al., 2007). The starters are prepared by grinding of softened rice with various parts of different plant species. The paste thus obtained is sometimes mixed with old powdered starters and made into dough out of which round flattened cakes of uniform sizes are made. These are fermented for some days and then dried using various methods to obtain the starter cake (SC) (Jeyaram et al., 2008; Shrestha et al., 2002). The fermentation is usually carried out in earthen pots at room temperature and takes about 5–7 days for completion of the entire process of preparation. The fermented mass is further diluted with water in appropriate ratio and strained with cloth to get the rice beer (RB) in liquid form. The methodology of fermentation carried out by different tribes is almost the same, except that the difference comes from the different types of plant species used in starter culture preparation (Tanti et al., 2010). Various plants have been reported to be used in the preparation of RB starter cultures in Northeast India by various authors. Some are *Albizia myriophylla* by the *Maiteis* in the state of Manipur (Singh and Singh, 2006), *Amomum aromaticum* by the

Jaintia tribe of Meghalaya (Samati and Begum, 2007), *Plumbago zeylanica, Buddleja asiatica, Vernonia cinerea,* and *Gingiber officinale* in the state of Sikkim (Tsuyoshi et al., 2005), *Glycyrrhiza glabra* by the *Dimasas* in Assam (Chakrabarty et al., 2009), *Ananas comosus, Artocarpus heterophyllus, Calotropis gigantea, Capsicum frutescens,* etc. by the Rabha tribe of Assam (Deka and Sarma, 2010), and sprouted rice grains by the *Angamis* in Naga-land (Bernal et al., 1996).

During the fermentation process, a succession of microbes with a delicate balance between different kinds is observed along with changes in biochemical parameters especially in sugar contents (Shrestha et al., 2002). SCs consist of a consortium of different groups of microflora like molds (Tamang et al., 2007), yeast (Jeyaram et al., 2008; Tsuyoshi et al., 2005), and lactic acid bacteria (LAB) (Tamang and Sarkar, 1995). The use of this kind of mixed cultures for fermentation also contributes to the synthesis of various esters and alcohols (Yoo et al., 2010). The amylolytic microbes *Mucor circinelloides, Rhizopus chinensis, Saccharomycopsis fibuligera, Saccharomycopsis capsularis,* and *Pichia burtonii* have been isolated from the starter culture *marcha* used in Sikkim, whereas ethanol production was shown by the isolated strains *Saccharomyces bayanus, Candida glabrata,* and *Pichia anomala* (Tsuyoshi et al., 2005; Tamang and Sarkar, 1995). The LAB *Lactobacillus plantarum, Lactobacillus brevis,* and *Pediococcus pentosaceus* have been isolated from samples of starter cultures used in the states of Sikkim and Manipur (Tamang et al., 2007).

The process of manufacturing RB consists of saccharification of the rice starch by fungal enzymes followed by alcoholic fermentation by yeasts supplied by the starters. This process is unique and the product differs from commercial malt beer or wine. Even though the methodology of production has resemblance with malt beer, however, there is difference in the sacchari-fication process of both. In malt beer, the enzymes for conversion of starch to sugars (α and β amylases) and proteases are produced during the germina-tion process of the barley grains, whereas in case of RB, these enzymes are produced by fungus supplied externally. The whole process of preparation involves saccharification of the starch present in steamed or boiled rice by fungal enzymes followed by alcoholic fermentation by yeasts. These caus-ative organisms are supplied by traditional starters that are usually in the form of dry powder or hard balls or cakes. Various plant materials are used in the preparation of these starters.

In this chapter, the key ingredients used in the preparation of RB starter cultures in Assam, India, and the fermentation technologies followed by the indigenous people were studied. The presence of various components that might contribute to the unique characteristics and nutritional aspects RB prepared in Assam was studied. The comparative evaluation has been done based on the composition of different organic acids, carbohydrates, amino acids, and aromatic compounds. It was also seen that detailed differentiation in between these SCs and the RB produced with these cakes has not been reported earlier. Hence, it was aimed to bring about a clear distinction in between these cakes and the RB produced from them by examining their physical, microbiological, and biochemical parameters.

5.2   MATERIALS AND METHODS

5.2.1   MATERIALS

Samples of RB and SCs were collected from four different states of Northeast India, namely, Assam, Nagaland, Arunachal Pradesh, and Meghalaya. Collection was made from the locations that were predominantly involved in the process of making RB, either for self-consumption or for commercial purposes. All the samples were collected in replicates of three in 500-mL sterile glass sample bottles (Borosil, India), marked according to the place of collection, brought to the laboratory under refrigerated condition on the same day, and stored at 4°C. Both the microbiological and biochemical examination of the samples were started within 24 h of storage. A nonglutinous variety of rice (*Oryza sativa*) named *Mahsuri*, collected from Assam Agricultural University, Jorhat, Assam, was used as a substrate for the preparation of RB in the laboratory. The chemicals and standards were obtained from HiMedia (India) and Sigma-Aldrich Corporation (USA).

5.2.2   FIELD SURVEY

A field survey was carried out in the villages and rural areas of the states of Assam, Nagaland, Arunachal Pradesh, and Meghalaya. The areas were selected based on the information available upon the prevalence of traditional methods of RB preparation. Information was collected from the producers predominantly involved in the process of making RB. The women in all the communities visited were mostly involved and they were inquired about their practices for preparation such as making of SCs along with plants and their parts added, fermentation procedure, duration, and uses of the beverage.

Some of the nearby fields and forests were visited along with local help and the available plant samples were collected and stored in plastic bags and sealed. Later on, these samples were dried and made into herbarium as per the guidelines given by Anderson (1999). Further identification of the collected plant species, the plant samples, and herbariums were done by Department of Agronomy, Assam Agricultural University, Jorhat, Assam, and Department of Botany, Darrang College, Tezpur, Assam.

5.2.3 BIOCHEMICAL ANALYSIS OF THE COLLECTED RICE BEER SAMPLES

5.2.3.1   COLOR MEASUREMENT

The color measurement of the SCs and RB was done by analyzing the samples in a Hunter Lab Color Quest (Ultrascan Vis, HunterLab, USA). The measurement was done without altering the original shape or size of the cakes. The results were expressed in Commission Internationale de l'Eclairage L, a, and b (CIELAB) systems in which L indicates the degree of lightness or darkness ($L = 0$ indicates perfect black and $L = 100$ indicates most perfect white), a indicates degree of redness (+) and greenness (−), and b indicates degree of yellowness (+) and blueness (−).

5.2.3.2   SAMPLE PREPARATION FOR BIOCHEMICAL ANALYSIS

Initially, 20 mL of each of the samples in three replicates were taken and made CO_2 free for carrying out high-performance liquid chromatography (HPLC) and biochemical analyses, except for the study of volatile compounds. This was done by transferring the test samples to a large flask and shaking, first gently and then vigorously in a refrigerated incubator shaker (Excella E24R, NBS, USA), maintaining the temperature at 20–25°C as per the AOAC Official Method 920.49 (AOAC, 2010; Patáková-Jůzlová et al., 1998).

5.2.3.3   pH, ACIDITY MEASUREMENT, AND ALCOHOL CONTENT ESTIMATION

The pH and acidity were determined according to AOAC Official Methods 945.10 and 950.07, respectively (AOAC, 2010). The undiluted test portions were tested in a digital pH meter (pH510, Eutech Instruments) and the indicator titration method was used to obtain the total acidity of the samples. The results were reported as percentage of lactic acid (1 mL of 0.1 M alkali = 0.0090 g lactic acid). The alcohol content (by weight) was measured by the specific gravity method under AOAC Official Method 935.22 (AOAC, 2010). A 100-mL calibrated pycnometer was used to find the specific gravity and then the corresponding percentage of alcohol by volume and weight was calculated.

5.2.3.4   HPLC ANALYSIS OF ORGANIC ACIDS

Extraction: The samples were first mixed with a mixture of acetonitrile and type I water (70:30) and then stirred continuously for 2 h in a shaker. The mixture was then centrifuged for 10 min at 10,000 rpm. The supernatant was then filtered through Whatman No. 4 filter paper and the filtrate was again subjected to solid phase extraction using Sep-Pak® C18 cartridge. This extract was used for analysis of organic acids (Gomis, 2000).

Analytical conditions: The analysis of organic acids was carried out in an HPLC system (Ultimate 3000, Dionex, Germany) equipped with an autosampler. The injection volume was 20 µL and the detector used was Ultimate 3000 Variable Wavelength detector at 210 nm (UV range). The column used was Acclaim OA® (5 µm beads size, 4.0 × 250 mm, Thermo Scientific). The mobile phase was 0.2 M sodium sulfate solution (pH adjusted to 2.68 with methane sulfonic acid). An isocratic run was used with a constant flow rate of 0.6 mL/min at a temperature of 30°C.

5.2.3.5   HPLC ANALYSIS OF CARBOHYDRATES

Extraction: Prior to the analysis of carbohydrates, the samples were once again degassed for 15 min in an ultrasonic bath (RZ 08892-26, Cole Parmer, USA) to remove any residual gases. It was then passed through 0.22 µm pore size organic syringe filter of 30 mm diameter (SF2-1, HiMedia) and then through a Sep-Pak® C18 cartridge which had previously been activated with 10 mL of methanol, followed by 10 mL of type I water (Bernal et al., 1996; Nogueira et al., 2005).

Analytical conditions: The analysis of carbohydrates was carried out in an HPLC system (Ultimate 3000, Dionex, Germany) equipped with an autosampler. The injection volume was 20 µL and Shodex RI-101® Refractive Index detector was used with plus polarity at 512 µRIU recorder and 500 µRIU/V integrator ranges. The column used was a Hamilton HC-75® Ca²⁺ column. The mobile phase used was type I water with an isocratic flow rate of 0.6 mL/min at a temperature of 80°C.

5.2.3.6   HPLC ANALYSIS OF AMINO ACIDS

Acid hydrolysis: The samples were taken in a hydrolysis tube, dried, mixed with 6 M HCl containing 0.1% phenol, sealed, and then hydrolyzed at 110°C for 24 h in vacuum. Following hydrolysis, the residual acid was dried off in a vacuum oven. The samples were then suspended in 100 mM HCl and passed through 0.22-µm size syringe driven filter (Blackburn, 1978; Brand-Williams et al., 1995).

Derivatization procedure: The derivatization of amino acids was done by modification of the method of Bank et al. (1996). Sample (100 µL), 900 µL of borate buffer (1 M, pH 6.2) and 1 mL of fluorenylmethyloxycarbonyl chloride (FMOC-Cl) were taken and vortexed. It was then kept for 2 min and then 4 mL of *n*-pentane was added, followed by vortexing for 45 min. The upper layer was discarded and the lower layer was used for injection after passing through 0.22 µm size syringe driven filter.

Analytical conditions: The analysis of amino acids was carried out in an HPLC system (Ultimate 3000, Dionex, Germany) equipped with an autosampler. The injection volume was 20 µL and the Ultimate 3000 Variable Wavelength detector was used at 265 nm (UV range). The column used was Acclaim 120® (C18, 5 µm beads size; 4.0 × 250 mm, Thermo Scientific) and the column oven temperature was maintained at 30°C. The mobile phase used was (A) acetate buffer:acetonitrile (9:1) and (B) acetate buffer:acetonitrile (1:9) with a flow rate of 1.0 mL/min. The flow gradient for the two solvents is shown in Figure 5.1.

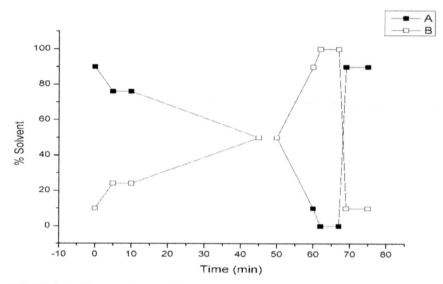

FIGURE 5.1 Flow gradient used in the analysis of amino acids by HPLC. (A) Acetate buffer:acetonitrile (9:1) and (B) acetate buffer:acetonitrile (1:9).

5.2.3.7 GC–MS ANALYSIS OF AROMATIC COMPONENTS

Extraction: The aromatic compounds present in the RB samples were extracted by slight modification of the method of Patáková-Jůzlová et al. (1998). Samples (100 mL) were first filtered through Whatman No. 4 filter paper and then distilled in a rotary evaporator (8763.RV0.000 Roteva, Equitron, India) until 20% of the total volume had been collected as distillate. This distillate was then extracted with 10% dichloromethane in a separating funnel by shaking for 10 min. The DCM extract was filtered through 0.2 μm pore size filter driven syringe and stored at 4°C until analysis.

GC–MS analysis: The GC–MS system (Clarus 600 Gas Chromatograph and Clarus 600C Mass Spectrometer, PerkinElmer, USA) was equipped with thermal conductivity detector (TCD) and mass spectrometer detector (Photomultiplier Tube Detector). The mass range selected was from m/z 50–500. A split-type injector was used. The injection volume was 1 μL. The carrier gas was helium (1 mL/min). The oven temperature was at 50°C for 2 min and then programmed to 250°C for 10 min at a range of 10°C/min and held for 10 min. The column used was Elite 5MS®. The stationary phase of the column was 5% phenyl and 95% methyl-polysiloxane.

5.2.3.8 MINERAL ELEMENTS ANALYSIS BY ATOMIC ABSORPTION SPECTROSCOPY

All glassware and digestion tubes were soaked in 10% nitric acid for 24 h and then rinsed several times with Milli-Q water prior to use. The whole digestion was carried out in an acid digestion system (KES 06L, Pelican Equipments, India) operated at 350°C. Five milliliter of sample was taken in a Kjeldahl flask and boiled down to a small bulk with HNO_3. Then, added few glass beads, 10 mL of H_2SO_4, and 10 mL of HNO_3, and heated gently until the liquid appreciably darkened in color. Then, HNO_3 in small proportions (1 mL) was added and continued heating until darkening again took place and continued addition of acid and heating to fuming for 10 min until the solution failed to darken. The solution was allowed to cool and 10 mL of Milli-Q water was added and boiled gently to fuming. Then, again allowed the solution to cool again and added 5 mL of Milli-Q water and boiled gently to fuming. Finally, it was cooled and made up the volume to 50 mL with Milli-Q water (Ranganna, 2008).

The quantitative analysis was carried out in an AAS system (iCE 3000, Thermo Scientific, USA) operated in a double-beam mode. The analysis was carried out in flame mode with a 100-mm burner and the fuel used was a mixture of acetylene and air. The measurement mode was absorbance and the lamps used were hollow cathode lamps, each with distinctive emission wavelengths. Background correction was accomplished with a deuterium lamp in analyses where emission wavelength was less than 300 nm. All the standards used were AAS grade and purchased from Sigma-Aldrich, USA. The software used for analysis of data SOLAAR.

5.2.4 EFFECT OF THE MICROBIAL STARTER CAKES ON SOME QUALITY ATTRIBUTES OF RICE BEER

5.2.4.1 PHYSICAL ANALYSIS

5.2.4.1.1 Volume and Density Measurement of the Starter Cakes

Certain volume of toluene was measured in a 1000-mL graduated measuring cylinder and the SC (whose mass had already been recorded) was placed in the cylinder and completely submerged. The difference between the measurements in the cylinder before and after placing of the starters gave the volume (cm³) of the SC. The true density was calculated by dividing the mass with the actual volume (g/cm³) (Webb, 2001).

5.2.4.1.2   Texture Analysis of the SC

This analysis was carried out in a texture analyzer (TA-HD Plus 5187, Stable Micro Systems, UK). A p75 probe was used with a 100-kg load cell and a heavy-duty platform (HDP/90). The pretest, test, and posttest speeds were set at 1.00, 0.50, and 5.00 mm/s, respectively. The trigger force used was 20 g.

5.2.4.1.3   Color Measurement

This was done according to Section 5.2.3.1.

5.2.4.2   PRODUCTION OF RICE BEER IN THE LABORATORY

The rice was first boiled in distilled water for 10 min. This was followed by cooling the rice to room temperature. The SC was powdered in a clean mortar and pestle and then mixed with the boiled rice at a ratio of 5 g/kg of rice. This mixture was transferred to sterile glass containers. Fermentation was allowed to take place at 30°C in an incubator for eight days. After the completion of fermentation, the produce was strained using a muslin cloth and the filtrate was further diluted with distilled water in 1:1 ratio. This procedure was adapted from the traditional methodology for preparation of RB followed by the indigenous people of Northeast India. Nine types of RBs were thus produced which were further used for analysis. The local names of the SCs and RBs and the different codes used for them are shown in Table 5.1.

5.2.4.3   BIOCHEMICAL ANALYSIS

5.2.4.3.1   Sample Preparation for Biochemical Analysis

This was done according to Section 5.2.3.2.

5.2.4.3.2   pH, Acidity, and Alcohol Content Estimation

This was done according to Section 5.2.3.3.

5.2.4.3.3   Total Soluble Solids Measurement

This estimation was carried out in a digital Abbe refractometer (DR-A1, Atago, Japan) at room temperature.

5.2.4.3.4   Estimation of Proximate Composition, Reducing Sugars, Starch, and Amylose

All of these were done according to standard AOAC official methods (AOAC, 2010).

5.2.4.3.5   Total Polyphenol Content Estimation

The concentration of total phenolic compounds was determined according to Bray and Thorpe (1954). The sample extracted was treated with Folin–Ciocalteu reagent and the absorbance was read at 650 nm in a UV–vis spectrophotometer (Spectrascan UV-2600, Thermo Scientific).

5.2.4.3.6   Radical Scavenging Activity Estimation

This experiment was performed according to 2,2-diphenyl-1-picrylhydrazyl (DPPH) cation free radical scavenging activity (RSA) method of Brand-Williams et al. (1995).

5.2.4.4   MICROBIAL ANALYSIS

Plate count agar (PCA) was used for general aerobes, potato dextrose agar (PDA) supplemented with tartaric acid and Rose Bengal chloramphenicol agar (RBCA) for yeasts and molds, respectively, media of deMan, Rogosa, and Sharpe (MRS) supplemented with $CaCO_3$ and bromocresol purple indicator for LAB, Salmonella–Shigella agar (SSA) for Salmonella and Shigella species, Baird Parker agar (BPA) for coagulase positive Staphylococcus species, eosin methylene blue (EMB) agar for enterobacteriaceae and modified MYP Agar for *Bacillus cereus*. The PDA and RBCA plates were maintained at 27°C, the PCA, SSA, and BPA plates at 37°C and the MRS and EMB plates in an anaerobic gas pack system at 37°C. The results obtained were expressed as log of colony forming units (CFU) per gram of sample.

5.2.5　STATISTICAL ANALYSIS

This was carried out using the software Origin Pro (Version 8.0). Values were taken as mean of three replicates and SD was calculated. The Fisher's least significant difference was taken at $p < 0.05$ and different superscripted alphabets has been used to represent the difference along a column.

5.3　RESULTS AND DISCUSSIONS

5.3.1　OBSERVATIONS ON THE METHODOLOGIES FOLLOWED BY DIFFERENT TRIBES IN THE PREPARATION OF RICE BEER

5.3.1.1　HOR-ALANK—KARBI TRIBE

The *Karbis* are one of the major tribes of Assam and are settled mostly in the districts of Karbi Anglong and North Cachar Hills. They prepare a traditional alcoholic beverage called *hor-alank.* This beverage is used as a refreshing drink and also bears significance in many social ceremonies and events. For preparation of *hor-alank*, the yeast starter culture called *thap* first needs to be prepared. For preparing *thap* (Fig. 5.2), rice is soaked in water for 1 day. The soaked rice is then mixed with leaves of *marthu* (*Croton joufra*), *janphong* (*A. heterophyllus*), *jockan* (*Phlogocanthus thysiflorus*), *hisou-kehou* (*Solanum viarum*), and barks of *themra* (*Acacia pennata*) plant. The mixture is grinded together in a wooden mortal called *long* with a pestle called *lingpum* to make a paste. This paste is then made into small flat-shaped cakes of about 6 cm in diameter and 0.5 cm in thickness. These are overlaid with powder of previous *thaps* and kept in a bamboo sieve called *ingkrung* and dried for about 3 days under the sun or above the fireplace. These can be stored for about 1 year for further use. For preparing beer, rice is first boiled, then spread, and allowed to cool. It is followed by mixing with powdered *thaps* (5 kg rice + 7 *thaps*). The whole mixture is kept in a large container and covered, first with plastic bags and then with sack. It is left to ferment for a period of 2 days at room temperature. After that, it is mixed with water and further fermented for 2 (summer) to 4 (winter) days.

FIGURE 5.2   A *thap.*

5.3.1.2   SUJEN—DEORI TRIBE

Being one of the oldest settlers of Assam, the Deoris are mostly inhabitant of Lakhimpur, Sibasagar, Dibrugarh, and Tinsukia districts of Assam, India. The indigenous RB of the Deoris is known as *sujen* (Fig. 5.3). The starter material is known as *perok kushi* (Fig. 5.4). The plant materials used for preparing *perok kushi* are leaves of *bhatar duamali* (*Jasminum sambac*), *thok thok* (*Cinnamomum byolghata*), *tesmuri* (*Zanthoxylum hamiltonianum*), *zing zing* (*Lygodium flexuosum*), *zuuro* (*Acanthus leucostychys*), *bhilongoni* (*Cyclosorus exlensa*), *sotiona* (*Alstonia scholaris*), roots of *dubusiring* (*Alpinia malaccensis*), and the stem and rhizome of the plant *jomlakhoti* (*Costus speciosus*). All these are washed and cut into small pieces. They are then grinded in a specialized wooden grinder called as *dheki*. The mixture is then soaked in water in a vessel until the water becomes colored. The whole mixture is added to grinded rice in a vessel to make dough. Round balls of about 4 cm diameter is made out of this and dried either in the sunlight or over the fire hearth by placing in a bamboo mat called as *aaphey*. After getting dried, they are placed in a bamboo container called as *kula* (Fig.

5.5), the inside of which is laid with *kher* (paddy straw). Its mouth is again covered with *kher* and is kept over the hearth for storage. They can be kept in this manner for many months and can be used as and when required. For fermentation of *sujen*, an earthen pot (*disoh*) is first sterilized by washing it with ash and placing it over the hearth for drying and fumigation. Rice is first boiled and then allowed to cool by spreading on banana leaves placed above an *aaphey*. This is followed by addition of powdered *perok kushi* to the cooled rice (1 starter per 3 kg of rice). The mixture is kept in a *disoh*, the mouth of which is sealed with *kol pat* (banana leaves) and left for fermentation to take place for about 4–5 days. It can then be diluted and filtered (Fig. 5.6). It is said that the fermented mass in the *disoh* can be stored for up to 1–2 months at room temperature.

FIGURE 5.3   *Sujen* served on a brass cup.

FIGURE 5.4   A *perok kushi.*

FIGURE 5.5   A *kula.*

FIGURE 5.6   A Deori woman filtering *sujen.*

5.3.1.3   *XAJ PANI—AHOM COMMUNITY*

The Ahoms or Tai-Ahoms are an ethnic group settled in Assam and are of Tai origin. They are a part of the Assamese society and are found all over Assam. The Ahoms prepare RB in their own traditional way and name it as *xaj pani* or *koloh pani.* The SC is known as *vekur pitha* (Fig. 5.7) and consists of various parts of several plant species. The mainly used are leaves of *banjaluk* (*Oldenlandia corymbosa*), *kopou lota* (*Lygodium* sp.), *horumini-muni* (*Hydrocotyle sibthorpioides*), *bormanmunii* (*Centella asiatica*), *tubuki lota* (*Cissampelos pareira*) and seeds of *jaluk* (*Piper nigrum*). All these are washed and dried well and then grinded in an *ural* (wooden mortar) with a pestle and mixed with grinded rice and a little water in a vessel and made into a paste. From this, oval shaped balls of about 4.5 cm × 3 cm are made and placed on *kol pat* [banana (*Musa* sp.) leaves] and dried either in the sun or over the fire place by taking care not to bring them not to close to the fire. After a period of about 5 days, they become hard and are ready to be used. This *vekur pitha* can be stored for up to a year and used when needed. For preparing *xaj pani*, rice (either glutinous or nonglutinous) is half cooked and spread on banana leaves to cool down. It is then mixed with powdered *vekur pitha* (1 per kg of rice) and again spread for some time. The mixture is kept on a *koloh* (earthen pot) and the mouth is sealed. This is kept in a closed room for a period of 3–5 days. After this, some amount of water is added to the fermented mass and left for about 10 min. Filtration is done by straining the mass by using a cloth (Fig. 5.8).

FIGURE 5.7   A *vekur pitha.*

FIGURE 5.8   An Ahom woman filtering *xaj pani.*

5.3.1.4   APONG—MISING TRIBE

Although inhabiting in many districts of Assam, the Misings are concentrated mostly in the districts of Dhemaji, Lakhimpur, and Jorhat. They are said to have migrated to Assam from the state of Arunachal Pradesh. The RB prepared by the Misings is known as *apong* and the SC is called as *apop pitha* (Fig. 5.9). The different leaves needed for preparing *apop pitha* are of the plants *bormanimuni* (*C. asiatica*), *horumanimuni* (*H. sibthorpioides*), *banjaluk* (*O. corymbosa*), *kuhiar* (*Saccharum officinarum*), *dhapat tita* (*Clerodendrum viscosum*), *bhilongoni* (*C. exlensa*), *bam kolmou* (*Ipoemea* sp.), *senikuthi* (*Scoparia dulcis*), *lai jabori* (*Drymeria cordata*), *jalokia* (*Capsicum annuum*), *anaras* (*A. comosus*), and *kopou dhekia* (*L. flexuosum*). All these leaves are cleaned and dried by placing on a bamboo mat called *opoh*. They can be either used freshly or dried in the sun before addition. Soaked rice and the leaves are grinded separately in a *kipar* (wooden grinder) and they are mixed together in a vessel with little water. From the dough, oval-shaped balls of about 6 cm × 3 cm are made and dried in the sun. Before starting the fermentation process, the *killing* (earthen pot) used for fermentation is first fumigated by placing it on a *torap* (a bamboo frame constructed over the fire place) until the pot turns blackish (Fig. 5.10). After that, boiled rice is spread on a *kol pat* (banana leaf) and allowed to cool. To this, powdered *apop pitha* is added (1 *apop pitha* for 1 kg of rice) and the whole mixture is kept inside the *killing* and the mouth of the pot is covered with banana leaves or leaves of *bhilongoni* (Fig. 5.11). This is left for fermentation to take place for a period of about 5 days. A little water is added to the fermented product and is filtered to get the *apong* (Fig. 5.12). The Misings also prepare another kind of RB and it is known by the name *sai mod*. In this method, hay and husk are half burned till they become black in color. This ash is mixed in equal amount with boiled rice and to it the *apop pitha* is added. In this case, the amount of *apop pitha* added in double quantity with respect to *apong* preparation. The mixture is compactly packed in a *killing* and fermented for about 15 days. It is filtered in the same way as *apong*.

FIGURE 5.9   An *apop pitha*.

FIGURE 5.10   A *killing* before fumigation.

FIGURE 5.11   Fermentation in a *killing*.

FIGURE 5.12   A Mising woman filtering *apong*.

5.3.1.5   JOU BISHI—BODO TRIBE

The *Bodos* are one of the largest linguistic groups in Northeast India and among the earliest settlers of Assam. They inhabit most of the regions in Assam but resides mostly in the Bodoland regions. The local RB prepared by the Bodos is known as *jou bishi* and the SCs are known as *ankur* (Fig. 5.13). For preparing *angkur*, different plant materials are said to be used based on their availability in different regions. However, the most common species are leaves of *agarsita* (*Xanthium strumarium*) and *dongphang rakhep* (*S. dulcis*) and either roots or leaves of *lokhunath* (*C. viscosum*). These plants are first washed properly and allowed to dry in the air. Rice grains are soaked for about 5 h in normal temperature water and allowed to soften. This is then mixed with the plants and grinded together in a wooden mortar with a pestle and this set of apparatus is called *wayal*. Dough is made by adding a little water to the mixture. They are then made into round cakes of about 5.5 cm diameter and 0.5–1 cm thickness and covered with powder of the mixture to which water is not added. This is followed by covering with *gigab* (paddy straw) and allowed to dry for a period of 3–4 days. These can be stored in moisture-free places for more than a year. For preparing the beer, either glutinous or nonglutinous rice can be used. When glutinous rice is used the product is known as *maibra jou bishi* and when nonglutinous rice is used, it is known as *matha jou bishi*. The rice is first boiled with care not to allow it to overcook. It is then cooled and allowed to dry. To this, powdered *ankur* is added (about one *ankur* for 1 kg of rice) and mixed well. This mixture is put inside a plastic bag and kept closed for one night. After this, a little water is added to it and left in a *baiphu* (earthen pot) covered with banana leaves for a period of at least 3 days. The fermented mass if further mixed with water and strained to get the liquid *jou bishi*.

FIGURE 5.13 An *ankur*.

5.3.1.6 JUDIMA—DIMASA TRIBE

The Dimasas are one of the earliest indigenous ethnic groups of Northeastern India. They are mostly found in the North Cachar Hills of Assam and Dimapur in Nagaland. The SC for preparing *judima* is called as *umhu* or *humao* (Fig. 5.14) and is a mixture of rice and bark of *thempra* (*Acacia pennata*) plant. The barks are cut into small pieces and dried in the sun. Rice is soaked in water until it is softened. It is then grinded in a wooden or metallic mortal pestle called *rimin* along with the barks of *thempra* plant. A little water is added to make a paste. They are then made into cakes of appropriate sizes and allowed to dry for a period of one week. They can be stored for many months. For preparing *judima*, rice is boiled and allowed to cool. It is mixed with powdered *humao* (one large-sized *humao* is sufficient for 5 kg of rice) and kept in a large container which is covered with jute gunny bags. After about a week, slightly yellowish juices come out of the mass which indicates the completion of fermentation. This can further be diluted with water and filtered for consumption (Fig. 5.15).

FIGURE 5.14 A *humao*.

FIGURE 5.15   A Dimasa woman serving *judima*.

5.3.1.7   *ZUTHO—ANGAMI TRIBE*

Nagaland is chiefly a mountainous state and is inhabited by many different Naga tribes. Each of these tribes has some common culture and traditions and they are all regarded as to having warrior background. The local brew prepared by the Angami tribe is known as *zutho* (Fig. 5.16). This starter material used in the preparation of *zutho* is known as *piazu*, which is basically sprouted rice. For preparing *piazu*, unhulled rice is first soaked in water for a period of about 3–4 days. After this, some of the water is drained out and the grains are allowed to germinate. This may sometimes take about a week depending on the prevailing temperature. After being dried in the air, the sprouted grains are pounded on a wooden mortar with a pestle. The powder obtained is known as *piazu*. For preparing *zutho*, rice is first boiled and then allowed to cool by spreading on a bamboo mat. To this rice, *piazu* (about 10 g for 1 kg of rice) is added and mixed well. The amount of *piazu* added is needed more (almost double) during the months of winter. The mixture is then left to ferment in a closed earthen or wooden vessel for about 4 days in summer and about a week in winter. After completion of fermentation, some amount of water is added to the rice and is filtered by using a bamboo or plastic mesh and usually served in bamboo cups (Fig. 5.17). The Ao tribe of Nagaland also prepared this brew in a similar manner and they call it as *litchumsu*.

FIGURE 5.16 *Zutho* after fermentation.

FIGURE 5.17 *Zutho* served in bamboo mugs.

5.3.1.8   OPO—ADI-GALO TRIBE

Located in the far Northeast India, Arunachal Pradesh is inhabited by many different tribes and each of these bears their own cultural resemblance. This study was done in Pasighat subdivision of East Siang district and the contribution came from the *Adi-Galo* tribe residing in that area. The local RB prepared by this tribe is called as *opo* and the SC is known as *siiyeh* (Fig. 5.18) or *opop*. For preparing *opop*, leaves and barks of the plants *dhapat* (*C. viscosum*) and *Lohpohi* (*Veronia* sp.) are washed, sun dried, and then made into powder. This is then mixed with powdered rice and a little bit of previously prepared *opo* to make a paste. From this flat cakes of about 10–11 cm diameter are made and placed upon bamboo mats. The mats are then kept in the hearth for about 3–4 days, when the cakes become hardened. These can be stored for many months. For preparing *opo*, rice husk called *ampe* is half burnt till they become black in color. After that, rice is boiled and then spread on a bamboo mat called as *peche*. After the rice gets cooled, it is mixed with the burnt husk in 1:1 ratio. To this, powdered *opop* is added (about 100 g of the starter for 10 kg of the mixture) and mixed well. This mixture is then put in a plastic container, the walls of which are covered with leaves of a locally available plant called as *oko* (Zingiberaceae family). The mouth is also sealed with *oko* leaves and is left undisturbed for about 5 days. After this, the contents are mixed well and are again left in the same manner for a longer duration. The product becomes ready after about 20 days of fermentation. It is also kept for longer durations for production of more alcohol. For filtration, a special type of funnel called as *perpur* is used where *oko* leaves are used as the filter. The fermented mass is first placed on the *perpur* and then hot water is poured over it slowly to obtain the *opo* as the filtrate (Fig. 5.19). The quantity of water poured depends on the desired concentration of the final product.

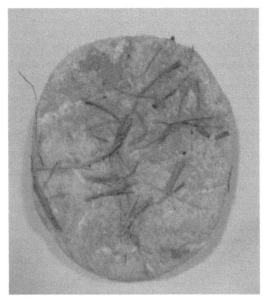

FIGURE 5.18   A *siiyeh*.

FIGURE 5.19   An Adi-Galo woman filtering *opo*.

5.3.1.9 SADHIAR—KHASI TRIBE

The Khasis are an indigenous group of tribal people, the majority of whom live in the state of Meghalaya. They are also found in small populations in Assam, and in parts of Bangladesh. This study was done among the Khasi community of Shillong in Meghalaya. The RB prepared by the Khasis is called as *sadhiar* and the starter is known as *thiat*. For making *thiat*, dried leaves of *khawiang* (*A. aromaticum*) are ground in a wooden mortar pestle. Also *khoso* (local variety of rice) is ground in the mortar pestle into a fine powder. These two are then mixed and a little spring water is added to the mixture to make dough from which the round flattened cakes are made. These are dried on a *malieng* (bamboo basket) which is kept over the hearth until the cakes get hardened. For brewing *sadhiar*, 4–5 kg of *khoso* is mixed with spring water and cooked in a metallic vessel with continuous stirring. The cooked rice is then spread on a *malieng* for cooling and drying. Then to this, 2–3 finely crushed cakes of *thiat* is mixed. The mixture is then put in a cone-shaped basket called *shang*. The whole basket is covered with a cloth and left for 2–3 days. The fermented mash known as *jyndem*, from which the off-whitish RB (*sadhiar*) is filtered out.

5.3.2 BIOCHEMICAL ANALYSIS OF THE RICE BEER SAMPLES

5.3.2.1 GENERAL CHARACTERISTICS OF RICE BEER SAMPLES

Details of RB samples collected across the Northeastern states of India including its origin of preparation are presented in Table 5.1. The alcohol content in all the samples was mild and ranged from 3.99% (sample AsB2) to 5.09% (sample AsB3). Acidic pH was recorded in all of the samples and varied from 4.16 in sample NaB1 to 4.81 in sample MeB1. Sample NaB1 showed the highest value of L (74.97) indicating it to be the whitest, while sample AsB5 showed the lowest value of L (1.11). The a values ranged from 0.43 (AsB5) to 0.88 (AsB4), and hence redness was more prevalent than greenness. The b values varied among the samples and the highest value was recorded in ArB1 (14.01) and lowest in AsB2 (1.77), showing more and less yellowness, respectively.

The biochemical examination of the samples revealed that there are significant differences in quality among them. One of the plausible reasons for this variation might be attributed to differences in the methodology of making each product, and especially the kind of starter cultures used.

TABLE 5.1 Details of the Nine Rice Beer Samples Collected from Various Regions of Northeast India.

Code	Local name	Community	Place of collection	Alcohol content (%)	pH	Color		
						L	a	b
AsB1	*Jou-bishi*	Bodo	Gossaigaon, Assam	4.33 ± 0.07	4.22 ± 0.04	4.28 ± 0.12	0.71 ± 0.13	2.01 ± 0.09
AsB2	*Sujen*	Deori	Lakhimpur, Assam	3.99 ± 0.03	4.56 ± 0.02	3.11 ± 0.21	0.44 ± 0.06	2.63 ± 0.09
AsB3	*Hor-alank*	Karbi	Diphu, Assam	5.09 ± 0.24	4.72 ± 0.01	2.56 ± 0.42	0.63 ± 0.09	1.77 ± 0.15
AsB4	*Xaj-pani*	Ahom	Sibsagar, Assam	4.12 ± 0.01	4.39 ± 0.01	1.98 ± 0.05	0.88 ± 0.04	4.63 ± 0.32
AsB5	*Apong*	Mising	Lakhimpur, Assam	4.77 ± 0.04	4.31 ± 0.08	1.11 ± 0.09	0.43 ± 0.21	2.59 ± 0.54
NaB1	*Judima*	Dimasa	Dimapur, Nagaland	4.13 ± 0.09	4.16 ± 0.04	56.35 ± 0.76	0.54 ± 0.09	7.34 ± 0.14
NaB2	*Jutho*	Angami	Chumukedima, Nagaland	4.68 ± 0.10	4.24 ± 0.05	13.77 ± 0.62	0.77 ± 0.11	2.65 ± 0.12
ArB1	*Opo*	Adi-Galo	Passighat, Arunachal Pradesh	4.59 ± 0.05	4.56 ± 0.03	22.58 ± 0.20	0.54 ± 0.22	14.01 ± 0.28
MeB1	*Sadhiar*	Khasi	Shillong, Meghalaya	4.45 ± 0.11	4.81 ± 0.03	6.20 ± 0.21	0.78 ± 0.06	2.56 ± 0.41

For instance, in the samples NaB1 and AsB3 collected from Dimapur and Diphu, respectively, the main component of the starter was the bark of the plant *Acacia pennata*. However, in case of the sample NaB2 collected from Chumukedima, the starter was only the powder of sprouted paddy, whereas the starters for all the other samples contained roots, leaves, or whole plant of various herbs and shrubs. In addition, the strain of microbes carrying out the fermentation process also differs from product to product (Jeyaram et al., 2008; Tamang et al., 2007; Tsuyoshi et al., 2005; Tamang and Sarkar, 1995).

5.3.2.2   ANALYSIS OF ORGANIC ACIDS

The result of HPLC analysis for the content of different organic acids present in the various samples is shown in Table 5.2. Lactic acid was found to be predominant among all the other acids and its concentration varied significantly. It was recorded the highest (9119.42 mg/L) in AsB4 and lowest in NaB2 (618.76 mg/L). Propionic acid was recorded in less amount in NaB2 (0.19 mg/L) and not found in rest of the samples. Oxalic acid was present in seven of the samples and there was no significant difference among the samples NaB2, AsB1, and MeB1, and the highest concentration was shown by NaB1 (727.04 mg/L). Three samples evinced the presence of citric acid; they were NaB1 (292.70 mg/L), AsB1 (492.34 mg/L), and AsB3 (361.96 mg/L). The highest concentration of tartaric acid was recorded in NaB2 (2618.37 mg/L) and lowest in MeB1 (37.32 mg/L). Out of all the tested samples, five of them showed the presence of tartaric acid. Succinic acid was found in the samples NaB2, AsB2, AsB3, and ArB1 in varying concentrations from 543.94 to 914.68 mg/L. Pyruvic acid was present in six of the samples and NaB1 (215.09 mg/L) had the highest concentration followed by AsB5 (33.84 mg/L) and rest of the samples did not reveal significant difference. Except NaB1 and MeB1, all the samples showed the presence of formic acid and the highest concentration was shown by AsB5 (500.92 mg/L). Acetic acid was found in AsB1 (1172.84 mg/L), AsB5 (531.70 mg/L), and MeB1 (1331.30 mg/L).

A variety of organic acids were present in the samples and most of them are natural products of microorganisms or intermediates in their major metabolic pathways (Sauer et al., 2008). The high concentration of lactic acid found in all the samples of the present study can be attributed to the lactic acid fermentation of sugars undertaken by the group of LAB present in all of the rice beers (Steinkraus, 1983). These organic acids contribute to the unique and distinctive tartness and flavors of beers and are also responsible

TABLE 5.2 Organic Acid Profile of the Rice Beer Samples.

Sample	Organic acid concentration (mg/L)								
	Lactic acid	Propionic acid	Oxalic acid	Citric acid	Tartaric acid	Succinic acid	Pyruvic acid	Formic acid	Acetic acid
NaB1	4430.79 ± 18.68[d]	0.0[a]	727.04 ± 11.99[e]	292.70 ± 7.61[b]	83.81 ± 4.52[c]	0.0[a]	215.09 ± 15.56[c]	0.0[a]	0.0[a]
NaB2	618.76 ± 6.28[a]	0.19 ± 0.06[b]	0.17 ± 0.04[a]	0.0[a]	2618.37 ± 10.07[e]	914.68 ± 10.24[d]	7.65 ± 1.12[a]	63.05 ± 6.24[b]	0.0[a]
AsB1	6431.44 ± 28.44[g]	0.0[a]	4.15 ± 0.15[a]	492.34 ± 7.20[d]	0.0[a]	0.0[a]	1.76 ± 1.15[a]	102.18 ± 8.14[c]	1172.84 ± 10.01[c]
AsB2	4529.34 ± 15.85[e]	0.0[a]	0.0[a]	0.0[a]	80.72 ± 4.31[c]	686.73 ± 17.03[c]	0.0[a]	54.70 ± 5.33[b]	0.0[a]
AsB3	4076.94 ± 14.03[c]	0.0[a]	0.0[a]	361.96 ± 13.39[c]	0.0[a]	543.94 ± 19.02[b]	0.0[a]	208.65 ± 13.63[d]	0.0[a]
AsB4	9119.42 ± 13.13[i]	0.0[a]	255.98 ± 6.95[d]	0.0[a]	0.0[a]	0.0[a]	3.02 ± 1.17[a]	92.56 ± 6.35[c]	0.0[a]
AsB5	6024.76 ± 7.48[f]	0.0[a]	130.08 ± 5.41[b]	0.0[a]	428.58 ± 14.28[d]	0.0[a]	33.84 ± 3.56[b]	500.92 ± 11.83[e]	531.70 ± 30.44[b]
ArB1	1811.65 ± 9.56[b]	0.0[a]	151.54 ± 8.11[c]	0.0[a]	0.0[a]	912.59 ± 17.35[d]	0.0[a]	91.32 ± 9.07[c]	0.0[a]
MeB1	7438.05 ± 5.46[h]	0.0[a]	5.36 ± 1.10[a]	0.0[a]	37.32 ± 2.47[b]	0.0[a]	2.88 ± 0.66[a]	0.0[a]	1331.30 ± 29.57[d]

Note: Values are mean of three replicates ± standard deviation (SD). Means followed by different superscripted alphabet differs significantly ($p < 0.05$).

for the organoleptic properties apart from aiding in preservation process. Analysis of the organic acid content of Italian lager beers using HPLC methodologies showed that the total content of organic acids was in between 451 and 712 mg/L which corroborate our results. The most common organic acid found in the Italian beers was lactic acid (128 mg/L), followed by citric (116 mg/L), acetic (108 mg/L), succinic (68 mg/L), malic (63 mg/L), and pyruvic and formic acid (44 mg/L) (Montanari et al., 1999). The occurrence of acetic, lactic, citric, malic, pyruvic, and succinic acid in 58 samples of lager beer representing 20 different brands from different parts of the world have also been reported by Nord et al. (2004). Lee et al. (2012) found that the static fermentation of brown rice produced high contents of acetic, oxalic, tartaric, and malic acids with increasing concentration of the starter *nuruk*. They also found that acetic, tartaric, and malic acid contents were higher in static culture than agitated culture. The presence of these organic acids helps in increasing the shelf-life of the products by inhibiting the growth of spoilage bacteria like *Escherichia coli* and *Salmonella* spp. (Salmond et al., 1984; Blocher and Busta, 1985). Such inhibition of microbial growth by organic acids is affected by several factors such as inhibition of essential metabolic reactions (Krebs et al., 1983), membrane disruption (Freese et al., 1973), stress on intracellular pH homeostasis (Cole and Keenan, 1987), and the accumulation of toxic anions (Booth and Kroll, 1989).

5.3.2.3   ANALYSIS OF CARBOHYDRATES

The HPLC analysis of carbohydrates profile of the nine RB samples (Table 5.3) revealed that glucose was present in six samples and NaB1 evinced the highest value (3675.88 mg/L) and concentration in all the samples differed significantly. On the other hand, fructose was present in AsB3 and ArB1 and its concentration did not differ significantly. Raffinose, a trisachharide of galactose, fructose, and glucose, was present in seven samples and AsB2 showed the highest concentration of 1413.00 mg/L. Trehalose was detected in two samples, namely AsB4 and AsB5. Another monosaccharide (5C), that is, arabinose was found in six samples and its concentrations ranged from 18.84 to 49.26 mg/L. Xylose which is also one of the important pentose monosaccharides was detected in AsB1 (1881.85 mg/L) and AsB3 (45.01 mg/L). The hexose monosaccharide galactose was present in five samples and differed significantly in its concentration, and the sample NaB1 evinced the highest value (3023.80 mg/L). A naturally occurring deoxy sugar rhamnose was found only in MeB1 at (2828.63 mg/L). The less known classical

TABLE 5.3 Carbohydrate Profile of the Rice Beer Samples.

Sample	Carbohydrate concentration (mg/L)												
	Glucose	Fructose	Sucrose	Raffinose	Maltose	Trehalose	Arabinose	Xylose	Galactose	Rhamnose	Melibiose	Lactose	Inositol
NaB1	3675.88 ± 17.08^g	0.0^a	0.0^a	230.16 ± 22.50^b	0.0^a	0.0^a	18.84 ± 0.51^b	0.0^a	3023.80 ± 9.01^f	0.0^a	0.0^a	0.0^a	0.0^a
NaB2	1744.91 ± 5.90^d	0.0^a	0.0^a	503.81 ± 7.03^f	0.0^a	0.0^a	49.26 ± 1.48f	0.0^a	0.0^a	0.0^a	0.0^a	0.0^a	20.02 ± 1.62^e
AsB1	0.0^a	0.0^a	0.0^a	450.75 ± 12.96^e	0.0^a	0.0^a	20.45 ± 0.78^b	1881.85 ± 9.57^c	0.0^a	0.0^a	0.0^a	0.0^a	23.87 ± 4.82^f
AsB2	319.36 ± 10.49^b	0.0^a	0.0^a	1413.00 ± 14.29^g	0.0^a	0.0^a	31.20 ± 1.72^c	0.0a	0.0^a	0.0^a	0.0^a	0.0^a	15.14 ± 3.45^d
AsB3	1855.23 ± 10.00^c	20.03 ± 0.08^b	0.0^a	395.79 ± 5.69c	0.0^a	0.0^a	0.0^a	45.01 ± 3.53^b	497.11 ± 13.58^c	0.0^a	0.0^a	0.0^a	13.57 ± 1.23^{cd}
AsB4	3647.89 ± 25.17^f	0.0^a	0.0^a	0.0^a	0.0^a	979.79 ± 13.62^c	46.84 ± 1.53^e	0.0^a	1540.12 ± 13.99^d	0.0^a	0.0^a	0.0^a	0.0^a
AsB5	0.0^a	0.0^a	0.0^a	0.0^a	0.0^a	608.98 ± 5.10^b	34.81 ± 0.74^d	0.0^a	2565.76 ± 4.92^e	0.0^a	0.0^a	0.0^a	0.0^a
ArB1	950.55 ± 2.11^c	21.83 ± 3.47^b	0.0^a	433.41 ± 4.62^{de}	0.0^a	0.0^a	0.0^a	0.0^a	253.20 ± 5.23^b	0.0^a	0.0^a	0.0^a	11.09 ± 1.4c
B1	0.0^a	0.0^a	0.0^a	431.01 ± 9.28^d	0.0^a	0.0^a	0.0^a	0.0^a	0.0^a	2828.63 ± 8.66^b	0.0^a	0.0^a	4.79 ± 0.29^b

Note: Values are mean of three replicates ± standard deviation (SD). Means followed by different superscripted alphabet differs significantly (p < 0.05).

sugar inositol (or cyclohexane-1,2,3,4,5,6-hexol) was also present in six samples in low concentration. Although most of the common carbohydrates were present, the disaccharides namely sucrose, maltose, and melibiose were not found in the present study.

Carbohydrates that were found in varying quantities in all the samples might attribute as the major sources of energy and the monosaccharides contribute to the sweetness of the product. The presence of simple sugars like glucose represents the easily metabolized carbohydrate, whereas other complex forms like raffinose can act as dietary fibers that have several health benefit effects such as prevention of heart diseases, diabetes, obesity, and certain gastrointestinal diseases (Anderson et al., 2009). The quantification of carbohydrates in pilsner beer (a type of pale lager beer) by HPLC revealed that fructose, glucose, maltose, maltotriose, and maltotetraose were present in 2.4–2.6, 4.2–406, 0.35–38.5, 1.3–8.1, and 1.1–2.1 g/L, respectively (Nogueira et al., 2005).

5.3.2.4   ANALYSIS OF AMINO ACIDS

The amino acid profile of the samples is presented in Table 5.4. All the samples contained arginine, serine, aspartic acid, glutamic acid, glycine, tyrosine, proline, valine, phenylalanine, isoleucine, leucine, histidine, and lysine and concentration differed significantly. Alanine was present in only NaB2 and AsB3 and methionine in only AsB1, AsB3, and AsB4. Aspartic acid was found in high concentration (1091.81–14,626.35 mg/L), and rest of the amino acids varied with samples. Results revealed that most of the essential amino acids were present in the tested RB samples. In addition, some of the conditionally essentially amino acids (amino acids which are to be supplied exogenously with diet for some specific population) (Reeds, 2000) were also found, for example, arginine, glycine, glutamic acid, histidine, proline, serine, and tyrosine.

The amino acids obtained in relatively high amounts in the present study signify that RB can be a good source of essential nutrients and energy for body metabolism. Nord et al. (2004) also reported the occurrence of all these amino acids in 58 samples of lager beer representing 20 different brands. Kabelová et al. (2008) also measured the content of 16 free amino acids in 35 beers commercially available in Czech Republic using HPLC method. Proline was found to be the most common amino acid with a concentration of 40–250 mg/L. Proline has been described as the chief amino acid present in wines by Ough (1968), who found that the average value of proline content

TABLE 5.4 Amino Acid Profile of the Rice Beer Samples.

Sample	Amino acid concentration (mg/L)														
	Arginine	Serine	Aspartic acid	Glutamic acid	Glycine	Alanine	Tyrosine	Proline	Methionine	Valine	Phenyl alanine	Isoleucine	Leucine	Histidine	Lysine
NaB1	1104.53 ± 25.04[f]	349.24 ± 24.69[a]	6855.74 ± 31.86[c]	1964.80 ± 37.99[c]	1059.51 ± 45.07[c]	0.0[a]	7354.06 ± 49.50[f]	463.01 ± 5.96[f]	0.0[a]	435.58 ± 22.03[c]	349.67 ± 13.51[c]	233.07 ± 3.76[cd]	434.88 ± 10.76[b]	1472.20 ± 78.14[a]	365.05 ± 20.64[b]
NaB2	473.02 ± 23.66[b]	519.62 ± 11.95[b]	3935.96 ± 31.57[b]	1171.45 ± 61.21[b]	866.29 ± 11.41[a]	2093.85 ± 6.18[c]	338.98 ± 5.58[a]	320.64 ± 9.35[d]	0.0[a]	351.07 ± 8.55[d]	381.97 ± 12.69[d]	247.42 ± 11.81[de]	386.57 ± 1.49[c]	3848.21 ± 33.40[d]	491.34 ± 6.13[d]
AsB1	847.51 ± 4.36[d]	1034.78 ± 33.17[e]	12,415.78 ± 96.10[f]	2890.68 ± 39.14[h]	1240.41 ± 37.56[d]	0.0[a]	399.78 ± 18.82[b]	515.58 ± 15.17[g]	108.22 ± 4.48[c]	454.12 ± 31.01[c]	572.33 ± 5.96[g]	431.80 ± 4.85[f]	692.46 ± 14.09[h]	6763.49 ± 11.49[h]	775.15 ± 18.04[f]
AsB2	453.33 ± 23.17[b]	572.75 ± 20.55[b]	11,273.79 ± 66.43[e]	2341.91 ± 42.16[f]	836.63 ± 33.29[a]	0.0[a]	489.17 ± 12.82[c]	375.23 ± 6.09[e]	0.0[a]	508.59 ± 14.47[f]	416.65 ± 6.19[e]	378.90 ± 21.48[e]	533.64 ± 9.75[f]	4729.79 ± 28.42[g]	575.56 ± 12.44[e]
AsB3	1147.56 ± 45.84[g]	760.85 ± 34.23[c]	9380.68 ± 55.96[d]	2558.65 ± 26.90[g]	1054.61 ± 49.72[bc]	1131.59 ± 27.25[b]	1113.79 ± 61.43[e]	450.40 ± 9.70[f]	96.45 ± 8.40[c]	512.25 ± 10.95[f]	472.09 ± 7.24[f]	356.07 ± 10.18[e]	566.17 ± 12.26[g]	2663.85 ± 10.37[c]	454.21 ± 28.44[c]
AsB4	977.10 ± 19.77[e]	1214.89 ± 79.71[f]	14,626.35 ± 355.97[h]	2540.44 ± 55.31[g]	1381.06 ± 24.83[e]	0.0[a]	721.19 ± 19.10[d]	380.81 ± 13.44[e]	95.85 ± 3.29[b]	251.94 ± 5.91[c]	466.69 ± 18.25[f]	264.25 ± 27.32[d]	464.91 ± 11.86[e]	3047.17 ± 32.59[e]	363.94 ± 6.89[b]
AsB5	519.39 ± 20.61[c]	788.55 ± 24.24[cd]	12,434.13 ± 53.70[f]	1341.38 ± 23.44[c]	886.43 ± 11.43[a]	0.0[a]	526.57 ± 15.50[c]	238.67 ± 15.14[b]	0.0[a]	157.05 ± 21.14[b]	172.44 ± 6.57[a]	163.73 ± 6.82[b]	284.91 ± 21.80[b]	1930.43 ± 49.65[b]	285.10 ± 1.72[a]
ArB1	281.36 ± 15.56[a]	849.01 ± 17.54[d]	12,948.99 ± 53.48[g]	1636.62 ± 38.65[d]	876.81 ± 17.99[a]	0.0[a]	494.86 ± 17.81[c]	266.54 ± 11.21[c]	0.0[a]	247.24 ± 12.14[a]	156.93 ± 14.63[a]	217.93 ± 17.13[c]	283.36 ± 14.46[a]	4019.48 ± 6.62[f]	427.57 ± 13.75[c]
MeB1	278.60 ± 11.77[a]	2120.16 ± 44.36[g]	1091.81 ± 12.15[a]	850.08 ± 33.94[a]	1005.41 ± 23.43[b]	0.0[a]	374.87 ± 4.78[ab]	173.04 ± 14.63[a]	0.0[a]	120.77 ± 8.36[a]	216.09 ± 16.42[b]	125.16 ± 3.78[a]	211.25 ± 2.88[a]	4761.02 ± 50.27[a]	357.55 ± 16.03[b]

Note: Values are mean of three replicates ± standard deviation (SD). Means followed by different superscripted alphabet differs significantly ($p < 0.05$).

in California wines was 869 mg/L. The proteins rich in proline are also responsible for producing the haze (turbidity) in beer by combining with the polyphenols (Siebert, 2010). The presence of these amino acids indicates the presence of low molecular weight peptides in RB with bioactive and sensory active properties as have been described by Han and Xu (2011).

5.3.2.5   ANALYSIS OF AROMATIC COMPOUNDS

GC–MS analysis is a very precise and time efficient method for analyzing the volatile and semivolatile compounds and has been used for this purpose by several other workers (Luo et al., 2008; Isogai et al., 2004, 2005; Mo et al., 2009; Chuenchomrat et al., 2008). The volatile and semivolatile compounds are considered the major components responsible for imparting the distinctive flavor to wines and beers (Luo et al., 2008; Isogai et al., 2004, 2005). The occurrence of different volatile organic compounds in the RB varieties, under study, is presented in Table 5.5. The relative peak areas of the compounds are listed as percentage of areas occupied by the individual peak out of the total area of all the detected peaks. Various alcohols and esters were detected in all the samples. It has been observed that phenylethyl alcohol was present in all the varieties of RB under investigation. All the samples contained significant amounts except NaB1 and ArB1. The compound 2,2'-oxybis-ethanol was detected in six RB varieties with maximum peak area up to 37.28%. In case of butanedioic acid diethyl ester, only four samples showed the presence of the compound with peak area in the range of 0.66–3.85%. The compound 3-methyl-1-butanol was detected only in AsB2, ArB1, and MeB1. However, the presence of the compound was negligible in AsB2. Ethyl acetate (in ArB1 and MeB1), 1,4-butanediol (in ArB1 and MeB1), and 6-methylheptyl ester-2-propenoic acid (in NaB1 and NaB2) were present only in two samples each. NaB2 contained the compounds namely (S)-(+)-5-methyl-1-heptanol, 2-(1,1-dimethylethyl)-cyclobutanone, and (S)-(+)-6-methyl-1-octanol, whereas NaB1 contained 2-methyl-dipropanoate-1,3-propanediol, 5-methyl-2-(1-methylethyl)-1-hexanol, and 2-isopropyl-5-methyl-1-heptanol. On the other hand, AsB1 (5-methyl-4-octanol and hexyl ester chloroacetic acid) and AsB3 (4-butoxy-butanoic acid and didodecyl phthalate) revealed the presence of two compounds each. Samples ArB1 and AsB5 contained 3,3-dimethyl-4-methylamino-butan-2-one and 3-hydroxy-butanal, respectively. The occurrences of these compounds was less than 8% except 2-(1,1-dimethylethyl)-cyclobutanone which showed 18.04%.

TABLE 5.5 Aromatic Compounds in the Rice Beer Samples Detected by GC–MS Analysis.

Organic compounds	Retention time (min)	Rice beer varieties/Sample								
		NaB1	NaB2	AsB1	AsB2	AsB3	AsB4	AsB5	ArB1	MeB1
Ethyl acetate	2.0	✗	✗	✗	✗	✗	✗	✗	✓ (25.12)	✓ (0.12)
1,4-Butanediol	2.1	✗	✗	✗	✗	✗	✗	✗	✓ (20.26)	✓ (11.01)
3-Methyl-1-butanol	2.9	✗	✗	✗	✓ (0.36)	✗	✗	✗	✓ (39.37)	✓ (32.46)
(S)-(+)-5-methyl-1-heptanol	3.0	✗	✓ (3.75)	✗	✗	✗	✗	✗	✗	✗
5-Methyl-4-octanol	3.2	✗	✗	✓ (3.31)	✗	✗	✗	✗	✗	✗
2-Methyl-dipropanoate-1,3-propanediol	3.5	✓ (16.46)	✗	✗	✗	✗	✗	✗	✗	✗
2,2'-Oxybis-ethanol	3.9	✓ (15.97)	✓ (1.25)	✓ (1.86)	✗	✓ (3.09)	✓ (37.28)	✗	✗	✓ (20.61)
3-Amino-2-methyl-butanoic acid	5.6	✗	✗	✗	✗	✗	✗	✗	✓	✗
2-(1,1-Dimethylethyl)-cyclobutanone	6.9	✓ (14.62)	✓ (18.04)	✗	✗	✗	✗	✗	✗	✗
5-Methyl-2-(1-methylethyl)-1-hexanol	7.90	✗	✗	✗	✗	✗	✗	✗	✗	✗
3,3-Dimethyl-4-methylamino-butan-2-one	7.92	✗	✗	✗	✗	✗	✗	✗	✓ (7.37)	✗
4-Butoxy-butanoic acid	7.93	✗	✗	✗	✗	✓ (2.31)	✗	✗	✗	✗
(S)-(+)-6-methyl-1-octanol	8.1	✗	✓ (2.95)	✗	✗	✗	✗	✗	✗	✗
Phenylethyl alcohol	8.9	✓ (7.67)	✓ (67.52)	✓ (93.03)	✓ (96.19)	✓ (86.91)	✓ (67.72)	✓ (71.96)	✓ (4.48)	✓ (35.79)
Diethyl ester butanedioic acid	9.8	✗	✓ (3.85)	✓ (0.66)	✓ (1.61)	✓ (1.86)	✗	✗	✗	✗
6-Methylheptyl ester-2-propenoic acid	10.5	✓ (39.67)	✓ (2.65)	✗	✗	✗	✗	✗	✗	✗
Hexyl ester chloroacetic acid	12.6	✗	✗	✓ (1.14)	✗	✗	✗	✗	✗	✗
2-Isopropyl-5-methyl-1-heptanol	12.9	✓ (5.62)	✗	✗	✗	✗	✗	✗	✗	✗
3-Hydroxy-butanal	13.1	✗	✗	✗	✗	✗	✗	✓ (6.92)	✗	✗
Didodecyl phthalate	19.2	✗	✗	✗	✗	✓ (3.27)	✗	✗	✗	✗

Note: "✗" denotes the absence of the compound and "✓" denotes the presence of the compound along with the relative peak area in percentage.

The present study revealed varying patterns in the content of volatile aromatic compounds. The production of similar groups of aromatic compounds by different strains of yeasts in fermentation mixture of *sochu* (a Japanese alcoholic drink) have also been reported by Yamamoto et al. (2012) who studied the fermentation process at different temperatures. Many volatile compounds have been characterized as odor-active compounds in RB and these are considered to provide alcohol-like, sweet, fruity, buttery, and pungent aromas (Chuenchomrat et al., 2008). A large number of volatile and aromatic compounds in RB have also been earlier identified by Isogai et al. (2005) using GC–MS methods. Many volatile and nonvolatile components contribute to the distinctive flavor of beer (Smogrovicova and Domeny, 1999) and a diverse group of volatile and semivolatile aromatic compounds were detected in the samples. However, the influence of these volatiles on the actual aroma profile of the RB will depend on their threshold values. Phenylethyl alcohol is the most abundant compound in the samples studied and is an important constituent in many essential oils, flavors, and perfumery and moreover it has antimicrobial properties (TGSC, 2012). Lee et al. (2012) also found high content of phenylethyl alcohol in agitated fermentation of brown rice to produce vinegar using the starter *nuruk.* The occurrence of the group of higher alcohol like 1-butanol, 1-hexanol, 2,3-butanediol, phenylethyl alcohol, 2-butanol, ethyl alcohol, and 3-ethoxy-1-propanol have also been reported in alcoholic beverage prepared from fruits and these may contribute distinctive flavor to the beverage (Butkhup et al., 2011).

5.3.2.6   PRESENCE OF DIFFERENT MINERAL ELEMENTS

Humans require more than 22 mineral elements, all of which can be supplied by an appropriate diet. However, improper diet plans very often lead to a deficiency of minerals such as iron, zinc, calcium, magnesium, copper, or selenium. Also, other trace elements such as copper and zinc are essential for and human and animal nutrition (He et al., 2013). The content of some of the mineral elements detected in the beers is shown in Table 5.6. The highest content in all the beers was that of Mg and it ranged in between 14.34 mg/L in MlB1 and 28.89 mg/L in AsB1. Na, K, and Ca were also detected in good amount in all the beers. The content of Cu, Fe, and Mn remained below 1 mg/L in all the beers, whereas content of Zn was in between 0.54 and 1.57 mg/L. The minerals in alcoholic beverages can be attributed to many factors, such natural sources including raw materials, soil, water, and yeast, and to contamination during the making process, transport, and storage. Their levels in beer can be a significant parameter affecting its consumption. They have certain beneficial effects on human body, for example, Fe is an essential constituent of hemoglobin, myoglobin, and other enzymes, Zn, Mg, and Cu are also essential for numerous enzymes and Cu is also a constituent of hair, bone, and other body organs (Pan et al., 2013). However, high intake of any mineral may have negative effect on human health. Hence, in this study, the amount of elements detected in all the RBs signifies a safe level for human consumption.

In similar studies, Pan et al. (2013) analyzed trace elements in Chinese rice wine by ICP–OES and found them to be in the range of 0.088–0.106 mg/L for Co, 0.097–0.164 mg/L for Cr, 0.085–0.126 mg/L for Cu, 0.181–0.308 mg/L for Fe, 0.179–0.311 mg/L for Mg, 0.102–0.184 mg/L for Mn, 0.219–0.349 mg/L for Se, and 0.090–0.139 mg/L for Zn. He et al. (2013) also studied the mineral content in North China rice wines and found that the amount of K was the largest (667.430 mg/kg) followed by Mg (305.578 mg/kg), Na (199.004 mg/kg), and Ca (167.231 mg/kg). Among the trace elements, Fe (9.925 mg/kg) was found in the highest content.

TABLE 5.6 Mineral Content in the Rice Beer Samples.

Sample	Concentration (mg/L)							
	Cu	Na	Zn	K	Ca	Fe	Mg	Mn
AsB1	0.24	10.48	0.98	13.36	12.30	0.82	28.89	0.29
AsB2	0.14	11.89	1.31	12.39	5.47	0.53	25.62	0.18
AsB3	0.14	8.88	0.93	11.30	3.42	0.39	24.05	0.10
AsB4	0.12	8.98	0.93	13.08	3.92	0.56	26.28	0.24
AsB5	0.05	10.31	1.06	12.89	4.06	0.48	26.04	0.23
NaB1	0.15	12.87	1.57	13.37	6.06	1.42	25.52	0.23
NaB2	0.17	13.38	1.16	12.75	3.07	0.46	23.74	0.19
MlB1	0.00	13.82	0.54	8.09	2.76	0.25	14.34	0.07
ArB1	0.09	11.06	0.69	17.13	13.78	0.26	27.63	2.50

5.3.3 EFFECT OF THE MICROBIAL STARTERS ON THE QUALITY OF RICE BEER

The various SCs used, their codes, and the code names for the RBs prepared in the laboratory are shown in Table 5.7.

TABLE 5.7 Various Codes for the Samples.

Microbial starter cake (SC)		Rice beer prepared in laboratory (RB)
Local name	Code name	Code name
Angkur	AsS1	AsR1
Perok-kushi	AsS2	AsR2
Thap	AsS3	AsR3
Vekur-pitha	AsS4	AsR4
Apop-pitha	AsS5	AsR5
Umhu	NaS1	NaR1
Piazu	NaS2	NaR2
Siiyeh	ArS1	ArR1
Thiat	MeS1	MeR1

5.3.3.1 PHYSICAL PROPERTIES OF THE SC

Some physical characteristics of the SC are shown in Table 5.8. The shape of six of the samples was similar, that is, round and flat, while three others were oval balls. Sample ArS1 was the largest, with a total volume of 142.15 cm^3 while sample AsS4 was the smallest with a volume of 13.65 cm^3. The highest densities were observed in the oval-shaped samples, that is, AsS5, AsS5, and MeS1 with values of 0.75, 0.74, and 0.70 g/cm^3, respectively, with no significant difference (at $p < 0.05$), while the densities of all the other samples varied significantly.

Significant differences in hardness, springiness, cohesiveness, and gumminess was observed among most of the SC. AsS4 was found to be the softest with a value of 7.2 kg force and AsS2 was the hardest with a value of 113.39 kg force. AsS4 also had the least values of springiness, cohesiveness, and gumminess with values of 0.23 mm, 0.02, and 0.15 kg force, respectively. NaS2 had the highest values of springiness (0.62 mm) and cohesiveness (0.30) and AsS2 had the highest values of gumminess (31.94 kg force).

TABLE 5.8 Some Physical Characteristics of the Microbial Starter Cakes.

Sample code	Shape	Total volume (cm³) ± SD	True density (g/cm³) ± SD	Parameters Hardness (kg) ± SD	Springiness (mm) ± SD	Cohesiveness ± SD	Gumminess (kg) ± SD
NaS1	Round flat	16.91 ± 3.50[ab]	0.53 ± 0.01[ab]	21.44 ± 0.31[c]	0.29 ± 0.01[b]	0.11 ± 0.01[d]	2.23 ± 0.09[c]
NaS2	Round flat	78.02 ± 22.9[e]	0.65 ± 0.11[cd]	36.57 ± 0.15[f]	0.62 ± 0.02[g]	0.30 ± 0.02[h]	11.02 ± 0.45[e]
AsS1	Round flat	35.66 ± 2.51[cd]	0.60 ± 0.03[bc]	73.34 ± 0.30[h]	0.34 ± 0.01[c]	0.19 ± 0.01[f]	13.38 ± 0.78[f]
AsS2	Round flat	33.63 ± 2.54[bcd]	0.55 ± 0.07[ab]	113.39 ± 0.34[i]	0.39 ± 0.004[d]	0.28 ± 0.004[g]	31.94 ± 0.6[g]
AsS3	Round flat	19.84 ± 2.29[abc]	0.74 ± 0.01[e]	72.44 ± 0.41[g]	0.43 ± 0.02[e]	0.09 ± 0.01[c]	7.04 ± 0.72[d]
AsS4	Oval	13.65 ± 4.34[a]	0.75 ± 0.03[e]	7.2 ± 0.06[a]	0.23 ± 0.01[a]	0.02 ± 0.004[a]	0.15 ± 0.01[a]
AsS5	Oval	24.89 ± 3.68[abcd]	0.74 ± 0.01[e]	27.83 ± 0.28[e]	0.40 ± 0.004[d]	0.02 ± 0.001[a]	0.57 ± 0.04[ab]
ArS1	Round flat	142.15 ± 19.08[f]	0.52 ± 0.02[a]	23.67 ± 0.29[d]	0.25 ± 0.004[a]	0.05 ± 0.004[b]	1.16 ± 0.07[b]
MeS1	Oval	40.08 ± 5.69[d]	0.70 ± 0.03[de]	18.31 ± 0.18[b]	0.47 ± 0.02[f]	0.13 ± 0.004[c]	2.38 ± 0.13[c]

Note: Values are mean of three replicates ± standard deviation (SD). Means followed by different superscripted alphabet differs significantly ($p < 0.05$).

5.3.3.2 COLOR IN CIELAB EXPRESSION

The results (Table 5.9) were expressed in Commission Internationale de l'Eclairage L, a, and b (CIELAB) systems. L indicates the degree of lightness or darkness ($L = 0$ indicates perfect black and $L = 100$ indicates most perfect white), a indicates degree of redness (+) and greenness (−), and b indicates degree of yellowness (+) and blueness (−). A direct correlation was observed in between the color of the SC and that of the respective RB. Starter NaS1 with the highest value of L (74.97) produced the beer with the highest value of L (78.58), while starter AsS4 with the lowest value of L (50.08) also produced the beer with the lowest L value (0.83). Similarly, the starter AsS4 with an a value of 5.10 produced the beer AsR4 with an a value of 0.13 and the starter AsS3 with an a value of 1.02 produced the beer AsR3 with an a value of 1.80. The same trend was also observed in case of b values with starter NaS1 ($b = 15.25$) producing beer NaR1 ($b = 23.47$) and starter AsS1 ($b = 10.99$) producing beer AsR1 ($b = 1.21$). Thus, it was observed that the color of the starters greatly influence the color of the final product, even if added at a very small ratio. Since the color further influences the sensory characteristics of the RB, a proper combination of SC and substrate is needed to be maintained.

TABLE 5.9 Color of the Microbial Starter Cakes and Rice Beer in CIELAB Expression.

Sample code	Color					
	$L \pm SD$		$a \pm SD$		$b \pm SD$	
	SC	RB	SC	RB	SC	RB
NaS1/ NaR1	74.97 ± 2.20^c	78.58 ± 0.26^h	2.29 ± 0.25^{cd}	0.55 ± 0.04^{bc}	15.25 ± 0.63^c	23.47 ± 0.35^f
NaS2/ NaR2	73.96 ± 6.65^c	2.07 ± 0.06^d	2.16 ± 0.69^{bcd}	0.70 ± 0.02^{bc}	14.82 ± 1.24^c	2.65 ± 0.06^{cd}
AsS1/ AsR1	59.62 ± 1.65^b	1.16 ± 0.04^b	1.91 ± 0.27^{bc}	0.57 ± 0.02^{bc}	10.99 ± 0.24^a	1.21 ± 0.17^a
AsS2/ AsR2	52.46 ± 1.76^a	5.38 ± 0.10^f	3.12 ± 0.26^f	0.40 ± 0.02^{ab}	13.14 ± 0.43^b	2.91 ± 0.61^d
AsS3/ AsR3	83.05 ± 0.92^d	1.99 ± 0.01^d	1.02 ± 0.16^a	0.13 ± 0.03^a	10.87 ± 0.43^a	2.26 ± 0.02^c
AsS4/ AsR4	50.08 ± 5.74^a	0.83 ± 0.03^a	5.10 ± 0.63^g	1.80 ± 0.54^d	15.20 ± 1.03^c	1.53 ± 0.02^{ab}
AsS5/ AsR5	51.18 ± 1.72^a	0.93 ± 0.05^a	1.45 ± 0.33^{ab}	0.67 ± 0.02^{bc}	13.28 ± 0.90^b	1.39 ± 0.01^a
ArS1/ ArR1	72.08 ± 1.97^c	14.50 ± 0.07^g	2.88 ± 0.16^{df}	1.74 ± 0.05^d	15.19 ± 0.20^c	10.19 ± 0.18^e
MeS1/ MeR1	62.67 ± 1.56^b	1.43 ± 0.04^c	2.14 ± 0.65^{bcd}	0.83 ± 0.03^c	12.38 ± 0.94^b	1.83 ± 0.03^b

5.3.3.3 PROXIMATE COMPOSITION

The proximate composition of the samples is given in Table 5.10. All the results have been expressed on wet basis. Low content of moisture was found in the SC and varied from 8.96% (AsS4) to 10.55% (NaS1). Except for NaS1, there were no significant differences among the other samples. Among the RBs, the moisture content varied from 90.30% (NaR1) to 98.50% (ArR1).

The ash content ranged in between 0.43% and 1.79% for the SC and in between 0.02% and 0.37% for the RB. These results tally with the ash content of *Ou* (Thai rice wine) samples which have also been found to range in between 0.1% and 0.3% (Chuenchomrat et al., 2008). Other reported values of ash content on a dry matter basis in RB are 5.1% (Thapa and Tamang, 2004) and 1.7% (Tamang and Thapa, 2006). Tamang and Sarkar (1995) also studied various characteristics of *marcha* cakes and found them to contain 13% w/w moisture and 0.7% w/w ash (dry weight basis). The SC with the highest content of ash, namely, AsS3, AsS4, and AsS5, was also responsible for producing the beer with the highest content of ash, namely, ASR3, AsR4, and AsR5.

Crude fiber was present in varying content in all the starter and beer samples, with the samples NaS1 (2.48%) and AsR5 (0.29%) having the highest content. The content of fats was minimal in all the SC samples, however, with significant differences. Among the RB samples, except AsR1 and MeR1, there was significant difference in fats content among all the other samples. NaR2 and AsR4 were found to have the highest content with 0.76% and 0.86%, respectively. The fat content of *kodo ko jaanr* (fermented finger millet beverage) (Thapa and Tamang, 2004) was however found to be higher (2% DM) than all the samples studied by us.

Protein was found to be present in all the SC and there was significant difference in its content among most of the samples. The highest content was found in MeS1 (4.68%) and AsS5 (1.65%) had the lowest content. In the RB samples, protein was found to be present in the range of 0.25% (MeR1) to 1.01% (AsR4). There was no significant difference among NaR2, AsR1, and AsR2, and among AsR5, ArR1, and MeR1. However, no correlation was observed in the protein content of the SC and the RB produced from them. The results can be corroborated to the protein content of *Ou* samples which vary from 0.45% to 0.99% (Chuenchomrat et al., 2008). The protein content of similar products reported on a dry basis was 9.5% (Tamang and Thapa, 2006) and 9.3% (Thapa and Tamang, 2004).

Since rice, a starch-rich substrate is the major constituent of both the SC and the RB, high content of carbohydrates was found in the SC as well as RB.

TABLE 5.10 Proximate Composition of the Microbial Starter Cakes and the Prepared Rice Beer on Wet Basis.

Sample code	Moisture (%) ± SD		Ash (%) ± SD		Crude fiber (%) ± SD		Fats (%) ± SD		Protein (%) ± SD		Carbohydrates (%) ± SD	
	SC	RB*	SC	RB	SC	RB	SC	RB	SC	RB	SC*	RB
NaS1/ NaR1	10.55 ± 1.42[a]	90.30	0.82 ± 0.02[b]	0.17 ± 0.003[c]	2.48 ± 0.34[de]	0.09 ± 0.01[ab]	1.27 ± 0.63[c]	0.06 ± 0.01[a]	4.35 ± 0.21[de]	0.90 ± 0.05[g]	80.53	8.47 ± 1.75[c]
NaS2/ NaR2	13.17 ± 1.06[b]	97.19	0.77 ± 0.05[ab]	0.02 ± 0.01[a]	2.47 ± 0.50[de]	0.23 ± 0.04[d]	2.75 ± 0.21[d]	0.76 ± 0.02[g]	4.48 ± 0.27[def]	0.51 ± 0.06[e]	76.36	1.29 ± 0.11[a]
AsS1/ AsR1	10.53 ± 1.66[a]	98.14	1.26 ± 0.09[c]	0.17 ± 0.01[c]	2.74 ± 0.04[e]	0.24 ± 0.02[d]	0.28 ± 0.01[ab]	0.36 ± 0.02[c]	4.26 ± 0.12[d]	0.45 ± 0.05[d]	80.93	0.63 ± 0.18[a]
AsS2/ AsR2	10.08 ± 0.77[a]	98.42	0.93 ± 0.52[bc]	0.16 ± 0.01[d]	1.89 ± 0.12[bc]	0.31 ± 0.03[f]	1.29 ± 0.07[c]	0.11 ± 0.02[b]	4.81 ± 0.34[f]	0.47 ± 0.04[d]	81.00	0.83 ± 0.05[a]
AsS3/ AsR3	9.83 ± 0.65[a]	97.84	1.79 ± 0.17[d]	0.24 ± 0.02[f]	2.23 ± 0.08[cd]	0.18 ± 0.01[c]	0.28 ± 0.07[ab]	0.16 ± 0.02[c]	2.59 ± 0.11[b]	0.77 ± 0.06[f]	83.28	0.80 ± 0.04[a]
AsS4/ AsR4	8.96 ± 1.89[a]	93.68	1.76 ± 0.16[d]	0.33 ± 0.01[g]	1.53 ± 0.09[b]	0.12 ± 0.03[b]	0.23 ± 0.02[a]	0.86 ± 0.02[h]	1.90 ± 0.27[a]	1.01 ± 0.09[h]	85.62	4.00 ± 0.09[b]
AsS5/ AsR5	9.56 ± 1.45[a]	97.58	1.78 ± 0.13[d]	0.37 ± 0.01[h]	1.46 ± 0.39[b]	0.29 ± 0.02[ef]	0.63 ± 0.04[b]	0.43 ± 0.02[f]	1.65 ± 0.07[a]	0.35 ± 0.03[c]	84.92	0.98 ± 0.08[a]
ArS1/ ArR1	9.91 ± 2.02[a]	98.50	0.43 ± 0.12[a]	0.14 ± 0.01[c]	0.79 ± 0.25[a]	0.06 ± 0.02[a]	0.24 ± 0.03[a]	0.32 ± 0.02[d]	3.68 ± 0.09[c]	0.28 ± 0.04[b]	84.95	0.70 ± 0.02[a]
MeS1/ MeR1	9.02 ± 0.91[a]	97.69	1.75 ± 0.12[d]	0.04 ± 0.01[b]	1.84 ± 0.18[bc]	0.27 ± 0.02[de]	0.47 ± 0.09[ab]	0.37 ± 0.01[e]	4.68 ± 0.02[ef]	0.25 ± 0.03[a]	82.24	1.37 ± 0.34[a]

*Calculated after subtracting the sum of other compositions from 100.

Carbohydrate was the major constituent in the starters and AsS4 (85.62%) was found to have the highest content, while NaS2 (76.36%) had the lowest content. In case of RB, NaR1 had the highest content of carbohydrates with 8.47%, followed by AsR4 with 4.0%. All the other RB samples had concentration in the range of 0.70–1.37% and showed no significant difference.

5.3.3.4 BIOCHEMICAL ATTRIBUTES OF THE RB

The values of some of the biochemical attributes of the RB samples are shown in Table 5.11. These readings were helpful in understanding the general quality aspects of the RB from this region.

The pH of all the samples was found to be low. AsR1, AsR3, and AsR4 showed no significant difference in their pH. MeR1 had the lowest value (pH 3.35), whereas ArR1 had the highest (pH 5.11). The low pH has been reported in other types of beer like *jutho* (pH 3.6) (Teramoto et al., 2002), *kodo ko jaanr* (pH 4.1) (Isogai et al., 2004), *poko* (pH 3.2–3.0) (Shrestha et al., 2002). The pH of different varieties of *tapuy* (Philippine RB) has also been found to range in between 4.6 and 5.0 (Perez, 1988). The total acidity of the samples was expressed as percentage of lactic acid. There was no significant difference among the samples NaR2, AsR1, and AsR5, in between NaR1 and AsR2 and in between NaR2 and AsR5. The highest value (0.77%) was found in MeR1 and lowest (0.33%) in ArR1. The acidity values were in line with *yakju* (Korean RB) brewed with different wild-type yeast strains (Kim et al., 2010). The values were however more than that of *bhaati jaanr* during whose fermentation the titrable acidity was found to increase from 0.01% to 0.2% till the 4th day, and remained at a level of 0.17% till the end (Tamang and Thapa, 2006).

All the samples had similar alcohol by weight content within the range of 3.93–4.39% and there was very little significant difference. The alcohol contents were found to be similar to that of *zutho* (Teramoto et al., 2002), *bhaati* jaanr (Tamang and Thapa, 2006), and *kodo ko jaanr* (Thapa and Tamang, 2004) which were 5.0%, 5.9%, and 4.8%, respectively, and more than that of *poko* (Shrestha et al., 2002) which had 1.0–1.6%. This content was less than that found in samples of *yakju* of Korea (Kim et al., 2010), *Ou* of Thailand (Chuenchomrat et al., 2008), and *tapuy* of Philippines (Perez, 1988). There was significant difference in the total soluble solids (TSS) value of all the samples except in between AsR2 and AsR3. The values of total titritable acidity,

TABLE 5.11 Biochemical Attributes of the Rice Beer Prepared under Laboratory Condition.

Sample code	Parameters								
	pH ± SD	Acidity (%) ± SD	Alcohol (%) ± SD	TSS (°Bx) ± SD	RS (g/100 g) ± SD	Starch (g/100 g) ± SD	Amylose (g/100 g) ± SD	TPC (mg/100 g) ± SD	RSA (%) ± SD
NaR1	4.06 ± 0.01c	0.47 ± 0.01c	3.93 ± 0.01a	16.27 ± 0.06i	3.47 ± 0.19c	0.85 ± 0.04bc	0.48 ± 0.05a	10.06 ± 0.18g	45.28 ± 0.61a
NaR2	4.63 ± 0.01f	0.50 ± 0.01d	4.24 ± 0.06b	3.20 ± 0.10b	0.21 ± 0.01a	1.38 ± 0.08e	0.52 ± 0.05a	1.83 ± 0.15b	81.11 ± 1.51e
AsR1	4.28 ± 0.01e	0.50 ± 0.01d	4.30 ± 0.02b	6.20 ± 0.50g	0.25 ± 0.07a	1.07 ± 0.02d	0.52 ± 0.05a	2.19 ± 0.20c	90.95 ± 0.39f
AsR2	4.72 ± 0.01g	0.40 ± 0.01b	4.26 ± 0.05b	4.03 ± 0.06d	0.23 ± 0.02a	1.07 ± 0.12d	0.55 ± 0.05ab	2.00 ± 0.02bc	90.29 ± 0.54f
AsR3	4.23 ± 0.01d	0.75 ± 0.01f	4.39 ± 0.36a	4.27 ± 0.12e	0.32 ± 0.13a	0.88 ± 0.03c	0.82 ± 0.06cd	5.05 ± 0.002f	69.93 ± 0.68d
AsR4	4.27 ± 0.01e	0.75 ± 0.01f	4.00 ± 0.01a	13.40 ± 0.10h	1.67 ± 0.07b	0.94 ± 0.05c	0.69 ± 0.07bc	4.71 ± 0.02e	63.70 ± 2.49c
AsR5	3.60 ± 0.01b	0.58 ± 0.01e	4.37 ± 0.02b	3.40 ± 0.23c	0.20 ± 0.05a	1.36 ± 0.05e	0.68 ± 0.07bc	0.93 ± 0.08a	47.21 ± 2.93a
ArR1	5.07 ± 0.01h	0.32 ± 0.01a	4.35 ± 0.02b	6.03 ± 0.06f	0.21 ± 0.01a	0.76 ± 0.04ab	1.21 ± 0.05e	2.62 ± 0.13d	56.89 ± 2.57b
MeR1	3.35 ± 0.01a	0.76 ± 0.02f	4.35 ± 0.03b	2.63 ± 0.06a	0.20 ± 0.004a	0.74 ± 0.05a	0.84 ± 0.04d	0.91 ± 0.04a	44.38 ± 0.41a

Note: Values are mean of three replicates ± standard deviation (SD). Means followed by different superscripted alphabet differs significantly (p < 0.05).

TSS, total soluble solids; RS, reducing sugars; RSA, radical scavenging activity.

pH, alcohol were similar with the findings of Akpinar-Bayizit et al. (2010), who studied the changes in total titritable acidity, pH, alcohol, organic acid profiles, and sensory properties during the fermentation of boza, an alcoholic beverage produced from rice, maize, millet, and wheat flours in Turkey.

The TSS value of the samples ranged from a minimum of 2.63°Bx in MeR1 to a maximum of 16.27°Bx in NaR1. NaR1 had the highest content of reducing sugars with 3.47%, followed by AsR4 (1.67%). All the other samples had concentration in the range of 0.20–0.32% and showed no significant difference. The concentration of reducing sugars in samples of *tapuy* has been reported to be in between 0.07% and 0.21% (Perez, 1988) and that in *zutho* to be 6.3 mg/mL) (Teramoto et al., 2002).

Starch content ranged in between 0.74 g/100 g (MeR1) and 1.38 g/100 g (NaR2) and there was less significant difference among the samples. The amylose content also varied depending on the content of starch. The partial sweetness of the products is explained by the presence of carbohydrates, especially the reducing sugars in them. The presence of starch and amylose in the final product indicates the partial conversion of starch to sugars and it is also due to the straining practice followed instead of proper filtration.

The total polyphenol content (TPC) and the antioxidant activity of the samples are presented in Table 5.11. The sample NaR1 had the highest content of polyphenols (10.06 mg/100 g) followed by AsR3 (5.05 mg/100 g) and AsR4 (4.71 mg/100 g), and there was significant difference among most of the samples. All the samples showed moderate-to-high DPPH free RSA. Both AsR1 and AsR2 showed the highest activity with values of 90.95% and 90.29% RSA, respectively. The RSA of other samples varied and remained in the range of 44.38% RSA (MeR1) to 81.11% RSA (NaR2). The presence of polyphenols may account for the high antioxidant activity exhibited by most of the samples, which in turn includes health benefits such as prevention of diseases like cancer and coronary heart disease. The antioxidant property may also be attributed to the various plants used in preparing the SC. High content of polyphenols and antioxidant activity was also observed by Zujko and Witkowska (2014) in different type of wines and beer and concluded that the antioxidant potential of the foods tested was related to the TPC.

5.3.3.5   MICROBIOLOGICAL PROFILE OF THE SAMPLES

The count of five different groups of microbes, namely, plate counts, LAB, yeasts, molds, and *Staphylococcus* sp. observed in the SC samples is given in Table 5.12. All the analyses were done in three replicates and the mean values were taken.

The plate count of heterotrophic bacteria includes all bacteria that use organic nutrients for growth. They are present in all types of water, food, soil, vegetation, and air. They were found in high numbers in both the SC and the RB samples. In case of SC, there was no much significant variation in their counts. NsS2 (8.58 log CFU/g) had the highest count, while AsS5 (5.03 log CFU/g) had the lowest count. However, significant variation in counts was observed in case of the RB samples, with AsR1 (10.28 log CFU/g) having the highest count and AsR3 (2.54 log CFU/g) with the lowest count. Similar count of mesophilic aerobes in fermented *poko*, a rice-based fermented food of Nepal have been reported by Shrestha et al. (2002). They found the count to start from 7.9×10^7 CFU/g on the first day of fermentation and decrease to a count of 1×10^7 CFU/g on the fifth day. Thapa and Tamang (2004) have also reported the total count of aerobes in *kodo ko jaanr*, a fermented finger millet beverage to be around 7.4 log CFU/g.

LAB are a group of fermentative bacteria and are abundant in nutrient rich environments where carbohydrates and proteins are usually present. They have remarkable selective advantages in diverse ecological niches due to the efficient use of nutrients and the production of lactic acid during growth (Mayo et al., 2010). The LAB were found to be present in all the samples in considerable high number and their count varied significantly in all the samples studied. Among the SC, NaS2 had the highest numbers with a count of 7.76 log CFU/g, while lowest numbers were found in AsS5 with a count of 3.71 log CFU/g. Tamang et al. (2007) also reported the average popula-tion of LAB in *hamei* (SC used in Manipur, India) to be log 6.9 CFU/g and *marcha* (SC used in Sikkim, India) to be log 7.1 CFU/g. The isolates from *hamei* were identified as *L. plantarum* and that from *marcha* as *L. brevis*. Among the RB samples, NaR2 had the maximum population with a count of 7.55 log CFU/g and AsR3 had the minimum with a count of 4.32 log CFU/g. Similar results were obtained by Thapa and Tamang (2004), who found the count of LAB in *kodo ko jaanr* (fermented finger millet beverage) to range from 4.1 to 6.5 log CFU/g. Shrestha et al. (2002) also found the population of LAB in the fermentation mixture of *poko* (fermented rice product) to increase from an initial value of 3.5×10^6 CFU/g on day 1 to a value of 5×10^7 CFU/g on day 5. The dominance of LAB in fermented products results in the inhibition of the growth of pathogens and spoilage microbes (Sriphochanart and Skolpap, 2010) and most of the LAB are also probiotic in nature (Ljungh and Wadström, 2006), which adds to the uniqueness of this type of beer.

TABLE 5.12   Microbiological Profile of the Samples.

Sample code	Log CFU/g ± SD									
	Plate counts		LAB		Yeasts		Molds		Staphylococcus sp.	
	SC	RB	SC	RB	SC	RB	SC	RB	SC	RB
NaS1/ NaR1	8.32 ± 0.16^{bc}	9.51 ± 0.03^{f}	7.56 ± 0.29^{e}	5.43 ± 0.03^{c}	7.14 ± 0.70^{c}	6.55 ± 0.06^{b}	5.72 ± 0.50^{bc}	0.0^{a}	0.0^{a}	0.0^{a}
NaS2/ NaR2	8.58 ± 1.19^{c}	7.28 ± 0.09^{c}	7.76 ± 0.14^{e}	7.55 ± 0.03^{i}	8.78 ± 0.76^{f}	8.34 ± 0.08^{h}	5.82 ± 0.81^{bc}	0.0^{a}	0.0^{a}	4.30 ± 0.21^{d}
AsS1/ AsR1	6.28 ± 0.75a^{b}	10.28 ± 0.12^{g}	7.29 ± 0.36^{e}	6.81 ± 0.03^{f}	8.39 ± 0.40^{ef}	7.73 ± 0.03^{f}	7.67 ± 0.26^{d}	0.0^{a}	0.0^{a}	2.27 ± 0.07^{b}
AsS2/ AsR2	6.83 ± 1.87^{abc}	9.36 ± 0.11^{f}	6.51 ± 0.29^{d}	7.15 ± 0.05^{h}	3.11 ± 0.10^{a}	8.32 ± 0.08^{h}	6.46 ± 0.35^{c}	0.0^{a}	0.0^{a}	4.81 ± 0.14^{e}
AsS3/ AsR3	6.82 ± 0.59^{abc}	2.54 ± 0.12^{a}	6.68 ± 0.39^{d}	4.32 ± 0.07^{a}	7.77 ± 0.52^{cde}	7.24 ± 0.05^{d}	7.58 ± 0.55^{d}	0.0^{a}	0.0^{a}	0.0^{a}
AsS4/ AsR4	6.43 ± 1.14^{abc}	8.49 ± 0.19^{e}	5.62 ± 0.45^{c}	4.61 ± 0.12^{b}	4.63 ± 0.42^{b}	6.68 ± 0.05^{c}	5.26 ± 0.35^{b}	0.0^{a}	0.0^{a}	0.0^{a}
AsS5/ AsR5	5.03 ± 0.85^{a}	7.40 ± 0.23^{c}	3.71 ± 0.09^{a}	6.92 ± 0.03^{g}	3.33 ± 0.34^{a}	7.56 ± 0.04^{e}	4.15 ± 0.26^{a}	0.0^{a}	0.0^{a}	3.19 ± 0.10^{c}
ArS1/ ArR1	7.21 ± 1.50^{abc}	6.72 ± 0.15^{b}	4.62 ± 0.28^{b}	6.06 ± 0.05^{d}	8.23 ± 0.40^{def}	6.43 ± 0.08^{a}	5.49 ± 0.33^{b}	0.0^{a}	0.0^{a}	0.0^{a}
MeS1/ MeR1	6.44 ± 0.3^{abc}	7.79 ± 0.02^{d}	6.62 ± 0.25^{d}	6.53 ± 0.06^{e}	7.47 ± 0.45^{cd}	8.05 ± 0.04^{g}	5.35 ± 0.93^{b}	0.0^{a}	0.0^{a}	0.0^{a}

Note: Values are mean of three replicates ± standard deviation (SD). Means followed by different superscripted alphabet differs significantly ($p < 0.05$).

Amylolytic yeasts and molds accomplish starch hydrolysis and fermentation in a wide range of traditional alcoholic foods and beverages (Steinkraus, 1996). Yeasts were the dominant microbes in all the samples. There was significant difference in their count among the SC samples and varied from 3.11 log CFU/g in AsS2 to 8.78 log CFU/g in NaS2. In case of the RB samples, there was significant difference in their count except NaR2 and AsR2. Their counts ranged from 6.43 log CFU/g in ArR1 to 8.34 log CFU/g in NaR2. The count of yeasts in *bhaati jaanr*, which is a type of RB made in the Eastern Himalayas was found to increase from 10^5 CFU/g on day 1 of fermentation to 10^8 CFU/g on day 2, and then gradually decreased to level of 10^5 CFU/g on day (Tamang and Thapa, 2006). Shrestha et al. (2002) have also found an increase in the population of yeasts from 1.8×10^6 to 1.3×10^8 CFU/g from the first to the fifth day of fermentation of *poko*. The presence of yeasts in considerable high counts in all the samples also confirms that they are the primary organisms responsible for the alcoholic fermentation of RB. The count of molds in the SC was more consistent among all and remained in the range of 4.15–7.67 log CFU/g. The molds were however found to be absent from all the RB samples. The mucorales have roles (amylolytic or proteolytic enzyme activities) in the initial phase of fermentation, mostly in saccharification and their disappearance from the final product have been reported by others (Thapa and Tamang, 2004; Tamang and Thapa, 2006).

Jeyaram et al. (2008) studied the fungal species associated with *hamei* and found yeasts in the range of 8–9 log CFU/g and molds from 5 to 7 log CFU/g. The population of LAB and yeasts in *makgeolli*, a naturally fermented Korean RB was studied by Yoon et al. (2012). They found that *Saccharomyces cerevisiae* was predominant in the samples with an average count of 4.6×10^7 CFU/mL, whereas *L. plantarum* and *Weissella cibaria* were the predominant LAB species with an average count of 1.7×10^7 CFU/ mL.

Staphylococcus species were absent from all the SC samples but were present in four of the RB samples, namely, NaR2, AsR1, AsR2, and AsR5 in counts ranging from 2.27 to 4.95 log CFU/g. Their count in all the four samples differed significantly. The most possible source for this group of microbes in four of the RB may be the air. However, the presence of *Staphylococcus* species is not of much concern as most of them are harmless and reside normally on the skin and mucous membranes of organisms (Ryan, 2004). Members of the common food contaminating groups, namely, Enterobacteriaceae, *Salmonella* and *Shigella* species, and *Bacillus* sp. were not detected in any of the samples. The absence of such contaminants may be attributed to the low pH and high acidity of the products and also the antagonistic effect of the LAB group (Mayo et al., 2010).

5.4   CONCLUSIONS

It was observed that the process of RB preparation followed by different ethnic tribes residing in different states of Northeast India is more or less similar. The only difference is the ingredient in the form of different parts of various plants species. The tribes in different regions use different plant species based on their availability. The knowledge of the indigenous people in the use of the starter cultures as a source of yeast is very interesting. The local brews such as RB bears very significant resemblance of the culture and traditions of the tribal people residing in this part of the country. Each of the beverages prepared is rooted with the sociocultural practices of the individual tribes and also on various environmental factors. It has been found that the preparation of RB is considered as sacred by all the tribes and it occupies special recognitions in many of the occasions like rituals, festivals, marriages, and communal gathering. The consumption of mild amount of alcohol in the form of RB gives some relaxation to the hard working population of these states and practically has no side effect on their health. Apart from imparting color, flavor, and sweetness to the beer, the various plants used in the starter culture are also said to have many medicinal properties. Also some of the plant extracts may also provide certain nutrients for the survival of the microflora present in the SCs. The quality of the starter culture is said to be dependent on the variety of plant parts used and also on the maintenance of proper sanitary conditions. The preference of the variety of rice used for fermentation also differs from communities to communities. However, it is seen that glutinous rice is preferred more than nonglutinous rice, owing to the taste and alcohol content of the product.

The study revealed that all the RB samples collected from the Northeastern states of India are a potential source of nutrition owing to their various biochemical compositions like carbohydrates, amino acids, organic acids, aromatic compounds, etc. at appreciable amounts. The average consumption of RB by tribal communities of Northeast India is around two glasses (400 mL approx.) in the evening for 3–4 days in a week and no any health complication related to the consumption of RB has been reported. The level for consumption of alcohol also varies among individuals. The results revealed that the average content of alcohol is around 4.5% which is less as compared to 5–10% (avg. 7.5%) in malt beers. Therefore, consumption of RB at this level could be considered as safe for human health. This level is also within the limit issued by the World Health Organization. Their guidelines define one unit of alcohol as the equivalent of 8 g of ethanol and the "responsible" or "low," level of risk for men as "3 units per day and 21 per week" and for women as "2 units per day and 14 per week" spread throughout the week (including 2 alcohol free days per week). The presence of alcoholic groups such as phenylethyl alcohol and some other esters contributed to the appealing flavor and smell of all studied varieties of RB. Out of all samples, the AsB4 variety (locally called *Xaj-pani*, collected from Sibsagar district of Assam) prepared by the Ahom community showed relatively higher content of lactic acid, carbohydrates like glucose, trehalose, and galactose and almost all the amino acids as compared to the other samples. In addition, the alcohol content was also in medium range and this sample might be considered relatively more nutritious as compared to the others.

It was also observed that even though used for the same purpose, differences in terms of physical, chemical, and microbial properties were observed among the various SCs. Under the same conditions of fermentation and substrate type, there was significant difference in quality among all the types of RB and their quality was found to be affected by the type of SC. The plausible reason for this variation might be attributed to differences in the plants and rice used in preparing the starters, their ratio and also the differences in the microbial consortium. A direct correlation between the color of the starters and the RB was observed. The low moisture in the starters contributes to their shelf-life and the presence of ash (namely, minerals), protein, and fats in minimal amount in the RB makes this kind of product a balanced nutritional drink. The low pH and high acidity may prove beneficial in controlling the growth of spoilage microbes. Also phenolic compounds were found in all the RBs and these may contribute to the high antioxidant activity exhibited by most of them. The load of microbes belonging to the LAB group was also high in both the starters and the RB, and these may act as potential probiotic organisms present in the drink.

KEYWORDS

- **Northeast Indian tribes**
- **indigenous alcoholic beverages**
- **fermented mass**
- **microbes**
- **lactic acid bacteria**

REFERENCES

Akpinar-Bayizit, A.; Yilmaz-Ersan, L.; Ozcan, T. Determination of *boza's* Organic Acid Composition as It Is Affected by Raw Material and Fermentation. *Int. J. Food Prop.* **2010,** *13* (3), 648–656.

Anderson, L. C. Collecting and Preparing Plant Specimens and Producing an Herbarium. In *Tested Studies for Laboratory Teaching*; Karcher, S. J., Ed.; Kendall/Hunt Publishing Company: Dubuque, IA, 1999; vol. 20, pp 295–300.

Anderson, J. W.; Baird, P.; Davis, R. H.; Ferreri, S.; Knudtson, M.; Koraym, A.; Waters, V.; Williams, C. L. Health Benefits of Dietary Fiber. *Nutr. Rev.* **2009**, *67*, 188–205.

AOAC. *Official Method of Analysis of AOAC International*, 18th ed.; Association of Official Analytical Communities, USA, 2010.

Bank, R. A.; Jansen, E. J.; Beekman, B.; te Koppele, J. M. Amino Acid Analysis by Reverse-Phase High-Performance Liquid Chromatography: Improved Derivatization and Detection Conditions with 9-Fluorenylmethyl Chloroformate. *Anal. Biochem.* **1996**, *240*, 167–176.

Bernal, J. L.; Del Nozal, M. J.; Toribio, L.; Del Alamo, M. HPLC Analysis of Carbohydrates in Wines and Instant Coffees Using Anion Exchange Chromatography Coupled to Pulsed Amperometric Detection. *J. Agric. Food Chem.* **1996**, *44*, 507–511.

Blackburn, S. Sample Preparation and Hydrolytic Methods. In *Amino Acid Determination Methods and Techniques*; Blackburn, S., Ed.; Marcel Dekker Inc.: New York, USA, 1978; pp 17–38.

Blocher, J. C.; Busta, F. F. Multiple Modes of Inhibition of Spore Germination and Outgrowth by Reduced pH and Sorbate. *J. Appl. Bacteriol.* **1985**, *59*, 467–478.

Booth, I. R.; Kroll, R. G. The Preservation of Foods by Low pH. In *Mechanisms of Action of Food Preservation Procedures*; Gould, G. W., Ed.; Elsevier: London, 1989; pp 119–160.

Brand-Williams, W.; Cuvelier, M. E.; Berset, C. L. W. T. Use of a Free Radical Method to Evaluate Antioxidant Activity. *LWT—Food Sci. Technol.* **1995**, *28* (1), 25–30.

Bray, H. G.; Thorpe, W. V. Analysis of Phenolic Compounds of Interest in Metabolism. *Method Biochem. Anal.* **1954**, *1* (1), 27–52.

Butkhup, L.; Jeenphakdee, M.; Jorjong, S.; Samappito, S.; Samappito, W.; Chowtivannakul, S. HS–SPME–GC–MS Analysis of Volatile Aromatic Compounds in Alcohol Related Beverages Made with Mulberry Fruits. *Food Sci. Biotechnol.* **2011**, *20*, 1021–1032.

Chakrabarty, J.; Sharma, G. D.; Tamang, J. P. Substrate Utilisation in Traditional Fermentation Technology Practiced by Tribes of North Cachar Hills District of Assam. *Assam Univ. J. Sci. Technol.* **2009**, *4* (1), 66–72.

Chuenchomrat, P.; Assavanig, A.; Lertsiri, S. Volatile Flavour Compounds Analysis of Solid State Fermented Thai Rice Wine (*ou*). *Sci. Asia* **2008**, *34*, 199–206.

Cole, M. B.; Keenan, M. H. J. Effects of Weak Acids and External pH on the Intracellular pH of *Zygosaccharomyces bailii*, and Its Implications in Weak-Acid Resistance. *Yeast* **1987**, *3*, 23–32.

Deka, D.; Sarma, G. C. Traditionally Used Herbs in the Preparation of Rice-Beer by the Rabha Tribe of Goalpara District, Assam. *Indian J. Trad. Knowl.* **2010**, *9* (3), 459–462.

Freese, E.; Sheu, C. W.; Galliers, E. Function of Lipophilic Acids as Antimicrobial Food Additives. *Nature* **1973**, *41*, 321–325.

Gomis, D. B. HPLC Analysis of Organic Acids. In *Food Analysis by HPLC*; Nollet, L. M. L.; Ed.; Marcel Dekker Inc.: New York, 2000; pp 477–492.

Han, F. L.; Xu, Y. Identification of Low Molecular Weight Peptides in Chinese Rice Wine (Huang Jiu) by UPLC–ESI-MS/MS. *J. Inst. Brew.* **2011**, *117*, 238–250.

He, S.; Mao, X.; Liu, P.; Lin, H.; Du, Z.; Lv, N.; Han, J.; Qiu, C. Research into the Functional Components and Antioxidant Activities of North China Rice Wine (Ji Mo Lao Jiu). *Food Sci. Nutr.* **2013**, *1* (4), 307–314.

Isogai, A.; Utsunomiya, H.; Iwata, H. Changes in the Concentrations of Sotolon and Furfural during the Maturation of Sake. *J. Brew. Soc. Jpn (Jpn)* **2004**, *99*, 374–380.

Isogai, A.; Utsunomiya, H.; Kanda, R.; Iwata, H. Changes in the Aroma Compounds of Sake during Aging. *J. Agric. Food Chem.* **2005**, *53*, 4118–4123.

Jeyaram, K.; Singh, W. M.; Capece, A.; Romano, P. Molecular Identification of Yeast Species Associated with 'Hamei'—A Traditional Starter Used for Rice Wine Production in Manipur, India. *Int. J. Food Microbiol.* **2008**, *124*, 115–125.

Kabelová, I.; Dvořáková, M.; Čížková, H.; Dostálek, P.; Melzoch, K. Determination of Free Amino Acids in Beers: A Comparison of Czech and Foreign Brands. *J. Food Compos. Anal.* **2008**, *21*, 736–741.

Kim, H. R.; Kim, J. H.; Bae, D. H.; Ahn, B. H. Characterization of *yakju* Brewed from Glutinous Rice and Wild-Type Yeast Strains Isolated from *nuruks*. *World J. Microbiol. Biotechnol.* **2010**, *20*, 1702–1710.

Krebs, H. A.; Wiggins, D.; Stubbs, M.; Sols, A.; Bedoya, F. Studies on the Mechanism of the Antifungal Action of Benzoate. *Biochem. J.* **1983**, *214*, 657–663.

Lee, S. W.; Yoon, S. R.; Kim, G. R.; Woo, S. M.; Jeong, Y. J.; Yeo, S. H.; Kim, K. S.; Kwon, J. H. Effect of *nuruk* and Fermentation Method on Organic Acid and Volatile Compounds in Brown Rice Vinegar. *Food Sci. Biotechnol.* **2012**, *21*, 453–460.

Ljungh, A.; Wadström, T. Lactic Acid Bacteria as Probiotics. *Curr. Issues Interest Microbiol.* **2006**, *7*, 73–89.

Luo, T.; Fan, W.; Xu, Y. Characterization of Volatile and Semi-Volatile Compounds in Chinese Rice Wines by Headspace Solid Phase Microextraction Followed by Gas Chromatography–Mass Spectrometry. *J. Inst. Brew.* **2008**, *114* (2), 172–179.

Mayo, B.; Aleksandrzak-Piekarczyk, T.; Fernández, M.; Kowalczyk, M.; Álvarez-Martín, P.; Bardowski, J. Updates in the Metabolism of Lactic Acid Bacteria. In *Biotechnology of Lactic Acid Bacteria: Novel Applications*; Mozzi, F., Raúl, R., Raya, R. R. M., Vignolo, G. M., Eds.; Wiley Blackwell: Malden, MA, 2010; pp 3–34.

Mo, X.; Fan, W.; Xu, Y. Changes in Volatile Compounds of Chinese Rice Wine Wheat qu during Fermentation and Storage. *J. Inst. Brew.* **2009**, *115* (4), 300–307.

Montanari, L.; Perretti, G.; Natella, F.; Guidi, A.; Fantozzi, P. Organic and Phenolic Acids in Beer. *LWT—Food Sci. Technol.* **1999**, *32*, 535–539.

Nogueira, L. C.; Silva, F.; Ferreira, I. M.; Trugo, L. C. Separation and Quantification of Beer Carbohydrates by High-Performance Liquid Chromatography with Evaporative Light Scattering Detection. *J. Chromatogr. A* **2005**, *1065*, 207–210.

Nord, L. I.; Vaag, P.; Duus, J. Ø. Quantification of Organic and Amino Acids in Beer by ¹H NMR Spectroscopy. *Anal. Chem.* **2004**, *76*, 4790–4798.

Ough, C. S. Proline Content of Grapes and Wines. *Vitis* **1968**, *7*, 321–331.

Pan, X. D.; Tang, J.; Chen, Q.; Wu, P. G.; Han, J. L. Evaluation of Direct Sampling Method for Trace Elements Analysis in Chinese Rice Wine by ICP–OES. *Eur. Food Res. Technol.* **2013**, *236* (3), 531–535.

Patáková-Jůzlová, P.; Řezanka, T.; Viden, I. Identification of Volatile Metabolites from rice Fermented by the Fungus *Monascus purpureus* (Ang-kak). *Fol. Microbiol.* **1998**, *43*, 407–410.

Perez, C. M. Nonwaxy Rice for *tapuy* (Rice Wine) Production. *Cer. Chem.* **1988**, *65*, 240–243.

Ranganna, S. *Handbook of Analysis and Quality Control for Fruit and Vegetable Products*, 2nd ed.; Tata McGraw-Hill Publishing Company Limited: New Delhi, India, 2008.

Reeds, P. J. Dispensable and Indispensable Amino Acids for Humans. *J. Nutr.* **2000**, *130*, 1835S–1840S.

Ryan, K. J. Staphylococci. In *Sherris Medical Microbiology: An Introduction to Infectious Diseases*; Ryan, K. J., Ray, C. G., Eds.; McGraw Hill: New York, 2004; pp 261–271.

Salmond, C. V.; Kroll, R. G.; Booth, I. R. The Effect of Food Preservatives on pH Homeostasis in *Escherichia coli. Microbiology* **1984**, *130*, 2845–2850.

Samati, H.; Begum, S. S. Kiad, a Popular Local Liquor of Pnar Tribe of Jaintia Hills District, Meghalaya. *Indian J. Trad. Knowl.* **2007**, *6* (1), 133–135.

Sauer, M.; Porro, D.; Mattanovich, D.; Branduardi, P. Microbial Production of Organic Acids: Expanding the Markets. *Trends in Biotechnol.* **2008**, *26*, 100–108.

Shrestha, H.; Nand, K.; Rati, E. R. Microbiological Profile of *murcha* Starters and Physio-Chemical Characteristics of *poko*, a Rice Based Traditional Fermented Food Product of Nepal. *Food Biotechnol.* **2002**, *16*, 1–15.

Siebert, K. J. *Haze and Foam*, 2010. Retrieved on 15th March 2012 from http://www.nysaes.cornell.edu/fst/faculty/siebert/haze.html.

Singh, P. K.; Singh, K. I. Traditional Alcoholic Beverage, Yu of Meitei Communities of Manipur, *Indian J. Tradition. Knowl.* **2006**, *5* (2), 184–190.

Smogrovicova, D.; Domeny, Z. Beer Volatile By-Product Formation at Different Fermentation Temperature Using Immobilized Yeast. *Process Biochem.* **1999**, *34*, 785–794.

Sriphochanart, W.; Skolpap, W. Characterization of Proteolytic Effect of Lactic Acid Bacteria Starter Cultures on Thai Fermented Sausages. *Food Biotechnol.* **2010**, *24*, 293–311.

Steinkraus, K. H. Lactic Acid Fermentation in the Production of Foods from Vegetables, Cereals and Legumes. *Anton. Leeuwenh.* **1983**, *49*, 337–348.

Steinkraus, K. H. Introduction to Indigenous Fermented Foods. In *Handbook of Indigenous Fermented Foods*; Steinkraus, K. H., Ed.; pp 1–5, Marcel Dekker: New York, 1996.

Tamang, J. P.; Sarkar, P. K. Microflora of *Marcha*: An Amylolytic Fermentation Starter. *Microbios* **1995**, *81*, 115–122.

Tamang, J. P.; Thapa, S. Fermentation Dynamics during Production of *bhaati jaanr*, a Traditional Fermented Rice Beverage of the Eastern Himalayas. *Food Biotechnol.* **2006**, *20*, 251–261.

Tamang, J. P.; Dewan, S.; Tamang, B.; Rai, A.; Schillinger, U.; Holzapfel, W. H. Lactic Acid Bacteria in Hamei and Marcha of North East India. *Indian J. Microbiol.* **2007**, *47*, 119–125.

Tanti, B.; Gurung, L.; Sarma, H. K.; Buragohain, A. K. Ethanobotany of Starter Culture Used in Alcohol Fermentation by a Few Ethnic Tribes of Northeast India. *Indian J. Trad. Knowl.* **2010**, *9* (3), 463–466.

Teramoto, Y.; Yoshida, S.; Ueda, S. Characteristics of a Rice Beer (*zutho*) and a Yeast Isolated from the Fermented Product in Nagaland, India. *World J. Microbiol. Biotechnol.* **2002**, *18*, 813–816.

TGSC. *The Good Scents Company, USA Homepage*, 2012. Retrieved on 11th October 2011 from http://www.thegoodscentscompany.com.

Thapa, S.; Tamang, J. P. Product Characterization of *kodo ko jaanr*: Fermented Finger Millet Beverage of the Himalayas. *Food Microbiol.* **2004**, *21*, 617–622.

Tsuyoshi, N.; Fudou, R.; Yamanaka, S.; Kozaki, M.; Tamang, N.; Thapa, S.; Tamang, J. P. Identification of Yeast Stains Isolated from *marcha* in Sikkim, a Microbial Starter for Amylolytic Fermentation. *Int. J. Food Microbiol.* **2005**, *99* (2), 135–146.

Webb, P. *Volume and Density Determinations for Particle Technologists*; Micromeretics Instrument Corp.: Norcross, GA, 2001.

Yamamoto, H.; Mizutani, M.; Yamada, K.; Iwaizono, H.; Takayama, K.; Hino, M.; Kudo, T.; Ohta, H.; Kida, K.; Morimura, S. Characteristics of Aromatic Compound Production Using New *shochu* Yeast MF062 Isolated from *shochu* Mash. *J. Inst. Brew.* **2012**, *118*, 406–411.

Yoo, K. S.; Kim, J. E.; Moon, J. S.; Jung, J. Y.; Kim, J. S.; Yoon, H. S.; Choi, H. S.; Kim, M. D.; Shin, C. S.; Han, N. S. Evaluation of a Volatile Aroma Preference of Commercial Red Wines in Korea: Sensory and Gas Chromatography Characterization. *Food Sci. Biotechnol.* **2010**, *19*, 43–49.

Yoon, S. S.; Choi, J. A.; Kim, K. H.; Song, T. S.; Park, Y. S. Populations and Potential Association of *Saccharomyces cerevisiae* with Lactic Acid Bacteria in Naturally Fermented Korean Rice Wine. *Food Sci. Biotechnol.* **2012**, *21* (2), 419–424.

Zujko, M. E.; Witkowska, A. M. Antioxidant Potential and Polyphenol Content of Beverages, Chocolates, Nuts and Seeds. *Int. J. Food Prop.* **2014**, *17*, 86–92.

CHAPTER 6

PHYSICOCHEMICAL AND FUNCTIONAL PROPERTIES OF CASSAVA OF MANIPUR, INDIA

SINGAMAYUM KHURSHIDA and SANKAR CHANDRA DEKA*

Department of Food Engineering and Technology, Tezpur University, Napaam 784028, Sonitpur, Assam, India

Corresponding author. E-mail: sankar@tezu.ernet.in

ABSTRACT

Manihot esculenta commonly called cassava, *umangra* in Manipur (a South Asian region that belongs to India), is the third largest source of food carbohydrates in the tropics, after rice and maize. Cassava resists to drought, low-fertility soil, pest, and can adapt marginal land with less maintenance of irrigation. The physicochemical and functional properties of dry chips flour (DCF) and fermented cassava flour (FCF) were studied. Microwave-assisted extraction of pectin was employed for pectin-yield estimation. Functional properties, namely, swelling power (SP), solubility, pasting, color, water, and oil-holding capacities and in addition thermal and structural properties were investigated. The results revealed highest SP 11.41 g/g in FCF and highest solubility 6.04% in DCF. Fermentation decreased the total cyanide content. Total cyanide content of FCF was found within the safe limit of World Health Organization. Fermentation improved the physicochemical and functional properties of cassava flour suitable for food applications; however, flour color was found less desirable due to less lightness in color. Cassava is a sustainable, cheap food source of carbohydrate and can substitute for traditional staple food of wheat and rice in Manipur. The findings can be used in various foods and feed industries.

HIGHLIGHTS

- Fermentation increased WHC and OHC.
- Fermented flour resulted higher swelling power and lower solubility compared to dry chips flour.
- Cheap and sustainable source of carbohydrates.
- Provides avenues for functional food ingredients.

6.1 INTRODUCTION

Manihot esculenta is commonly called cassava umangra in Manipur (a South Asian region which belongs to India). The cassava root is rich in carbohydrate and can be an important food source for people at a large scale. It is an important food in many countries including Africa, Latin America, and some Asian countries. According to the United Nations Statistics Database, 2013, under the Food and Agriculture Organization, cassava is one of the most important food crops which ranks sixth globally in terms of annual production. For approximately 800 million people, cassava is a staple food throughout the world. The major nutrient present in it is carbohydrates that provides the majority of calories in the tropics. It is also generally consumed by tribal or poor people of Manipur, a North-Eastern state of India, and is an underutilized and neglected root which is usually used for pig feed. Several foods can be derived and processed from it (Pereira and Leonel, 2014). Gluten-free nontraditional flours are gaining popularity nowadays, especially flours obtained from root and tuber including cassava (Gularte et al., 2012). Cassava could also be a good source of dietary fiber. Traditional wheat flour for making various foodstuffs can be substituted by cassava flour. Development of gluten-free food has a great demand as it is helpful against celiac disease. There is wide range of potential health benefits of gluten-free products (Matos and Rosell, 2015). Manipur has a tough landscape with 90% hilly areas with improper road connectivity, and the bulky nature of cassava finds difficulty in transportation and therefore the processing of it by the simple techniques like drying reduces its size and prevents postharvest loss. However, cassava is also known to contain toxic cyanide compounds which have disastrous effects to human health. It is a threat to these bulk reserves of carbohydrates stored in cassava roots. Therefore, the crop needs extra attention from researchers as well as policy makers as it is an economically important crop in many parts of North-Eastern region of India including Manipur. The present study highlights the physicochemical and functional properties and cyanide content of cassava tuber flour. This scientific evidence will be helpful for better utilization of cassava of Manipur cultivars, and its importance to the tribal people of North-Eastern region of India, and will also help in fostering its applications in food and feed industry.

6.2   MATERIAL AND METHODS

6.2.1   COLLECTION OF RAW MATERIALS

Fresh cassavas were procured from Silent Tangkhul Naga village in Senapati district of Manipur, India.

6.2.2   PREPARATION OF SAMPLES

6.2.2.1   DRY CHIPS FLOUR

The fresh root was cleaned to remove soil and defective parts were washed with running tap water and peeled to remove the outer cover and also chopped into thin slices (thickness: 6 mm). The slices were spread over trays and then sun dried until moisture dropped to 10–12% and the dried chips were pulverized and kept for further analysis.

6.2.2.2   FERMENTED CASSAVA FLOUR

Peeled cassava slices were steeped in water in an earthen pot and left for fermentation for 48 h. It was then dewatered and dried in hot air oven at 45 ± 1°C for 24 h until the moisture dropped to 10–12% and the dry samples were grinded into powder.

6.2.3   CHEMICAL ANALYSIS

Standard procedure (AOAC, 2012) was used to estimate pH, moisture, ash, crude fiber, carbohydrates of dry chips flour (DCF), and fermented cassava flour (FCF). All the chemicals used were of reagent grade.

6.2.4 PECTIN EXTRACTION USING MICROWAVE-ASSISTED EXTRACTION TECHNIQUE

Pectin extraction was done using the method of Prakash et al. (2013) with a slight modification. One gram of cassava powder was kept into Pyrex 250-mL beaker (Prakash et al., 2013). It was thoroughly mixed with 16.9 mL distilled water having solid–liquid 1:16.9 g/mL and maintained pH 1.4 using 0.016 mol/L sulfuric acid. The mixture was kept inside a microwave oven (LG, MH-3948) at power of 360 W, with adjustable power for irradiation time 180 s. After microwave treatment, it was kept for some time until it reached room temperature and filtered using Whatman no. 1 filter paper. The filtrate was treated with an equal volume of 95% ethanol (v/v). It was kept for 1.5 h and the obtained wet sediment was washed with 95% ethanol (v/v) for 3 times. Most of the foods contain lesser amounts of pectin; care needs to be taken to draw calibration curves from spectrophotometric reading. The whole procedure for pectin extraction and quantification on the dietary fiber of food is very complex and time-consuming. Microwave-assisted extraction of pectin is an alternative extraction method with shorter time, higher extraction rate, quality products, and less solvent (Bailoni et al., 2005).

6.2.4.1 PECTIN YIELD DETERMINATION

The final sediment was extracted as wet pectin and kept for drying at 50 ± 1°C until it maintained a constant weight. The pectin yield (PY) was determined using following equation given by the method of Li et al. (2012) with slight modification.

$$PY = \left(\frac{M_0}{M} \right) \times 100 \tag{6.1}$$

where M_0 is the dried pectin weight; M is the dried cassava flour weight.

6.2.5 SWELLING POWER AND SOLUBILITY

Swelling power (SP) and solubility (S) were determined using Wang's method with slight modifications (Wang et al., 2010). Centrifuge tube (50 mL) was preweighed and 0.4 g of starch was added to it with 20 mL distilled water, and then the centrifuge tubes were kept in hot water bath for 30 min at

80°C. Centrifugation was done for 15 min at 6000g and the supernatant was collected to a preweighed petri dish and dried at 105°C in a hot-air oven until it reached a constant weight. The swollen starch sediment was weighed and the SP (g/g dry weight basis) and S (%) were calculated using the following relationships:

$$SP\ (g/g) = \frac{\text{Sediment weight}}{\left[\text{Weight of dry starch} \times (1-S)\right]} \tag{6.2}$$

$$S\ (\%) = \left(\frac{\text{Dried supernatant weight}}{\text{Weight of dry starch}}\right) \times 100 \tag{6.3}$$

6.2.6 PASTING PROPERTIES

Using Rapid Visco Analyzer (RVA) (Model Starch Master 2 from Newport Scientific, Australia), pasting properties were determined. The RVA sample in the aluminum canister was filled with 2 g of starch sample (corrected to 14% moisture basis) and mixed with 25 mL of distilled water. The temperature was held for 1 min at 50°C and then increased to 95°C for 3.75 min and kept for 2.5 min and cooled to 50°C for 3.75 min and then again held for 5 min. For the first 10 s, the paddle speed was kept at 960 rpm to homogenize the starch slurry and then maintained at 160 rpm till the end of the experiment; the viscosity was expressed in rapid visco units.

6.2.7 CYANIDE ESTIMATION

The total cyanide content in cassava flour was estimated by using spectrophotometric method (Nambisan and Shanavas, 2013).

6.2.8 COLOR

Color estimation was carried out using colorimeter (Ultrascan VIS, Hunterlab, USA), and L^*, a^*, and b^* values were noted. L-scale indicates lightness versus darkness, darkness ranges include (0–50) while that of lightness ranges (51–100); a^* scale indicates redness versus greenness and it has both negative and positive values. A positive value indicates redness while negative value indicates greenness. Similarly, *b** scale indicates yellowness versus blueness and have both positive and negative value. Yellowness is for positive value while blueness for negative value (CIE Commission Internationale de l'Eclairage, 2004).

6.2.9   WATER AND OIL-HOLDING CAPACITY

Water and oil absorption capacities were determined using Soluski's (1976) method. One gram of cassava flour sample was mixed with 10 mL of distilled water in preweighed centrifuged tube for measuring water-holding capacities (WHCs). In the case of measuring oil-holding capacity (OHC), instead of distilled water, refine soybean oil was used. The sample was kept for 10 min and centrifuged at 2000*g* for 10 min. WHC or OHC was expressed as the percentage of water or oil bound per gram of sample.

6.2.10   X-RAY DIFFRACTION

X-ray diffraction (XRD) analysis of flour samples was performed with an X-ray diffractometer (D/MAX 2500 V, Rigaku Corporation, Japan) operating at 30 kV and 15 mA with Cu $K\alpha$ radiation ($\lambda = 0.15418$ nm) at room temperature. The scans adjusted were in the range of 5–500 on a 2θ scale. Prior to X-ray scanning, sample of 50 mg was added into the slide for packing and the sample was dried at 50°C.

6.2.11   THERMAL ANALYSIS

Thermogravimetric analysis (TGA) and differential scanning calorimetry (DSC) (SHIMADZU, Model-TGA-50 and DSC-60, Software-TA-60WS) were used for thermal analysis. Thermal analysis of cassava flour samples was evaluated with the thermal analyzer (TGA-DSC). The measurements of TGA cassava flour samples were done using a thermogravimetric analyzer in nitrogen atmosphere at a heating rate of 10°C/min over a temperature range of 30–600°C. DSC measurements were performed. Cassava flour was weighed and placed in an aluminum pan with lid cover and sealed tightly. It was placed in the DSC instrument, with an empty aluminum pan used as reference. The samples were analyzed over a temperature range of 0–300°C at a heating rate of 10°C/min.

6.2.12 MORPHOLOGICAL STUDIES

Scanning electron microscopy-energy dispersive X-ray analysis (SEM-EDX) was performed using SEM-EDX absorption spectroscopy (Oxford Instrument, INCA X sight, model 7582). Cassava flour samples were taken and placed onto double-sided tape on a microscope stud using a sputter gold coat. SEM (JOEL JSM-6390LV, SEM, Oxford) was the instrument used for taking samples images. It was observed under 500×, 550× magnifying power with 20 kV voltage.

6.3   RESULT AND DISCUSSION

6.3.1 CHEMICAL ANALYSIS

The chemical composition of FCF and DCF is shown in Table 6.1. FCF found $10.58 \pm 0.5\%$, $0.69 \pm 0.21\%$, $32.90 \pm 0.2\%$, $2.27 \pm 0.02\%$, $63.94 \pm 0.1\%$, 6.94 ± 0.1 moisture, ash, PY, crude fiber carbohydrates, and pH, respectively. pH increased in FCF, whereas PY, crude fiber, ash, dry matter decreased in DCF. FCF had lesser carbohydrates content. These could be due to a possible transformation of some of the carbohydrate used by organisms as its carbon source to some other metabolites, such as protein or fat (Lehninger, 1987). The pH findings of the samples are within the range of 5–6, and these values are supported by the literature range of 5.5–8.5 (Muzanila et al., 2000; Ingram, 1975). pH is one important criterion for baking and the low level of pH (acidity) limits the substitution level of flour for composite flour baking (Aryee et al., 2006). In the case of pectin content in the samples, cassava has low pectin content as compared to citrus fruits. However, certain literature cited that 38.2% pectin content is found in cassava by-products meal as protopectin. The total pectin content found is 49.8 g/kg DM in dried beet pulp, and at an average of 33.8 ± 0.3 g/kg DM in fruits and vegetables, 13.2 ± 8.4 g/kg DM in legumes and tubers, and only 2.8±0.5 g/kg DM in cereals (Bailoni et al., 2005).

TABLE 6.1   Chemical Composition of Fermented Chips Flour (FCF) and Dry Chips Flour (DCF).

Samples	Moisture (%)	Ash (%)	Pectin yield (%)	Crude fiber (%)	Carbohydrates (%)	pH
FCF	10.58 ± 0.5	0.69 ± 0.21	32.90 ± 0.2	2.27 ± 0.02	54.75 ± 0.1	6.94 ± 0.1
DCF	7.62 ± 0.8	1.78 ± 0.01	38.00 ± 0.01	2.97 ± 0.05	63.94 ± 0.9	5.68 ± 0.03

Mean value of three replications with standard deviation.

6.3.2 TOTAL CYANIDE CONTENT

Total cyanide content in cassava flour is presented in Table 6.2. The antinutritional factor of cassava is from glycosides, mainly linamarin, present in the cell vacuoles. Disruption of the cells leads to contact with the enzyme linamarase, which is located in the cell walls, and leads to subsequent hydrolysis into glucose and cyanohydrins. By inhalation or ingestion, cyanide can poison a person (Jack et al., 2010). The total cyanide content of FCF and DCF was recorded at 9.11 ± 0.27 and 15.3 ± 0.23 ppm, respectively. The findings of the present investigation are within normal range, that is, the normal range of cyanogen content falls between 15 and 400 ppm fresh weight. Total cyanogen content is 6.8 (0.4 SD) mg HCN/kg for sun-drying chips in Kagera region (Muzanila et al., 2000). The variation can be ascribed to the age of the plant, particularly growing conditions (i.e., soil type, humidity, and temperature) and cultivars (Cardoso et al., 2005). The World Health Organization (WHO) has set the safe level of cyanogen in cassava flour at 10 ppm and the acceptable limit in Indonesia is 40 ppm (Emmanuel, 2002). FCF was found within the safe limit of WHO and both FCF and DCF were found within the acceptable limit of Indonesia. The difference in finding in FCF and DCF may be attributed to different processing techniques. Sun-drying of cassava root pieces result in decrease of toxic linamarin and is also associated with the low level of moisture; however, the stabilization takes place at 15% moisture. The cyanogen leaching takes place in fermentation processing technique which physically separates cyanogen from cassava tissue (Mkpong et al., 1990); the natural fermentation is one of the most efficient method for cyanogen removal from cassava (Westby and Choo, 1994).

6.3.3 SWELLING POWER AND SOLUBILITY

SP and solubility (S) of cassava flours are presented in Table 6.2. The SP and solubility (S) as the solubilization of starch components were quantified (Leach et al., 1959). SP was found higher (11.40 ± 0.2 g/g) in FCF, but in the case of solubility, DCF had higher than FCF. The SP showed the difference between FCF and DCF and the same with solubility. The different processing methods, one in dry form and another in fermented form, are attributed to the above findings. Processing parameters and environmental factors also had impact in swelling and solubilization properties (Moorthy and Ramanujam, 1986). Several findings had claimed a probable range of SP (0.8–15.5%)

and solubility (0.2–16.6%) at 75°C among ~4050 genotypes from a world collection (Sánchez et al., 2009) and our results fall in line with the above-mentioned ranges. At water-stress condition during initial and later stage, SP of starch was found to decrease (Santisopasri et al., 2001). As compared to potato and cereal starches, medium SP was found in cassava starch (Rickard et al., 1991).

TABLE 6.2   Functional Properties of Cassava.

Sample	SP (g/g)	S (%)	WHC (%)	OHC (%)	Total cyanide content (ppm)
FCF	11.40 ± 0.2	2.67 ± 0.21	208.6 ± 0.11	179.11 ± 0.11	9.11 ± 0.27
DCF	8.06 ± 0.7	6.04 ± 0.23	170.03 ± 0.54	160.54 ± 0.26	15.3 ± 0.23

Mean value of three replications with standard deviation.
SP, swelling power; S, solubility; WHC, water-holding capacity; OHC, oil-holding capacity.

6.3.4   WATER AND OIL ABSORPTION CAPACITY

WHC and OHC of cassava flour are presented in Table 6.2. WHC was found almost similar in both FCF and DCF. FCF showed higher oil absorption capacity (208.6 ± 0.21%) compared to DCF. The mechanism of OHC is the physical entrapment of oil by proteins (Kinsella, 1979). Fermentation may increase the protein content which may contribute in increasing OHC in FCF.

6.3.5   PASTING PROPERTIES

Pasting profile and properties of cassava flour obtained from RVA for DCF and FCF are illustrated in Figure 6.1a and b and Table 6.3. FCF has higher pasting temperature (PT), peak viscosity (PV), hold viscosity, final viscosity (FV), breakdown (BD), and set back viscosity than DCF. Fermentation brought changes in cassava-pasting properties. Various changes took place in fermentation and are responsible to increase all pasting properties. A paste containing the majority of swollen and intact granules showed highest paste viscosity (Thomas and Atwell, 1997). FCF showed higher swollen and intact granule than DCF as FCF has higher paste viscosity. The processing conditions on later and initial stages increased PT and decreased PV of pasting (Santisopasri et al., 2001). It has been observed that there is a little difference in starch pasting from dry chips either through wet or dry milling (Abera and Rakshit, 2003) which allows managing the crop supply in the industry during off seasons and glut periods. FV is the ability of the sample to form a gel which takes place after cooking and followed by cooling. BD is another starch property for application in food. The starch granule swells when subjected to heat and shear and undergoes fragmentation. It leads to reduction in viscosity which indicates the BD for the starch which is an undesirable quality as it leads to cohesive nature and uneven viscosity (Moorthy, 2004).

TABLE 6.3 Pasting Properties of Cassava Dry Chips Flour (DCF), Fermented Chips Flour (FCF) Obtained from RVA Viscogram.

Sample	PT (°C)	PV (cP)	HV (cP)	FV (cP)	BD (cP)	Setback 1 (cP)
DCF	38	815	804	1119	11	315
FCF	116	1961	1541	2104	420	563

PT, pasting temperature; PV, peak viscosity; HV, hold viscosity; FV, final viscosity; BD, breakdown.

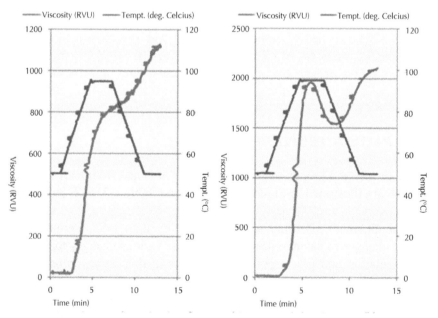

FIGURE 6.1 (See color insert.) Pasting profile of (a) dry chips flour (DCF) and (b) fermented chips flour (FCF) of cassava.

6.3.6 COLOR

L, *a*, and *b* values are presented in Table 6.4 and the criterion of color is an important factor for selection of flour. Usually, cassava root is white in color. When it is processed into flour or starch, it also possesses white color. The Hunter *L*, *a*, and *b* values were used to describe completely the white color intensity of the native and fermented flour. DCF has higher values of *L* compared to FCF which signifies that DCF is whiter in color than FCF. *L* value is one of the most important entities to describe the final flour color property, while *a* and *b* of both the samples give a negative value which indicates more closeness to green and blue, respectively. One can derive processing effect on the color of cassava flour. Fermentation makes the cassava flour darker in color. The difference may be ascribed to different processing techniques which give quantitative and qualitative differences of the pigments present in native and fermented cassava flour.

TABLE 6.4 *L*, *a*, and *b* Values of Cassava Dry Chips Flour (DCF) and Fermented Chips Flour (FCF).

Sample	*L**	*a**	*b**
FCF	78.98	−66.97	−26.12
DCF	79.02	−68.58	−24.62

6.3.7 X-RAY DIFFRACTION PATTERN

XRD patterns of DCF and FCF are given in Figure 6.2. DCF showed strong reflections at 15.3°, 17.6°, and 22.75° 2θ angles and FCF showed major sharp peaks at 11.3°, 15.2°, and 17.55° 2θ angles. The diffractogram revealed that both DCF and FCF flour is the mixture of A and B-type polymorphs. The diffractogram showed almost similar peak pattern except lesser peak of 11.3° found in FCF and higher peak of 22.75° in DCF. The difference in peak pattern may be due to the difference in processing technique like fermentation or difference in the moisture content of the sample. The above finding is supported by findings like the sample treatment can give rise to discrepancies. For example, the appearance of the C-type pattern from A-type polymorph can be due to increased moisture content of starch (Francisco et al., 1996). The mixture of A- and B types is C-type (Pérez and Bertoft, 2010). Various workers assigned cassava starch "A" and "C" XRD patterns (Moorthy, 2002). It is also found that cassava starch so far tested is

A or Ca-type (Ca-type denotes the portion of A-type is dominant over B-type in the polymorphic composition) (Zhu, 2015).

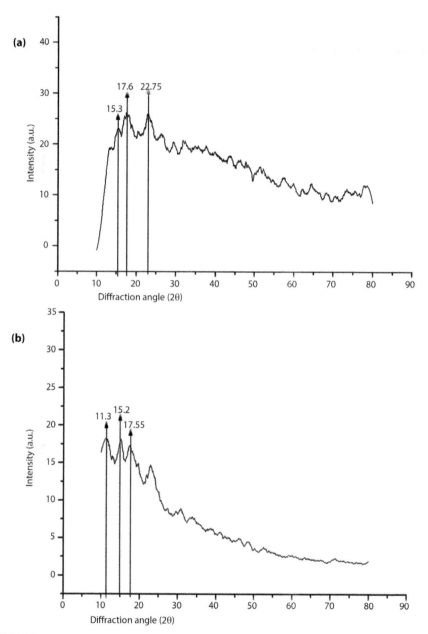

FIGURE 6.2   XRD graph of (a) dry chips flour (DCF) and (b) fermented chips flour (FCF).

6.3.8   STRUCTURAL ANALYSIS

Figure 6.3 illustrates the microphotographs of DCF and FCF. Both samples showed similarity in shape, but FCF showed slight lighter images compared to FCF. Both samples found maximum round granules with smooth surfaces. It also showed a small population of granules with truncated end. These findings are supported by the various literatures. Cassava starch has oval-to-round shape granules, with a small population of granules having truncated end (Charoenkul et al., 2011; Aggarwall and Dollimore, 1998). Different processing techniques on cassava flour showed a little effect on the morphology of cassava granule.

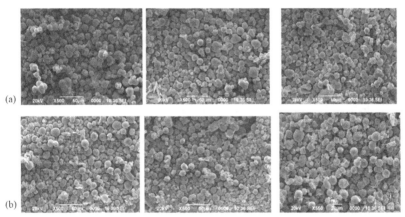

(a)

(b)

FIGURE 6.3   Scanning electron micrograph of (a) dry chips flour (DCF) and (b) fermented chips flour (FCF).

6.3.9   THERMAL PROPERTIES

The DSC thermographs of DCF and FCF are shown in Figure 6.4a and b and Table 6.5. DCF showed higher peak temperature of 95.73°C compared to FCF, which means that gelatinization of DCF is more than FCF. DCF will require more energy to start gelatinization. Gelatinization temperature of cassava flour of different varieties was reported within the range of 72–80°C (Charoenkul et al., 2011). Gelatinization temperature of FCF and DCF was found higher as compared to earlier findings. It could also be derived that the sample of DCF and FCF are from waxy genotypes as the literature noted. Waxy genotypes had higher gelatinization temperatures due to natural and artificial mutations (Cebalos et al., 2007; Raemakers et al., 2005). The melting temperature range (ΔT) was higher in LF. ΔT indicates the quality index and heterogeneity of the recrystallized amylopectin.

The TGA graph is illustrated in Figure 6.5. The TG curve of FCF and DCF also gives three stages of mass loss. The first-stage loss occurs in between 25°C and 106°C for FCF and 21°C and 103°C for DCF. The losses occurred due to the loss of moisture content through vaporization. The second-stage loss occurs at 251.26–339°C and 237.05–341.65°C for FCF and DCF, respectively. It is due to decomposition of hemicellulose and cellulose bond pyrolysis of starch. The third weight loss is also observed at 462–588°C for DCF and 385–598°C for FCF which is due to the formation of ash and decomposition of lignin (Gomand et al., 2010; Rudnik et al., 2006). The above finding was supported by LeVan's (1998) study and reported that there are three loss stages in TG curve (LeVan, 1998). The first-stage loss is from 30°C to 150°C, the loss is due dehydration and removal of moisture. The second-stage loss is between 200°C and 400°C, which occurs due to the decomposition of cellulose and hemicellulose and the third-stage loss occurs above 400°C which represents the decomposition of lignin and ash formation. The hemicellulose is less stable than the cellulose because of its side chains and it degrades before cellulose. The lignin degrades through hemicellulose and cellulose-degradation processes (Shafizadeh and DeGroot, 1976).

TABLE 6.5   Gelatinization Temperature Measured by Differential Scanning Calorimetry (DSC).

Sample	T_0	T_p	T_{end}	ΔT
FCF	83	94.05	124	41
DCF	91	95.73	117	26

T_0 = onset temperature; T_p = peak temperature; T_{end} = end temperature; and $\Delta T = T_{end} - T_0$.

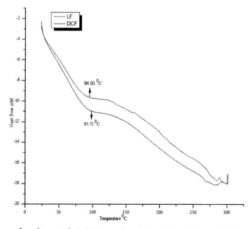

FIGURE 6.4   (See color insert.) DSC graph of dry chips flour (DCF) and fermented chips flour (FCF).

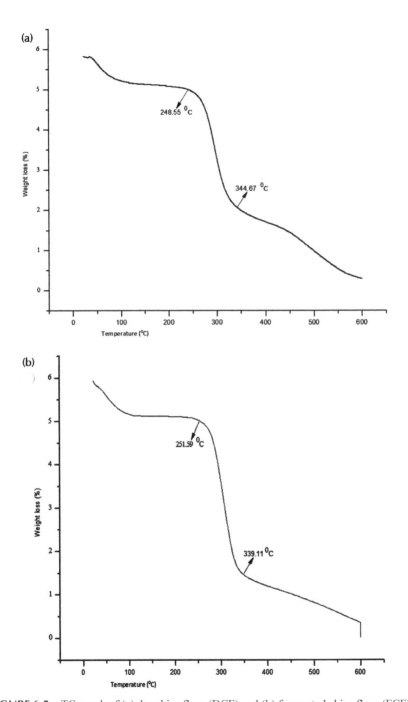

FIGURE 6.5   TG graph of (a) dry chips flour (DCF) and (b) fermented chips flour (FCF).

6.4   CONCLUSION

The North-Eastern region of India, Manipur, comprises mostly of hilly areas. Manipur has a tough geographical landscape inhabited by tribal people where cassava cultivation is carried out in Jhum areas. Cassava plays a crucial role in the food and nutritional security of the tribal population of North-Eastern region of India. The present findings can help designing value-added products in bakery and confectionery industries for utilizing cassava at a large scale. The novel technology of pectin extraction with assisted microwave can further look forward to extracting pectin from cassava meal and other by-products. The method was also less time consuming, easy, and sustainable. The roots present the antinutritional factor cyanogens. The maximum benefit out of cassava cannot be achieved until antinutritional factors of cyanogen are high. Scientists and researchers can further look forward searching effective method to reduce antinutritional factor to safe limit so that the rich carbohydrates can be fully utilized without posting harmful effect to health.

ACKNOWLEDGMENT

Financial help received from University Grants Commission (UGC), New Delhi (NET-JRF No. 1474/NET-DEC.2013) is duly acknowledged.

KEYWORDS

- physicochemical
- flour
- cassava
- cyanogens
- fermentation

REFERENCES

Abera, S.; Rakshit, S. K. Processing Technology Comparison of Physicochemical and Functional Properties of Cassava Starch Extracted from the Fresh Root and Dry Chips. *Starch—Stärke* **2003,** *55* (7), 287–296.

Aggarwall, P.; Dollimore, D. A. Thermal Analysis Investigation of Partially Hydrolyzed Starch. *Thermochim. Acta* **1998,** *319,* 17–25.

AOAC (Association of Official Analytical Chemists). In *Official Methods of Analysis,* 16th ed.; Horowitz, W.; Ed.; AOAC: Washington DC, 2012.

Aryee, F. N.; Oduro, I.; Ellis, W. O.; Afuakwa, J. J. The Physicochemical Properties of Flour Samples from the Roots of 31 Varieties of Cassava. *Food Control* **2006,** *17* (11), 916–922.

Bailoni, L.; Schiavon, S.; Pagnin, G.; Tagliapietra, F.; Bonsembiante, M. Quanti-Qualitative Evaluation of Pectins in the Dietary Fibre of 24 Foods. *Italian J. Anim. Sci.* **2005,** *4* (1), 49–58.

Cardoso, A. P.; Mirione, E.; Ernesto, M.; Massaza, F.; Cliff, J.; Rezaul, H. M.; Bradbury, J. H. Processing of Cassava Roots to Remove Cyanogens. *J. Food Compos. Anal.* **2005,** *18* (5), 451–460.

Cebalos, H.; Sánchez, T.; Morante, N.; Fregene, M.; Dufour, D.; Smith, A. M.; et al. Discovery of an Amylose-Free Starch Mutant in Cassava (*Manihot esculenta* Crantz). *J. Agric. Food Chem.* **2007,** *55,* 7469–7476.

Charoenkul, N.; Uttapap, D.; Pathipanawat, W.; Takeda, Y. Physicochemical Characteristics of Starches and Flours from Cassava Varieties Having Different Cooked Root Textures. *LWT—Food Sci. Technol.* **2011,** *44* (8), 1774–1781.

CIE. International Commission on Illumination, Colorimetry. *CIE, Technical Report,* 3rd ed.; Issue 15, Commission internationale de l'Eclairage, CIE Central Bureau, 2004.

Emmanuel, N. M. Cyanide Content of Gari. *Toxicol. Lett.* **2002,** *3* (1), 21–24.

Francisco, J. D.; Silverio, J.; Eliasson, A. C.; Larsson, K. A Comparative Study of Gelatinization of Cassava and Potato Starch in an Aqueous Lipid Phase (l2) Compared to Water. *Food Hydrocolloids* **1996,** *10,* 317–322.

Gomand, S. V.; Lamberts, L.; Derde, L. J.; Goesaert, H.; Vandeputte, G. E.; Goderis, B.; Visser, R. G. F.; Delcour, J. A. Structural Properties and Gelatinization Characteristics of Potato and Cassava Starches and Mutants Thereof. *Food Hydrocolloids* **2010,** *24* (4), 307–317.

Gularte, M. A.; Gomez, M.; Rosell, C. M. Impact of Legume Flours on Quality and In Vitro Digestibility of Starch and Protein from Gluten-Free Cakes. *Food Bioprocess Technol.* **2012,** *5* (8), 3142–3150.

Ingram, J. S. *Standards, Specifications and Quality Requirements for Processed Cassava Products*; TDRI: Lane, London, 1975.

Jack, B.; France, B.; Venkatesh, M.; Peter, R. Food Fortification—The Debate Continues. *Sci. Afr.* **2010,** *75.* Retrieved on August 23, 2010, from http://scienceinafrica.com/old/2002/june/food.htm.

Kinsella, J. E. Functional Properties of Soy Proteins. *J. Am. Oil Chem. Soc.* **1979,** *56* (3), 242–258.

Leach, H. W.; McCowen, L. D.; Schoch, T. J. Structure of the Starch Granule. 1. Swelling and Solubility Patterns of Various Starches. *Cer. Chem.* **1959,** *36,* 534–544.

Lehninger, A. L. *Principles of Biochemistry*; Worth Publishers: New York, 1987.

LeVan, S. L. Thermal Degradation. In *Concise Encylopedia of Wood and Wood-Based Materials*; Schniewind, A. P., Eds.; Pergamon Press: Elmsford, NY, 1998; pp 271–273.

Li, D.Q; Jia, X.; Wei, Z; Liu, Z. Y. Box–Behnken Experimental Design for Investigation of Microwave-Assisted Extracted Sugar Beet Pulp Pectin. *Carbohydr. Polym.* **2012,** *88* (1), 342–346.

Matos, M. E.; Rosell, C. M. Understanding Gluten-Free Dough for Reaching Bread with Physical Quality and Nutritional Balance. *J. Sci. Food Agric.* **2015**, *95* (4), 653–661.

Mkpong, O. E.; Yan, H.; Chism, G.; Sayre, R. T. Purification, Characterization, and Localization of Linamarase in Cassava. *Plant Physiol.* **1990**, *93* (1), 176–181.

Moorthy, S. N. Physicochemical and Functional Properties of Tropical Tuber Starches: A Review. *Starch—Stärke* **2002**, *54* (12), 559–592.

Moorthy, S. N. Tropical Sources of Starch, Starch in Food: Structure, Function, and Applications. *Food Science and Technology*; Woodhead Publishing: Sawston, Cambridge, United Kingdom, 2004; pp 321–359.

Moorthy, S. N.; Ramanujam, T. Variation in Properties of Starch in Cassava Varieties in Relation to Age of the Crop. *Starch—Stärke* **1986**, *38* (2), 58–61.

Muzanila, Y. C.; Brennan, J. G.; King, R. D. Residual Cyanogens, Chemical Composition, and Aflatoxins in Cassava Flour from Tanzanian Villages. *Food Chem.* **2000**, *70*, 45–49.

Nambisan, B.; Shanavas, S. Determination of Cyanogenic Potential by Spectrophotometric Method. *Assay for Determination of the Cyanide Potential of Cassava and Cassava Product*; Central Tuber Crops Research Institute: Sreekaryam, Trivandrum, 2013; pp 14–16.

Pereira, B. L. B.; Leonel, M. Resistant Starch in Cassava Products. *Food Sci. Technol.* **2014**, *34* (2), 298–302.

Pérez, S.; Bertoft, E. The Molecular Structures of Starch Components and Their Contribution to the Architecture of Starch Granules: A Comprehensive Review. *Starch—Stärke* **2010**, *62*, 389–420.

Prakash, M. J.; Sivakumar, V.; Thirugnanasambandham, K.; Sridhar, R. Optimization of Microwave Assisted Extraction of Pectin from Orange Peel. *Carbohydr. Polym.* **2013**, *97* (2), 703–709.

Raemakers, K.; Schreuder, M.; Suurs, L.; Furrer-Verhort, H.; Vinvken, J. P.; de Vetten, N. Improved Cassava Starch by Antisense Inhibition of Granule-Bound Starch Synthase I. *Mol. Breed.* **2005**, *16*, 163–172.

Rickard, J. E.; Asaoka, M.; Blanshard, J. M. V. The Physicochemical Properties of Cassava Starch. *Trop. Sci.* **1991**, *31*, 189–207.

Rudnik, E.; Matuschek, G.; Milanov, N.; Kettrup, A. Thermal Stability and Degradation of Starch Derivatives. *J. Therm. Anal. Calorimetry* **2006**, *85*, 267–270.

Sánchez, T.; Salcedo, E.; Ceballos, H.; Dufour, D.; Mafla, G.; Morante, N.; Calle, F.; Perez, J. C.; Debouck, D.; Jaramillo, G.; Moreno, I. X. Screening of Starch Quality Traits in Cassava (*Manihot esculenta* Crantz). *Starch–Stärke* **2009**, *61*, 12–19.

Santisopasri, V.; Kurotjanawong, K.; Chotineeranat, S.; Piyachomkwan, K.; Sriroth. K.; Oates, C. G. Impact of Water Stress on Yield and Quality of Cassava Starch. *Ind. Crops Prod.* **2001**, *13*, 115–129.

Shafizadeh, F; DeGroot, W. F. In *Thermal Uses and Properties of Carbohydrates and Lignins*; Shafizadeh, F.; Sarkanen, K. V.; Tilman, D. A.; Academic Press: New York, 1976; pp 1–18.

Soluski, F. W.; Garratt, M. O.; Slinkard, A. E. Functional Properties of Ten Legume Flours. *Can. Inst. Food Sci. Technol. J.* **1976**, *9*, 66–69.

Thomas, D. J.; Atwell, W. A. Gelatinization, Pasting, and Retrogradation. In *Starches Practical Guides for the Food Industry*; Eagen Press: St. Paul, MN, 1997; pp 25–30.

Wang, L.; Xie, B.; Shi, J.; Xue, S.; Deng, Q.; Wei, Y.; Tian, B. Physicochemical Properties and Structure of Starches from Chinese Rice Cultivars. *Food Hydrocolloids* **2010**, *24* (2), 208–216.

Westby, W.; Choo, B. K. Cyanogen Reduction during Lactic Fermentation of Cassava. *Acta Hortic.* **1994,** *375*, 209–215.

Zhu, F. Composition, Structure, Physicochemical Properties, and Modifications of Cassava Starch. *Carbohydr. Polym.* **2015,** *122*, 456–480.

CHAPTER 7

BIOFLAVONOIDS FROM *ALBIZIA MYRIOPHYLLA*: THEIR IMMUNOMODULATORY EFFECTS

KHWAIRAKPAM CHANU SALAILENBI MANGANG and SANKAR CHANDRA DEKA*

Department of Food Engineering and Technology, Tezpur University, Napaam, Tezpur 784028, Assam, India

Corresponding author. E-mail: sankar@tezu.ernet.in

ABSTRACT

Albizia myriophylla is therapeutically used in traditional Indian, Chinese, and Thai medicine, exhibiting a broad pharmacological effect. In one of our previous study, the ethanolic extract of *A. myriophylla* bark (AME) was used for its estimation of phytochemical, antioxidant, and antimicrobial properties. AME contained polyphenols, soluble sugars, terpenoids, anthraquinone, flavonoids, and tannins. The extract further exhibited antioxidant activity and inhibited lipid peroxidation and was effective against the spoilage organisms, namely, *Escherichia coli* MTCC 40 and *Staphylococcus aureus* MTCC 3160, whereas no activity was observed against *Sacharromyces cerevisae* ATCC 9763, *Lactobacillus plantarum* ATCC 8014 and *Lactobacillus sakei* ATCC 15521. It indicated a positive significance of the use of this extract along with probiotic bacterial strain for therapeutical applications. Also, high-performance liquid chromatography analysis of AME was done and 12 different phenolic compounds were detected. Among the phenolic compounds, three flavonoids namely, naringin, apigenin, and quercitin were found. Bioflavonoids are known to display health-beneficial effects and are termed as potential immunomodulatory agents. In our another study, *Pediococcus pentosaceus*, a probiotic strain, was found and isolated from *hamei*, in which *A. myriophylla* is one of the key ingredient. Thereby, a correlated activity of both AME constituting bioflavonoids and *P. pentosaceus* strains can be observed. *P. pentosaceus* is known to possess potential probiotic properties, stimulating immune activities effectively. Therefore, the synergistic effects of bioflavonoids from *A. myriophylla* and *P. pentosaceaus* strain may be further studied to observe better efficacy on immunomodulation.

7.1 INTRODUCTION

Albizia myriophylla is a native leguminous plant of Manipur belonging to fabaceae family and grows scattteredly in the wet and semi-evergreen forest of the hills. *Albizia* bark has been found to provide an ecological niche to a mold strain identified as *Mucor indicus*, which is responsible for the alcoholic fermentation of rice (Panmei et al., 2007). *A. myriophylla* bark (AME) is commonly utilized as an important constituent for the preparation of starter/catalyst called *"hamei"* which is added during the fermentation process of different fermented alcoholic beverage. In the preparation of *"hamei,"* finely chopped or powdered dried bark of *A. myriophylla* plant is mixed with water and the filtrate is used (Devi et al., 2013). Many experimental evidences of this genus have exhibited a broad pharmacological effects, including immunomodulatory (Barua et al., 2000), anticancer (Liang et al., 2005; Zhang et al., 2011), antimalarial (Rukunga et al., 2007), antioxidant (Jung et al., 2004), antimicrobial (Lam and Ng, 2011), anthelmintic (Egualea et al., 2011), and anti-inflammatory (Venkatesh et al., 2010) activities. In traditional Indian and Chinese medicine, *Albizia* plants are used therapeutically for insomnia, irritability, wounds, as antidysentric, antiseptic, antitubercular, etc.

In many food systems, plant phenolics act as antioxidants and prevent food deterioration. On the other hand, phenolics are antimutagenic and functional food components possessing health benefits or being able to prevent disease in human beings. Phenolics in alcoholic beverage are important in furnishing the color, producing the astringent taste and offering a reservoir for oxygen reduction. Some phenolics like caffeic, chlorogenic, ferulic, sinapic, and *p*-coumaric acids appear to be more active antioxidants, food flavor precursors, and considered to be an important part of the general defense mechanisms (Panmei et al., 2007).

Recently, plant extracts have attracted a great deal of attention mainly concentrated on their role in preventing diseases (Kaur and Kapoor, 2001). It is well accepted that oxidative stress which releases free oxygen radicals in the body is involved in the number of disorders including cardiovascular diseases, diabetes, and cancer which can be prevented by photochemicals in plant extracts. Oxidation caused by free radicals sets reduced capabilities to combat cancer, kidney damage, atherosclerosis, and heart diseases (Ames et al., 1993). Different phytochemicals in herbal products are safer than synthetic medicine and beneficial in the treatment of diseases caused by free radicals. Multiple biological effects of them have been described, such as antioxidants, cellular signals, cardioprotective effects, antibiotics, anti-inflammation, anti-allergic, anticoagulation, antineoplastic, antimutagenesis, and anticarcinogenesis (Ames et al., 1993; Ziegler, 1991; Halliwell and Gutteridge, 2001; Nakamura, 1997; Cook and Samma, 1996; Chen and Yen, 1997; Lin and Liang, 2000).

Immunomodulation is the alteration of immune responses to a desired level with monoclonal antibodies, cytokines, glucocorticoids, immunoglobulins, ultraviolet light, plasma pheresis, or related agents known to alter cellular or humoral immunity. The active agents responsible for inducing, enhancing, or suppressing an immune response are collectively called immune modulators. The prevalence of inflammatory diseases like allergies and autoimmune diseases is increasing. These diseases are often associated with a disturbed immune balance and a lack of immunological tolerance. Immunomodulatory regimens offer an attractive approach as they often have fewer side effects than existing drugs, including less potential for creating resistance in microbial diseases. Various flavonoids show a significant pharmacological and biochemical activity that affect the normal functions of immune cells such as B cells, T cells, macrophages, neutrophils, basophils, and mast cells (Middleton et al., 2000).

Moreover, modulation of the immune system in many different experimental models by different phytoconstituents of plants, such as lectins, peptides, and flavonoids, has also been reported (Shukla et al., 2009). Also, plant-derived products are of active interest in therapies for the modulation of immune cell activities being designed to combat inflammation, autoimmunity, and cancer. In several in vitro and in vivo studies, polyphenols, such as flavonoids, have shown to display immunomodulating properties. Further, some flavonoids have been found to exhibit anti-oxidant, antilipoperoxidant, antitumoral, antiplatelet, anti-ischemic, anti-allergic, and anti-inflammatory activities (Cao et al., 1997). Thereby, it is well established that there is a relationship between flavonoid intake and reduced risk of disease (Sanbongi et al., 1997). Henceforth, the impact of flavonoids on the immune system by modulating mechanisms involved in immune defense has been proposed in addition to their effects on inflammation against cancer, and in ameliorating oxidative stress (Benavente-Garcia and Castillo, 2008).

Also in our previous study (Mangang et al., 2017), the occurrence of *Pediococcus pentosaceus* strain in *hamei* has been observed. *P. pentosaceus* falls under the group of probiotic lactic acid bacteria. There are reports describing the probiotic properties of *Pediococcus* spp. (Yuksekdag and Aslim, 2010). This group of organisms can interact with mucosal immune cells or epithelial cells lining the mucosa to modulate specific functions of the mucosal immune system. Binding of microbe-associated molecular patterns with these innate pattern recognition receptors can activate antigen-presenting cells and modulate their function through the expression of surface receptors, secreted cytokines and chemokines, modulation of barrier functions, defense in production, and inflammatory signaling.

7.2 ALBIZIA myriophylla

Albizia species (family: Mimosaceae) were investigated to tap out the shelf of bioactive constituents that possess pharmacological properties. These species are used in folk medicine for the treatment of rheumatism, stomachache, cough, diarrhea, wounds, anthelmintic, etc. In traditional Indian and Chinese medicine, *Albizia* plants are used therapeutically for insomnia, irritability, wounds, as antidysentric, antiseptic, antitubercular, etc. Phytochemical studies on the genus *Albizia* have inferred them as a source of different group of natural product active against cytotoxicity and many other diseases. The narrower approach to reveal phytochemical, pharmacological, antioxidant, antidiabetic, anthelmentic, antibacterial, hepatoprotective, anti-inflammatory, and cytotoxic properties accompanied with possible number of bioactive constituents isolated from this species is discussed with a detailed description. This piece of report would promote these species for extensive research, to fetch the optimistic utility of phytoconstituents for its therapeutic applications (Kokila et al., 2013).

A. *myriophylla* has been found to be used as an important ingredient herb in various Thai herbal formulas for long by Thai traditional healers. The occurrence of different phenolic compounds has been observed in the ethanolic extract of AME in our previous study and some related images are being depicted in Figure 7.1 (Mangang et al., 2016). Three different flavonoids, namely, lupinifolin, 8-methoxy-7,3',4'-trihydroxyflavone, 7,8,3',4'-tetrahydroxyflavone, were found to be isolated from A. *myriophylla* by Joycharat et al. (2013). Some immunomodulatory activities of flavonoids isolated from plants have been reported earlier.

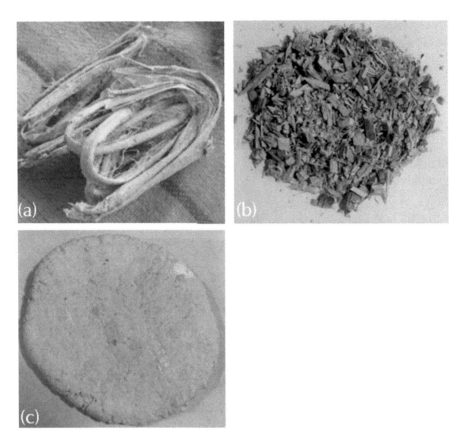

FIGURE 7.1 Different samples collected from Manipur: (a) *A. myriophylla* bark, (b) chopped *A. myriophylla* bark, and (c) *Hamei* from Andro.

Extraction, identification, and characterization of the flavonoids from different *Albizia* species were performed. Direct and sequential Soxhlet extraction showed maximum yield of the flavonoid from ethanolic extract and characterization of isolated flavonoid was done by infrared spectroscopy, ¹H nuclear magnetic resonance (NMR) spectroscopy, and mass spectrometry (Sharma and Janmeda, 2014). The flavonoid compounds were isolated and characterized by using thin-layer chromatography (TLC), purified by preparative TLC, and was identified using high-performance liquid chromatography (HPLC). Their structures and chemical bonds were analyzed using ultraviolet–visible spectrophotometery, Fourier transform-infrared spectroscopy, and NMR (¹³C and ¹H) techniques. Two flavonoids were identified as rutin and quercetin which showed a potent antioxidant radical-scavenging activity. The isolation of the above-characterized flavonoids would be useful to prepare plant-based pharmaceutical preparation to treat various complications linked with human diseases (Selvaraj et al., 2013).

7.3   PHYTOCHEMICALS ESTIMATION OF AME

This study was carried out on AME by Mangang et al. (2016). It comprises the estimation of different phytochemicals, namely, total phenolic compounds, total flavonoids, total soluble sugars, tannins, anthraquinone, terpenoids, alkaloids, and gylcosides.

TABLE 7.1   Concentration of Different Phytochemicals in AME.

Phytochemicals	Concentration (μg/mL) ± SD
Total soluble sugars	90.09 ± 9.58
Polyphenols	199.04 ± 9.63
Tannins	9.00 ± 0.71
Flavonoids	18.44 ± 1.64
Anthraquinone	16.87 ± 1.78
Terpenoids	60.33 ± 0.09
Alkaloids	ND
Glycosides	ND

Note: ND, not detected.

The phytochemicals study of AME revealed high content of polyphenols (199.04 mg/mL) followed by total soluble sugars and terpenoids content (Table 7.1). But alkaloids and glycosides contents were beyond the detectable range. Previous phytochemical studies of *A. myriophylla* have also observed the presence of phenolic acids, iminosugars, triterpene saponins, lignan glycosides, and alkaloids (Ito et al., 1994; Yoshikawa et al., 2002; Asano et al., 2005; Panmei et al., 2007).

7.4   HPLC ANALYSIS OF AME

It was performed for the identification of different phenolic compounds (Mangang et al., 2016). The HPLC chromatograms obtained for phenolic compounds are illustrated in Figure 7.2 and the results are presented in Table 7.2. In this analysis of AME, 12 different phenolic compounds were identified and indicated the presence of phenolic compounds in high propor-tionality. The highest amount was shown by salicylic acid (26,371.6 ppm) followed by naringin (17,698.12 ppm) and ferulic acid (2829.44 ppm) and the lowest amount by syringic acid (36.3 ppm). Desai et al. (2014), in the HPLC analysis of *Albizia julibrissin* foliage methanolic extract, confirmed the presence of three compounds: an unknown quercetin derivative with mass of 610 Da, hyperoside (quercetin-3-*O*-galactoside), and quercitrin (quercetin-3-*O*-rhamnoside). Lau et al. (2007) isolated catechin from *Albizia lebbeck* bark was confirmed with comparison with standard catechin by TLC and HPLC studies. Purity of isolated catechin was found to be 97.12% by HPLC. The HPLC method developed for quantification of catechin in the methanol extracts of *A. lebbeck* bark detected catechin peak at 1.79 min with the total run time 9 min. Linearity was obtained over a range of 10–60 mg/mL with correlation coefficient of 0.999. The detection limit and quantitation limit were found to be 1.0 and 3.0 mg/mL, respectively (Tatke et al., 2013).

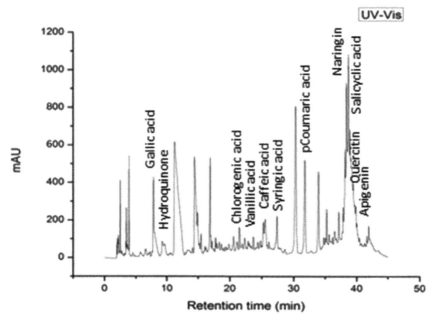

FIGURE 7.2 HPLC analysis for phenolic compounds.

TABLE 7.2   Phenolic Compounds Identified in AME by HPLC Analysis.

Phenolic compounds	Retention time (min)	Amount (ppm)
Gallic acid	7.09	339.81
Hydroquinone	8.97	49.84
Chlorogenic acid	22.85	1721.50
Vanillic acid	25.25	2685.73
Caffeic acid	26.12	780.48
Syringic acid	26.47	36.32
p-Coumaric acid	32.81	895.61
Ferulic acid	35.64	2829.44
Naringin	38.14	17,698.12
Salicylic acid	38.35	26,371.63
Quercitin	39.67	203.84
Apigenin	40.02	432.11

7.5   DPPH FREE RADICAL SCAVENGING ACTIVITY ASSAY OF AME

The antioxidant activity of AME was estimated by DPPH free radical scavenging activity assay (Mangang et al., 2016). The results shown in Figure 7.3 and Table 7.3 revealed that maximum antioxidant activity of 232.9% was observed with the highest concentration of AME. The minimum antioxidant activity of 27.8% was found in the least concentration of AME. The IC_{50} of AME was recorded to be 58 mg/mL. It was found that the antioxidant activity of the extract increased with increase in the concentration. The linearity in the curve for percentage inhibition was obtained after a concentration of 100 mg/mL, thus indicating the optimal AME concentration for scavenging activity. In the study of antioxidant activity of fruit exudate by Mathiesen et al. (1995), the reaction of DPPH with free radical scavenger was reported to cause a decline in the absorbance value.

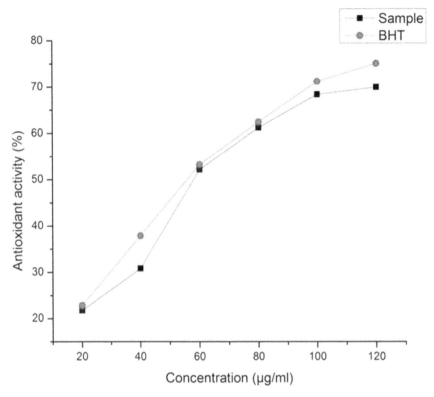

FIGURE 7.3 (**See color insert.**) Antioxidant activity of the *A. myriophylla* extract by DPPH free radical scavenging activity assay.

TABLE 7.3 Antioxidant Activity of the *A. myriophylla* Extract by DPPH Free Radical Scavenging Activity Assay.

Concentration (µg/mL)	Antioxidant activity (%)	
	Sample	BHT
20	21.74	22.81
40	30.83	37.84
60	52.17	53.26
80	61.26	62.45
100	68.38	71.19
120	69.96	75.07

7.6   THIOBARBITURIC ACID ASSAY OF AME

In the antioxidant activity determination of the extract by thiobarbituric acid (TBA) assay, the analysis was carried at six different concentrations of the extract (Mangang et al., 2016). The results are illustrated in Figure 7.4 and Table 7.4, where the maximum antioxidant activity of 94.5% was obtained at the highest concentration in 72 h. In the TBA assay, the inhibition of lipid peroxidation due to the presence of various concentrations of AME was evaluated. A linear correlation in between AME concentration and inhibition activity was noted. The highest inhibition activity was observed during the period of 72 h storage, and beyond that, there was a gradual decline. Sivakrishnan and Kottai (2013) observed that the concentration of thiobarbituric acid reactive substances was significantly reduced by ethanolic extract of *Albizia procera* with paracetamol-treated rats.

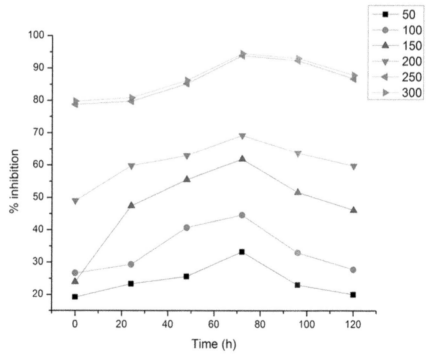

FIGURE 7.4   (See color insert.) Inhibition of lipid peroxidation (%) of the extract by thiobarbituric acid (TBA) assay.

TABLE 7.4 Inhibition of Lipid Peroxidation (%) of the Extract by Thiobarbituric Acid (TBA) Assay.

Concentration (µg/mL)	0 h	24 h	48 h	72 h	96 h	120 h
50	19.21	23.32	25.61	33.21	23.02	20.04
100	26.64	29.30	40.73	44.63	32.94	27.73
150	23.92	47.44	55.52	61.94	51.61	46.11
200	49.01	59.83	63.04	69.22	63.83	59.83
250	78.72	79.71	85.21	93.91	92.34	86.72
300	79.73	80.82	86.23	94.53	93.02	87.84

7.7 ANTIMICROBIAL ACTIVITY ANALYSIS OF AME

The study of antimicrobial activity of the extract against different microbes by Resazurin-based microtiter dilution assay (Mangang et al., 2016). The microtiter plates used for the study of antimicrobial activity of the extract against different microbes are depicted in Figure 7.5. The results for antimicrobial activity of AME against various microbes are presented in Table 7.5 which showed that there was almost no effect on the growth of *Sacharromyces cerevisae* ATCC 9763, *Lactobacillus plantarum* ATCC. Growth of *Staphylococcus aureus* MTCC 3160 and *Escherichia coli* MTCC 40 was not visible initially, but with the decrease in the concentration of AME, the spoilage microbes were detected. It was also observed that high concentrations (MIC of 10,937.5 mg/mL for *S. aureus* and 2734.4 mg/mL for *E. coli*) inhibited the growth of spoilage microbes but not on the growth of probiotic and fermenting microbes. *A. myriophylla*, which is a freely growing angiosperm, is a vigorous climbing plant and distributed abundantly in evergreen forests and other habitats in many East Asian countries. Only the barks of the plant are used in making of the AME and crude recovery is about 5%. Even though the MIC values are high against the pathogenic organisms and might be required in higher concentration while applying in food; yet economically, it can be considered as feasible. Some of the chemical groups such as phenolics and alkaloids have been established to be the major phytochemical classes of antibacterial metabolites (Samy and Gopalakrishnakone, 2010). The methanolic extract of *Ocimum sanctum* leaf has been reported previously to possess antibacterial activity against *S. aureus* (Rahman et al., 2010). This same species is known as a rich source of essential oils which were shown previously to be effective against both Gram-positive and Gram-negative bacteria. In addition, oil, alkaloids, glycosides, saponins, and tannins have also been reported from this plant species (Mondal et al., 2009).

TABLE 7.5   The Effect of the Extract against Different Microbes.

Concentration (μg/mL)	S. aureus	E. coli	S. cerevisae	L. plantarum	L. sakei
500,000.0	−ve	−ve	+ve	−ve	+ve
250,000.0	−ve	−ve	+ve	+ve	+ve
175,000.0	−ve	−ve	+ve	+ve	+ve
87,500.0	−ve	−ve	+ve	+ve	+ve
43,750.0	−ve	−ve	+ve	+ve	+ve
21,875.0	−ve	−ve	+ve	+ve	+ve
10,937.5	−ve	−ve	+ve	+ve	+ve
5468.8	+ve	−ve	+ve	+ve	+ve
2734.4	+ve	−ve	+ve	+ve	+ve
1367.2	+ve	+ve	+ve	+ve	+ve
683.6	+ve	+ve	+ve	+ve	+ve
341.8	+ve	+ve	+ve	+ve	+ve
170.9	+ve	+ve	+ve	+ve	+ve
85.5	+ve	+ve	+ve	+ve	+ve
42.8	+ve	+ve	+ve	+ve	+ve
21.4	+ve	+ve	+ve	+ve	+ve
10.7	+ve	+ve	+ve	+ve	+ve

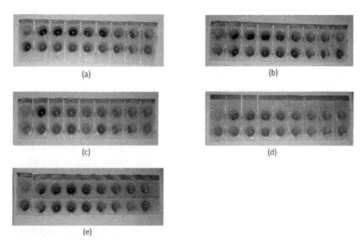

(a)   (b)

(c)   (d)

(e)

FIGURE 7.5   (See color insert.) Microtiter plates used for the study of antimicrobial activity of the extract against different microbes: (a) *S. aureus*, (b) *E. coli*, (c) *L. plantarum*, (d) *L. sakei*, and (e) *S. cerevisae*.

7.8 BIOFLAVONOIDS IN *A. myriophylla* AND ITS IMMUNOMODULATORY EFFECTS

In our previous study, the presence of three flavonoids, namely, naringin, apigenin, and quercitin in AME have been revealed (Mangang et al., 2016). Three different flavonoids, namely, lupinifolin, 8-methoxy-7,3′,4′-trihydroxyflavone, 7,8,3′,4′-tetrahydroxyflavone were also reported to be isolated from *A. myriophylla* by Joycharat et al. (2013). Flavonoids were found to impart a variety of biological activities, including anti-oxidant, anti-inflammatory, and antigenotoxic effects. Also, several in vitro and in vivo studies have revealed that polyphenols, such as flavonoids, display immunomodulating properties.

The effects of flavone luteolin and apigenin on immune cell functions, including proliferation, natural killer (NK) cell activity, and cytotoxic T-lymphocyte (CTL) activity of isolated murine splenocytes were investigated by Kilani-Jaziri et al. (2016). Luteolin and apigenin were found to significantly promote lipopolysaccharide (LPS)-stimulated splenocyte proliferation and enhance humoral immune responses. When compared to apigenin, luteolin induced a weak-cell proliferation of lectin-stimulated splenic T cells and both flavones significantly enhanced NK cell and CTL activities. Furthermore, both flavones were demonstrated to inhibit lysosomal enzyme activity, suggesting a potential anti-inflammatory effect. Henceforth, it was concluded that flavones exhibited an immunomodulatory effect that could be ascribed, in part, to its cytoprotective capacity via its anti-oxidant activity.

Flavonoids showing significant antitumor activity were indicated to enhance the immune function (Dai et al., 2008). Flavonoids including scutellarin, naringin, apigenin, luteolin, and wogonin significantly inhibited the proliferation and invasion of MHCC97H cells in a dose-dependent manner (Dai et al., 2013). Gong et al. (2015) evaluated the inhibition of tumor growth and immunomodulatory effects of flavonoids and scutebarbatines in Lewis-bearing C57BL/6 mice and indicated a positive effect, modulating the immune function. Experimental evidence was provided for the application in the treatment of lung cancer. Also, in an in vivo two-stage carcinogenesis test, lupinifolin exhibited a marked inhibitory effect on mouse skin tumor promotion (Itoigawa et al., 2002).

Lupinifolin was observed to significantly inhibit peristaltic index, intestinal fluid volume, and PGE2-induced enteropooling. They were also found to restore alterations in biochemical parameters such as nitric oxide, total carbohydrates, protein, DNA, superoxide dismutase, catalase, and lipid peroxidation. Lupinifolin indicated a significant recovery from Na^+ and K^+ loss and a pronounced antibacterial activity against bacterial strains mainly implicated in diarrhea (Prasad et al., 2013). In addition, it demonstrated selective cytotoxicity toward breast cancer with IC_{50} values of 3.3 µg/mL and expressed weakly cytotoxic activity toward KB and NCI-H187. Moreover, moderate antimycobacterial activity with MIC at 12.5 µg/mL was also expressed (Soonthornchareonnon et al., 2004).

In a study by Lyu and Park (2005), flavonoids were examined for their effects on LPS-induced NO production in RAW 264.7 macrophages and showed that all compounds were not strongly cytotoxic at the tested concentrations on human peripheral blood mononuclear cells and RAW 264.7 macrophages. On immunomodulatory properties, catechin, epigallo-catechin (EGC), naringenin, and fisetin were shown to repress NO production and TNF-a secretion. Furthermore, catechin, EGC gallate, epicatechin, luteolin, chrysin, quercetin, and galangin elevated IL-2 secretion while EGC, apigenin, and fisetin inhibited the secretion. These results revealed that flavonoids have the ability to modulate the immune response and have a potential anti-inflammatory activity. Moreover, no obvious structure–activity relationship regard to the chemical composition of the flavonoids and their cell biological effects was indicated.

The hydrolysis of naringin toward naringenin with immobilized naringinase was observed and their anti-inflammatory activity was evaluated (Ribeiro et al., 2008). The contribution of antioxidant properties to a possible anti-inflammatory effect was analyzed using an acute local inflammation model (rat paw edema induced by λ-carrageenan). The results revealed that rats ($n = 9$) pretreated with a solution of naringin, rats ($n = 9$) pretreated with a solution of naringenin, and rats ($n = 9$) treated with a solution of naringenin and naringin indicated a significant reduction on edema formation, 6 h after λ-carrageenan injection. Also, naringenin demonstrated a high in vivo anti-inflammatory activity and only 8% of paw edema ($p < 0.001$) was observed in rats pretreated with a solution of naringenin.

The neuroprotective effects of naringin were evidenced by upregulation of antioxidant enzymes, and attenuation of lipid peroxides formation (Mani and Sadiq, 2015). Thereby, naringin supplementation promoted the Nrf2 dissociation of Keap-1 and was likely contributed to its nuclear translocation in the brain. Naringin supplementation was also observed to control the abnormalities of deltamethrin-induced microglial activation by significantly attenuating the glial fibrillary acidic protein expression. Altogether, an activation of Nrf2/ARE pathways in brain by naringin was clearly indicated, protecting brain from DLM-induced neurotoxicity.

7.9 SYNERGISTIC EFFECTS OF BIOFLAVONOIDS AND *P. pentosaceus* STRAINS ON IMMUNOMODULATION

In our previous study, *P. pentosaceus* strains were detected in *hamei* and were isolated and identified (Mangang et al., 2017). *Hamei* is a traditional starter of fermentation in which AME is one of the key ingredients. Thereby, a correlated activity of both AME-constituting bioflavonoids and *P. pentosaceus* strains can be observed. *Pediococcus pentosaceus*, a novel strain of lactic acid, was found to display potential probiotic properties such as hydrophobicity, auto-aggregation, co-aggregation, and in vitro cell adhesion ability (Shukla and Goyal, 2014). *P. pentosaceus* isolated from Japanese traditional vegetable pickles was observed to stimulate immune activities effectively, showing allergic inhibitory effects and suggested the possibility of its use as functional foods (Jonganurakkun et al., 2008). Kekkonen (2008) investigated the immunomodulatory properties of probiotic strains by systematical screening in primary cell culture using human peripheral blood mononuclear cells and to evaluate the effects of probiotics on the immune system in healthy adults in randomized, double-blind, placebo-controlled intervention trials.

The modulation of gut microbial population by phenolics was reviewed to understand the two-way phenolic–microbiota interaction. Human intestinally isolated bacteria were shown to hydrolyze isoflavone C- and O-glycosides, as well as apigetrin, flavone O-glycoside (Kim et al., 2015). The intestinal microbial flora and its consistency apparently play a very central role in the metabolism of plant phenolics. The flavonoids are hydrolyzed by gut microbial enzymes to their aglycons, which represent the bioavailable and bioactive form (Puupponen-Pimia et al., 2001). Films with green tea and probiotics were able to extend shelf life of fish at least for a week and increase the beneficial lactic acid bacteria (Lopez-Lacey et al., 2014). Moreover, antimicrobial activity of flavonoids and phenolic acids from common Finnish berries was observed against selected bacterial species, including probiotic bacteria. Different sensitivities toward phenolics were exhibited by different bacterial species and these properties can be further utilized in the development of functional foods and in food preservative purposes (Puupponen-Pimia et al., 2001).

7.10   CONCLUSION

A. myriophylla were investigated to possess pharmacological properties with the presence of bioactive constituents. Phytochemicals study of AME revealed the presence of a wide array of phytochemicals, such as total polyphenols, total soluble sugars, and more. The HPLC study also depicted the presence of 12 numbers of phenolic compounds in high proportionality including three flavonoids, namely, naringin, apegenin, and quercitin. This signifies the use of this plant for therapeutical purposes. Natural flavonoids are known for their beneficial effects on health and were determined as potential immuno-modulatory agents, acting as direct and indirect antioxidants. Antioxidant activity of the extract was evident from the result of DPPH activity in which a linear correlation was observed in between the concentration of the extract and radical scavenging activity. Inhibition of lipid peroxidation was also shown by the extract, as apparent from the results of thiobarbituric assay. In the antimicrobial assay, it was seen that extract was effective against the spoilage organisms, namely, *E. coli* and *S. aureus*, whereas no activity was observed against *S. cerevisae*, *L. plantarum*, and *Lactobacillus sakei*. This indicated a positive significance of the use of this extract in the fermentation process as it was noneffective against the organisms responsible for fermentation. Also, there arises the possibility of its use in collaboration with some known probiotic bacteria to impact more enhanced beneficial health effects. Concisely, *A. myriophylla* was reported for the presence of bioactive constituents like flavonoids that possess pharmacological properties and natural flavonoids were determined as potential immunomodulatory agents. Therefore, the identification of potential bioflavonoids from *A. myriophylla* exhibiting distinctive immunomodulatory effects may be carried out. Also, both the natural flavonoids and *P. pentosaceus* were shown to efficiently stimulate the immune activities in healthy adults. Henceforth, the synergistic effects of bioflavonoids from *A. myriophylla* and *P. pentosaceaus* strain may be further studied to observe better efficacy on immunomodulation.

KEYWORDS

- *Albizia myriophylla*
- bioflavonoids
- *Pediococcus pentosaceus*
- synergistic effects
- immunomodulation

REFERENCES

Ames, B.; Shigena, M.; Hagen, T. Oxidants, Antioxidants and the Degenerative Diseases of Aging. *Proc. Nat. Acad. Sci. U.S.A.* **1993,** *90,* 7915–7922.

Asano, N.; Yamauchi, T.; Kagamifuchi, K.; Shimizu, N.; Takahashi, S.; Takatsuka, H.; Ikeda, K.; Kizu, H.; Chuakul, W.; Kettawan, A.; Okamoto, T. Iminosugar-Producing Thai Medicinal Plants. *J. Nat. Prod.* **2005,** *68,* 1238–1242.

Barua, C. C.; Gupta, P. P.; Patnaik, G. K.; Misra-Bhattacharya, S.; Goel, R. K.; Kulshrestha, D. K.; Dubey, M. P.; Dhawan, B. N. Immunomodulatory Effect of *Albizzia lebbeck*. *Pharm. Biol.* **2000,** *38,* 161–166.

Benavente-Garcia, O.; Castillo, J. Update on Uses and Properties of Citrus Flavonoids: New Findings in Anti-Cancer, Cardiovascular, and Anti-Inflammatory Activity. *J. Agric. Food Chem.* **2008,** *56,* 6185–6205.

Cao, G.; Sofic, E.; Prior, R. L. Anti-Oxidant and Pro-Oxidant Behavior of Flavonoids: Structure–Activity Relationships. *Free Radic. Biol. Med.* **1997,** *22* (5), 749–760.

Chen, H.; Yen, G. Possible Mechanisms of Antimutagen by Various Teas as Judged by their Effect on Mutagenesis by 2-Amino-3-Methylmidazo [4,5-*f*] Quinoline and Benzo(*a*) pyrine. *Mut. Res.* **1997,** *393,* 115–122.

Cook, N.; Samma, S. Flavonoids—Chemistry, Metabolism, Cardioprotective Effects and Dietary Sources. *J. Nutr. Biochem.* **1996,** *7,* 66–76.

Dai, Z. J.; Liu, X. X.; Tang, W.; Xue, Q.; Wang, X. J.; Ji, Z. Z.; Kang, H. F.; Diao, Y. Antitumor and Immune Modulating Effects of *Scutellaria barbata* Extract in Mice Bearing Hepato Carcinoma H22 Cells Derived Tumor. *Nan Fang Yi Ke Da Xue Xue Bao* **2008,** *28* (10), 1835–1837.

Dai, Z. J.; Wang, B. F.; Lu, W. F.; Wang, Z. D.; Ma, X. B.; Min, W. L.; Kang, H. F.; Wang, X. J.; Wu, W. Y. Total Flavonoids of *Scutellaria barbata* Inhibit Invasion of Hepatocarcinoma via MMP/TIMP In Vitro. *Molecules* **2013,** *18,* 934–950.

Desai, S.; Tatke, P.; Gabhe, S. Y. Isolation of Catechin from Stem Bark of *Albizia lebbeck*. *Int. J. Anal., Pharm. Biomed. Sci.* **2014,** *3,* 31–35.

Devi, M. R.; Bawari, M.; Paul, S. B. Neurotoxic Effect of *Albizia myriophylla* Benth, a Medicinal Plant in Male Mice. *Int. J. Pharm. Pharm. Sci.* **2013,** *5,* 243–248.

Egualea, T.; Tadesseb, D.; Giday, M. In vitro Anthelmintic Activity of Crude Extracts of Five Plants against Egg-Hatching and Larval Development of *Haemonchus contortus*. *J. Ethnopharmacol.* **2011,** *137,* 108–113.

Gong, T.; Wang, C. F.; Yuan, J. R.; Li, Y.; Gu, J. F.; Zhao, B. J.; Zhang, L.; Jia, X. B.; Feng, L.; Liu, S. L. Inhibition of Tumor Growth and Immunomodulatory Effects of Flavonoids and Scutebarbatines of *Scutellaria barbata* D. Don in Lewis-Bearing C57BL/6 Mice. *Evid.-Based Complement. Altern. Med.* **2015,** *2015,* 1–11.

Halliwell, B.; Gutteridge, J. *Free Radicals in Biology and Medicine*; Oxford Science Publication, Thomson Press Ltd.: India, 2001; pp 617–854.

Ito, A.; Kasai, R.; Yamasaki, K.; Duc, N. M.; Nham, N. T. Lignan Glycosides from Bark of *Albizzia myriophylla*. *Phytochemistry* **1994,** *37,* 1455–1458.

Itoigawa, M.; Ito, C.; Ju-ichi, M.; Nobukuni, T.; Ichiishi, E.; Tokuda, H.; Nishino, H.; Furukawa, H. Cancer Chemopreventive Activity of Flavanones on Epstein–Barr Virus Activation and Two-Stage Mouse Skin Carcinogenesis. *Cancer Lett.* **2002,** *176,* 25–2926.

Jonganurakkun, B.; Wang, Q.; Xu, S. H.; Tada, Y.; Minamida, K.; Yasokawa, D.; Sugi, M.; Hara, H.; Asano, K. *Pediococcus pentosaceus* NB-17 for Probiotic Use. *J. Biosci. Bioeng.* **2008,** *106* (1), 69–73.

Joycharat, N.; Thammavong, S.; Limsuwan, S.; Homlaead, S.; Voravuthikunchai, S. P.; Yingyongnarongkul, B. Antibacterial Substance from *Albizia myriophylla* Wood against Cariogenic *Streptococcus mutans*. *Arch. Pharm. Res.* **2013,** *36*, 723–730.

Jung, M. J.; Kang, S. S.; Jung, Y. J.; Choi, J. S. Phenolic Glycosides from the Stem Bark of *Albizzia julibrissin*. *Chem. Pharm. Bull.* **2004,** *52*, 1501–1503.

Kaur, C.; Kapoor, H. C. Antioxidants in Fruits and Vegetables the Millennoim's Health. *Int. J. Agric. Food Sci. Technol.* **2001,** *36*, 703–725.

Kekkonen, R. *Immunomodulatory Effects of Probiotic Bacteria in Healthy Adults*. Thesis submitted to Medical Faculty of the University of Helsinki for Public Examination in Lecture Hall 2, Biomedicum Helsinki: Haartmaninkatu, 8, 2008.

Kilani-Jaziri, S.; Mustapha, N.; Mokdad-Bzeouich, I.; Gueder, D. E.; Ghedira, K.; Ghedira-Chekir, L. Flavones Induce Immunomodulatory and Anti-Inflammatory Effects by Activating Cellular Anti-Oxidant Activity: A Structure–Activity Relationship Study. *Tumor Biol.* **2016,** *37*, 6571–6579.

Kim, M.; Lee, J.; Han, J. Deglycosylation of Isoflavone *C*-Glycosides by Newly Isolated Human Intestinal Bacteria. *J. Sci. Food Agric.* **2015,** *95*, 1925–1931.

Kokila, K.; Priyadharshini, S. D.; Sujatha, V. Phytopharmacological Properties of *Albizia* Species: A Review. *Int. J. Pharm. Pharm. Sci.* **2013,** *5*, 70–73.

Lam, S. K.; Ng, T. B. First Report of an Anti-Tumor, Antifugal, Anti-Yeast and Anti-Bacterial Hemolysis from *Albizia lebbeck* Seeds. *Phytomedicine* **2011,** *18*, 601–608.

Lau, C. S.; Carrier, D. J.; Beitle, R. R.; Bransby, D. I.; Howard, L. R.; Junior-Lay, J. O.; Liyanage, R.; Clausen, E. C. Identification and Quantification of Glycoside Flavonoids in the Energy Crop *Albizia julibrissin*. *Bioresour. Technol.* **2007,** *98*, 429–435.

Liang, H.; Tong, W. Y.; Zhao, Y. Y.; Cui, J. R.; Tu, G. Z. An Antitumor Compound Julibroside J28 from *Albizia julibrissin*. *Bioorg. Med. Chem. Lett.* **2005,** *15*, 4493–4495.

Lin, J.; Liang, Y. Cancer Chemoprevention by Tea Polyphenols. *Proc. Nat. Sci. Council, Republ. China, B* **2000,** *24*, 1–13.

Lopez-Lacey, A. M.; Lopez-Caballero, M. E.; Montero, P. Agar Films Containing Green Tea Extract and Probiotic Bacteria for Extending Fish Shelf-Life. *LWT—Food Sci. Technol.* **2014,** *55* (2), 559–564.

Lyu, S. Y.; Park, W. B. Production of Cytokine and NO by Raw 264.7 Macrophages and PBMC In Vitro Incubation with Flavonoids. *Arch. Pharm. Res.* **2005,** *28*, 573.

Mangang, K. C. S.; Das, A. J.; Deka, S. C. Shelf Life Improvement of Rice Beer by Incorporation of *Albizia myriophylla* Extracts. *J. Food Process. Preserv.* **2016.** DOI:10.1111/jfpp.12990.

Mangang, K. C. S.; Das, A. J.; Deka, S. C. Comparative Shelf Life Study of Two Different Rice Beers Prepared Using Wild-Type and Established Microbial Starters. *J. Inst. Brew.* **2017.** DOI:10.1002/jib.446.

Mani, V. M.; Sadiq, A. M. M. Flavonoid Naringin Inhibits Microglial Activation and Exerts Neuroprotection against Deltamethrin Induced Neurotoxicity through Nrf2/ARE Signaling in the Cortex and Hippocampus of Rats. *World J. Pharm. Sci.* **2015,** *3* (12), 2410–2426.

Mathiesen, L.; Malterud, K. E.; Sund, R. B. Antioxidant Activity of Fruit Exudate and C Methylated Dihydrochalcones from *Myrica gale*. *Plant. Med.* **1995,** *61*, 515–518.

Middleton, E.; Kandaswami, C.; Theoharides, T. C. The Effects of Plant Flavonoids on Mammalian Cells: Implications for Inflammation, Heart Disease and Cancer. *Pharm. Rev.* **2000**, *52*, 673–751.

Mondal, S.; Mirdha, B. R.; Mahapatra, S. C. The Science behind Sacredness of Tulsi (*Ocimum sanctum* Linn.). *Indian J. Physiol. Pharmacol.* **2009**, *53*, 291–306.

Nakamura, H. Effect of Tea Flavonoid Supplementation on the Susceptibility of Low Density Lipoprotein to Oxidative Modification. *Am. J. Clin. Nutr.* **1997**, *66*, 261–266.

Panmei, C.; Singh, P. K.; Gautam, S.; Variyar, P. S.; Devi, G. A. S.; Sharma, A. Phenolic Acids in *Albizia* Bark Used as a Starter for Rice Fermentation in *Zou* Preparation. *J. Food, Agric. Environ.* **2007**, *5*, 147–150.

Prasad, S. K.; Laloo, D.; Kumar, M.; Hemalatha, S. Antidiarrhoeal Evaluation of Root Extract, Its Bioactive Fraction, and Lupinifolin Isolated from *Eriosema chinense*. *Plant. Med.* **2013**, *79*, 1620–1627.

Puupponen-Pimia, R.; Nohynek, L.; Meier, C.; Kahkonen, M.; Heinonen. M.; Hopia, A.; Oksman-Caldentey, K. M. Antimicrobial Properties of Phenolic Compounds from Berries. *J. Appl. Microbiol.* **2001**, *90*, 494–507.

Rahman, M. S.; Khan, M. M. H.; Jamal, M. A. H. M. Anti-Bacterial Evaluation and Minimum Inhibitory Concentration Analysis of *Oxalis corniculata* and *Ocimum santum* against Bacterial Pathogens. *Biotechnology* **2010**, *9*, 533–536.

Ribeiro, I. A.; Rocha, J.; Sepodes, B.; Mota-Filipe, H.; Ribeiro, M. H. Effect of Naringin Enzymatic Hydrolysis towards Naringenin on the Anti-Inflammatory Activity of Both Compounds. *J. Mol. Catal., B: Enzym.* **2008**, *52–53*, 13–18.

Rukunga, G. M.; Muregi, F. W.; Tolo, F. M.; Omar, S. A.; Mwitari, P.; Muthaura, C. N.; Omlin, F.; Lwande, W.; Hassanali, A.; Githure, J.; Iraqi, F. W.; Mungai, G. M.; Kraus, W.; Kofi-Tsekpo, W. M. The Antiplasmodial Activity of Spermine Alkaloids Isolated from *Albizia gummifera*. *Fitoterapia* **2007**, *78*, 455–459.

Samy, R. P.; Gopalakrishnakone, P. Therapeutic Potential of Plants as Anti-Microbials for Drug Discovery. *Evid.-Based Complement. Altern. Med.* **2010**, *7*, 283–294.

Sanbongi, C.; Suzuki, N.; Sakane, T. Polyphenols in Chocolate, Which Have Anti-Oxidant Activity, Modulate Immune Functions in Humans In Vitro. *Cell. Immunol.* **1997**, *177*, 129–136.

Selvaraj, K.; Chowdhury, R.; Bhattacharjee, C. Isolation and Structural Elucidation of Flavonoids from Aquatic Fern *Azolla microphylla* and Evaluation of Free Radical Scavenging Activity. *Int. J. Pharm. Pharm. Sci.* **2013**, *5*, 743–749.

Sharma, V.; Janmeda, P. Extraction, Isolation and Identification of Flavonoid from *Euphorbia neriifolia* Leaves. *Arab. J. Chem.* **2014**. DOI:10.1016/j.arabjc.2014.08.019.

Shukla, R.; Goyal, A. Probiotic Potential of *Pediococcus pentosaceus* CRAG3: A New Isolate from Fermented Cucumber. *Probiot. Antimicrob. Proteins* **2014**, *6*, 11–21.

Shukla, S.; Mehta, A.; John, J.; Mehta, P.; Vyas, S. P.; Shukla, S. Immunomodulatory Activities of Ethanolic Extract of *Caesalpinia bonducella* Seeds. *J. Ethnopharmacol.* **2009**, *125*, 252–256.

Sivakrishnan, S.; Kottai, M. A. In vivo Antioxidant Activity of Ethanolic Extract of Aerial Parts of *Albizia procera roxb* (benth.) against Paracetamol Induced Liver Toxicity on Wistar Rats. *J. Pharm. Sci. Res.* **2013**, *5*, 174–177.

Soonthornchareonnon, N.; Ubonopas, L.; Kaewsuwan, S.; Wuttiudomlert, M. Lupinifolin, a Bioactive Flavanone from *Myriopteron extensum* (Wight) K. Schum. Stem. *Thai J. Phytopharm.* **2004**, *11* (2), 19–28.

Tatke, P.; Desai, S.; Gabhe, S. Y.; Gaitonde, V. Development of Rapid HPLC Method for Estimation of Catechin from *Albizia lebbeck* Bark Extracts and Its Validation. *J. Altern. Integr. Med.* **2013,** *2,* 248.

Venkatesh, P.; Mukherjee, P. K.; Kumar, N. S.; Bandyopadhyay, A.; Fukui, H.; Mizuguchi, H.; Islam, N. Anti-Allergic Activity of Standardized Extract of *Albizia lebbeck* with Reference to Catechin as a Phytomarker. *Immunopharmacol. Immunotoxicol.* **2010,** *32,* 272–276.

Yoshikawa, M.; Morikawa, T.; Nakano, K.; Pongpiriyadacha, Y.; Murakami, T.; Matsuda, H. Characterization of New Sweet Triterpene Saponins from *Albizia myriophylla. J. Nat. Prod.* **2002,** *65,* 1638–1642.

Yuksekdag, Z. N.; Aslim, B. Assessment of Potential Probiotic and Starter Properties of *Pediococcus* spp. Isolated from Turkish-Type Fermented Sausages (Sucuk). *J. Microbiol. Biotechnol.* **2010,** *20* (1), 161–168.

Zhang, H.; Samadi, A. K.; Rao, K. V.; Cohen, M. S.; Timmermann, B. N. Cytotoxic Oleanane-Type Saponins from *Albizia inundata. J. Nat. Prod.* **2011,** *74,* 477–482.

Ziegler, R. Vegetables, Fruits and Carotenoids and Risk of Cancer. *Am. J. Clin. Nutr.* **1991,** *53,* 251–259.

CHAPTER 8

PROCESS TECHNOLOGY OF SWEETENED YOGURT POWDER

DIBYAKANTA SETH, HARI NIWAS MISHRA, and
SANKAR CHANDRA DEKA[*]

Department of Food Engineering and Technology, Tezpur University, Napaam 784028, Sonitpur, Assam, India

[*]*Corresponding author. E-mail: sankar@tezu.ernet.in*

ABSTRACT

Sweetened yoghurt or misti dahi is a sweet fermented dairy product popular in eastern part of India. The shelf life of sweetened yoghurt is very less. So, an attempt was made to prepare a shelf-stable yoghurt powder using spray drying. This study aimed at optimizing the spray-drying process conditions in developing a sweetened yoghurt powder (SYP) using response-surface methodology. Process variables were 140–180°C inlet air temperature, 0.3–0.6 L/h feed rate, and 500–1000 kPa atomization pressure; and responses studied were moisture content (% db), bacteria survival ratio (N/N_0), acetaldehyde retention (%), and overall sensory score. Under optimal conditions (inlet air temperature 148°C, feed rate 0.54 L/h, and atomization pressure 898 kPa), the response variables including the moisture content, bacteria survival ratio, acetaldehyde retention, and overall sensory score were predicted and validated as 3.69% (db), 2.37×10^{-3}, 76.39% and 8.0, respectively. Furthermore, the influence of spray-drying conditions on functional and reconstitution properties of SYP was studied. Analysis of experimental data of solubility and dispersibility of SYP revealed negative correlation with inlet air temperature and atomization pressure. Solubility of powder was significantly ($p < 0.05$) affected by feed rate. Wetting time of powder increased significantly ($p < 0.001$) with the increase of spray-drying temperature. The bulk, tapped, and particle densities were in the range of 344.8–475.7, 551.7–782.5, and 1187.5–1666.7 kg/m^3, respectively. A significant quadratic effect ($p < 0.001$) of inlet air temperature and atomization pressure was observed on bulk density. Particle density increased with the increase in inlet air temperature and atomization pressure ($p < 0.001$). The results of the present investigation have the credible evidence to support that the processing conditions, namely, inlet air temperature and atomization pressure decreased the water activity and flow property of the SYP, whereas feed rate showed positive effect. Morphology of the optimally developed SYP showed a smooth and uniform texture with round particles. It was concluded that a reasonably good-quality shelf-stable powder can be developed from sweetened yoghurt having adequate bacterial count and acetaldehyde content by spray drying.

8.1 INTRODUCTION

Dahi is one of the oldest fermented milk products, which occupies a pivotal position in daily diet of Indian. Ancient scriptures of *Veda* and *Upanishads* describe the use of dahi. It is a fermented dairy product of yoghurt counterpart. About 7% of the total annual milk produced in India is utilized for dahi making for direct consumption (ICMR-DBT, 2011). Both in organized and unorganized sectors, dahi accounts for around 90% of the total cultured milk products produced in India (Barak and Mudgil, 2014). Large portion of dahi is still made in households. According to Food Safety and Standard Authority of India, it is a product obtained from pasteurized or boiled milk by souring, natural or otherwise, by a harmless lactic acid culture, or other harmless bacterial culture in conjunction with lactic cultures. *Dahi* may contain added cane sugar. *Dahi* shall have the same minimum percentage of milk fat and solid not fat as the milk from which it is prepared (FSSAI, 2011). *Dahi* is consumed as plain, salted, or sugared in India. The sweetened variety of *dahi*, known as *misti dahi* or *misthi doi* or *payodhi*, is mostly consumed in eastern states like West Bengal, Odisha, Bihar, and Assam. However, it has attracted consumers from all over India and nearby countries because of its pleasant aroma and flavor. It is characterized by light brown color, firm consistency, smooth texture, and a distinct aroma profile (Aneja et al., 2002; De, 1991). Of late, technology for industrial development of *misti dahi* or sweetened yoghurt was reported by Ghosh and Rajorhia (1990). With the technological advancement in the processing and packaging in dairy sector, sweetened yoghurt is now manufactured and marketed in different parts of the country

(Rao and Solanki, 2007). It can be seen in the retail market produced by leading brand owners of India with single-serve disposable cups.

One major hurdle in the expansion of sweetened yoghurt market has been its shelf life. It is susceptible to yeast and mold spoilage due to high water activity. The shelf life of sweetened yoghurt packed in glass cups was 2 days at 30°C, whereas at refrigerated condition (5 ± 1°C), sweetened yoghurt could be kept for 12 days (Ghosh, 1986) Attempts were also made to increase the shelf life of sweetened yoghurt by using permitted class II preservatives. Sarkar et al. (1992) used thermization process to increase the shelf life of sweetened yoghurt. They found that sweetened yoghurt treated at 60°C for 10 min increased the shelf life to 3 weeks stored at ambient temperature.

Many researchers tried to make shelf-stable fermented products from yoghurt, adopting freeze drying (Baisya and Bose, 1974; Rathi et al., 1990), microwave vacuum drying (Kim et al., 1997), or spray-drying technique (Koc et al., 2010). One of the advantages of powdered yoghurt is its convenience in adopting novel packaging technologies, which further increases its shelf life by checking oxidative degradation. Removal of moisture from yoghurt and conversion into powder not only increases its shelf life but also results in reduction of packing, transportation, and storage costs because of reduced bulk volume. Convenience to processors and users is an added advantage.

Spray drying is a technique used for producing food powders by atomizing liquid food to small droplets and exposing them to a very high temperature for short time. Because of short exposure time, many valuable components which otherwise would lose their functionality, remain intact. Koc et al. (2010) dried plain yoghurt by optimizing the inlet air temperature 171°C, outlet air temperature 60.5°C, and feed temperature 15°C. The thermal damage to the constituent microorganisms is less due to short residence time at high temperature compared to other thermal-drying processes. It was reported that at outlet temperature between 70°C and 75°C, the survival rate of *Lactobacillus delbrueckii* subsp. *bulgaricus* and *Streptococcus thermophilus* was better (Bielecka and Majkowska, 2000). Spray-dried yoghurt powder gives a firm body and texture after rehydration compared to powders manufactured by other methods. This is attributed to minimum damage to casein which forms a matrix in the yoghurt to give a firm consistency. Rehydration capacity of powder is very good compared to other drying processes. Rehydration is related to the particle size of the powder. Spray drying provides a uniform small particle size, thereby increases the reconstitution properties (Koc et al., 2014).

Stickiness in dairy powders is a common problem which has drawn much attention (Lloyd et al., 1996; Rennie, 1999). The stickiness is attributed to many factors like presence of low molecular weight sugars, moisture content, viscosity of product, etc. (Adhikari et al., 2001; Bhandari et al., 1997; Downtown et al., 1982). Sucrose present in sweetened yoghurt could cause the product stick to the wall of spray-drying chamber as a result reduce the yield. Higher molecular weight substances such as maltodextrin with low dextrose-equivalence (DE) are the well-established drying aid in production of sugar-rich food powders (Adhikari et al., 2009).

Numerous studies concerning the spray drying of yoghurt and other fermented products have been reported by various investigators (Izadi et al., 2014; Jaya, 2009; Kim and Bhowmik, 1990; Koc et al., 2010). However, studies concerning the spray drying of sugar-rich yoghurt and data on physicochemical properties are limited.

This study is aimed at optimizing spray-drying process parameters in terms of inlet air temperature, feed rate, and atomization pressure resulting in a process that minimizes moisture content, maintain satisfactory level of acetaldehyde, and cell survival in sweetened yoghurt powder (SYP).

8.2   MATERIALS AND METHODS

8.2.1   YOGHURT PREPARATION

A standard method was followed for preparation of sweetened yoghurt (Seth et al., 2016). The fat and SNF were standardized to 3% and 15%, respectively, and sugar was added at 10% (w/v). It was presumed that the same amount of water, which was present in yoghurt, could not be added to yoghurt powder for reconstitution, because it would resemble more as *Lassi* rather than yoghurt. Therefore, the sugar in yoghurt was reduced from 14% to 10% to maintain the sugar level below 15% in the reconstituted yoghurt even if the proportion of water is reduced. The standardized milk was inoculated with 1–2% working starter culture (*NCDC 263*) at 42 ± 1°C and incubated at that temperature in an incubator for about 6 h till the titratable acidity reached 1% lactic acid. Set sweetened yoghurt was kept in refrigeration temperature (5–7°C) till further use and analysis.

8.2.2 SPRAY DRYING

Sweetened yoghurt was tempered for 1 h at ambient temperature, then maltodextrin (20 *DE*) at 50% and SiO_2 at 0.01% of total solids were added to yoghurt blend properly. SiO_2 acts as a glidant which improves the flow-ability of powder (Rattes and Oliveira, 2007). A laboratory-scale spray dryer (Make: *LSD 01*, Advance Drying Systems, Mumbai) was used to carry out the spray-drying process (0.5–1 L/h water evaporation rate). Preheated sweetened yoghurt was pumped into a toughened glass drying chamber (0.2 m diameter and 0.4 m height) using a peristaltic pump and atomization was performed using a two fluid nozzle (inside nozzle diameter 0.5 mm). The drying air flow rate was constant at 100 m^3/min. Inlet air temperature (140–180°C), feed flow rate (0.3–0.6 L/h), and nozzle atomization pressure (500–1000 kPa) were adjusted according to the proposed experimental design using the Design Expert 7.0.0 software (Stat-Ease Inc., Minneapolis, MN, USA).

8.2.3 ANALYSIS OF SYP

8.2.3.1 PHYSICOCHEMICAL PROPERTIES

The acetaldehyde content of SYP was analyzed using GC–MS. The volatile compounds were separated on a capillary column (30 m long and 0.25 mm internal diameter) under the following conditions: injector temperature 200°C, carrier gas helium at a flow rate of 1.4 mL/min, oven temperature initially held at 50°C for 6 min, then raised to 180°C at 8°C/min and finally held at that temperature for 5 min. The MS was operated in an electron impact mode at an energy level of 70 eV.

A particle size analyzer (NanoPlus zeta potential and particle size analyzer, Particulate Systems, Atlanta, GA) was used to determine particle-size distribution. The particles under the Brownian motion scatter lights of different intensity when laser beams are irradiated to them. The fluctuation of scattered light is observed using a pinhole-type photon detector, from which particle size and particle size distributions are calculated. One gram of powder sample was suspended in water and analyzed at 25°C. The particle size (μm) was calculated based on its intensity (%) which represents a mean volumetric diameter and particle size distribution was represented by span and calculated using the following equation:

$$\text{Span} = \frac{D_{90} - D_{10}}{D_{50}} \qquad (8.1)$$

where D_{90}, D_{50}, and D_{10} are the diameters of sample at the 10th, 50th, and 90th percentile.

8.2.3.2 BULK, TAPPED, AND PARTICLE DENSITIES

A known quantity of yoghurt powder was taken in a 10-mL graduated cylinder and volume occupied was observed to calculate bulk density (ρ_B) (weight per volume). The same sample was tapped for 5 min (32 taps per minute approximately) and final volume was recorded and tapped density (ρ_T) was calculated (weight per volume). Particle density was measured using the method described by Jinapong et al. (2008). One gram of spray-dried yoghurt powder was transferred to a 10-mL graduated cylinder. A total of 5 mL petroleum ether was then added to the sample and shaken for some time so as to ensure all the particles are suspended. Finally, the wall of the cylinder was rinsed with 1 mL of petroleum ether to mix the attached lumps of powder if any to the sample. The final reading of the volume was taken and the tapped density was calculated as follows:

$$\rho_P = \frac{\text{Weight of the powder (g)}}{\text{Total volume of petroleum ether and suspended particles (mL)} - 6} \qquad (8.2)$$

8.2.3.3 FLOWABILITY AND COHESIVENESS

The spray-dried SYP was evaluated for its flowability and cohesiveness in terms of Carr's Index (CI) and Hausner ratio (HR), respectively. According to the method suggested by Jinapong et al. (2008), both the CI and HR were calculated from the bulk and tapped densities.

$$CI = \frac{\rho_T - \rho_B}{\rho_T} \times 100 \qquad (8.3)$$

$$HR = \frac{\rho_T}{\rho_B} \qquad (8.4)$$

8.2.3.3.1 Enumeration of Lactic Acid Bacteria

Plate count method was followed to enumerate the starter bacteria. *S. thermophilus* was enumerated on M17 agar (**HiMedia**) after being incubated aerobically at 37°C for 48 h (Dave and Shah, 1998). *L. delbrueckii bulgaricus* was enumerated by culturing in MRS agar (**HiMedia**) adjusted to pH 5.2 using 1 N HCl solution and incubating anaerobically at 37°C for 72 h (Kim et al., 1997). The survival ratio (SR) was calculated as follows:

$$\text{SR} = \frac{N}{N_0} \tag{8.5}$$

where N_0 and N represent the number of bacteria in the sweetened yoghurt before and after spray drying, respectively.

8.2.3.4 RECONSTITUTION PROPERTIES

8.2.3.4.1 Solubility

A total of 1 g of yoghurt powder was mixed with 100 mL distilled water and blended in a hand blender. The solution was transferred to 50-mL centrifuge tubes and centrifuged at 3000 rpm for 5 min. The solution was allowed to settle for 30 min before 25 mL of the supernatant was transferred to preweighed petri plates which were oven dried at 105°C for 5 h. The solubility (%) was calculated as the weight difference.

8.2.3.4.2 Dispersibility

A total of 10 mL distilled water at 25°C was taken in a 50-mL beaker. To the beaker, 1 g of sample was added. The sample was stirred vigorously for 15 s making 25 complete movements back and forth across the whole diameter of the beaker. The reconstituted powder was poured through a 212-μm sieve into a preweighed aluminum pan. The pan with sieved sample was dried at 105°C for 4 h. The dispersibility was calculated according to the formula given by Jinapong et al. (2008) as given in the following equation:

$$\text{Dispersibility (\%)} = \frac{(10+a) \times \%\text{TS}}{a \times (100-b)/100} \tag{8.6}$$

where a is the amount of yoghurt powder (g) taken, b is the moisture content in the powder, and %TS is the dry matter in percentage in the reconstituted powder after it has been passed through the sieve.

8.2.3.4.3　Wettability

Wettability of spray-dried SYP was evaluated according to the method described by Jinapong et al. (2008). According to the method, wettability is the time required for 1 g of powder deposited on the liquid surface to be completely submerged in 400 mL of distilled water at 25°C.

8.2.3.4.4　Morphology

Optimized SYP samples morphology was examined through micrograph to evaluate the structural changes occurred due to spray drying. The powder sample, prior to analysis by scanning electron microscope (SEM; JSM-6390LV; JEOL, Japan), was mounted on stubs with double-sided adhesive tape coated with a thin layer of gold and the SEM images were obtained (15 kV at 100–300× magnification).

8.3　RESULTS AND DISCUSSION

8.3.1　EFFECT OF SPRAY-DRYING PROCESS PARAMETERS ON SYP

8.3.1.1　MOISTURE CONTENT

The moisture content is a vital response parameter that influences the keeping quality of SYP. The moisture content of the spray-dried SYP varied from 1.38% to 5.26% (db). Koc et al. (2010) indicated a moisture level of 3.98–6.88% (wb) in yoghurt powder while varying the inlet temperature between 150 and 180°C. The results of predicted regression coefficient terms demonstrated that inlet air temperature had a negative effect and feed rate had positive effect on moisture content (Table 8.1). Increased inlet air temperature increases the temperature gradient between the atomized particles and drying air that provides a higher driving force for moisture removal and hence a lower residual moisture in the powder (Telang and

Thorat, 2010). On the other hand, increasing feed rate increases the mass density of feed inside the drying chamber and bigger droplet formation which reduces the evaporation rate (Alkahtani and Hassan, 1990). Hence, at high inlet temperature with low feed rate, the residual moisture found was less since the rate of evaporation was less (Fig. 8.1a) (Gong et al., 2008; Silva et al., 1997). In general, higher input energy in atomization produces finer droplets of powder and hence surface area for heat transfer increases. At a fixed feed flow rate, increased atomization pressure would produce powder of reduced moisture content because of finer droplet formation. However, in this study, the set energy level of atomization could not form varied droplet size which might be due to higher viscosity of sweetened yoghurt, and as a result, the effect of atomization pressure on moisture content of SYP could not be observed ($p > 0.05$).

8.3.1.2 BACTERIA SURVIVAL RATIO

Survival of lactic acid bacteria is highly dependent on spray-drying process variables, and among these, drying air temperature is more lethal to bacterial cells (Sarkar et al., 1992). The initial total plate count of sweetened yoghurt sample varied from 2.62×10^8 to 1.16×10^8 cfu/g (combining the cell counts of *S. thermophilus* and *L. bulgaricus*). The results demonstrated that all the independent variables had significant effect on SR at linear level. The coefficients of regression models indicated that with increase in inlet air temperature and atomization pressure, SR decreased, whereas it increased with increase in feed rate (Table 8.1). The SR in the present study varied from -5.987 to -2.650 log values. Quadratic effect of inlet air temperature and atomization pressure can be observed from Figure 8.1b which produced a curved surface plot. It was interpreted from the surface plot that a high SR can be achieved if drying of yoghurt is done at low inlet air temperature and high feed rate. Izadi et al. (2014) reported maximum survival rate of *S. thermophilus* in probiotic yoghurt at 150°C inlet air temperature, 500 m³/h blower air flow rate, and 2 L/h feed flow rate. Silva et al. (2011) however mentioned that survival of bacterial cell is mostly affected by outlet air temperature rather than the inlet air temperature. This was explained by the evaporative cooling effect inside the drying chamber. Teixeira et al. (1995) explained the death phenomenon in cells in terms of fatigue caused to cell membrane, cell wall, and DNA.

TABLE 8.1 Analysis of Variance and Regression Coefficients of the Second-order Polynomial Equations Obtained from Experimental Data

Source	df	Moisture content (%db)		SR ($\log_{10}$)		% Acetaldehyde retention		Overall acceptability	
		Coefficient	p value	Coefficient	p value	Coefficient	p value	Coefficient	p value
Model	9	2.98	<0.0001	-3.75	<0.0001	67.08	<0.0001	7.39	<0.0001
Linear									
x_1	1	-0.66	<0.0001	-0.90	<0.0001	-6.06	<0.0001	-0.46	<0.0001
x_2	1	0.72	<0.0001	0.25	0.0001	0.85	0.0054	0.15	0.0002
x_3	1	0.03	0.6006	-0.31	<0.0001	-0.12	0.6170	0.18	<0.0001
Interaction									
x_{12}	1	0.75	<0.0001	0.19	0.0076	0.025	0.9381	-0.15	0.0014
x_{13}	1	0.50	<0.0001	-0.43	<0.0001	-1.10	0.0050	-0.15	0.0014
x_{23}	1	-0.02	0.7618	0.03	0.6347	-0.42	0.2030	-0.15	0.0014
Quadratic									
x_1^2	1	0.42	<0.0001	-0.14	0.0064	0.84	0.0048	-0.14	0.0003
x_2^2	1	0.30	0.0002	0.06	0.1844	0.55	0.0416	-0.052	0.0773
x_3^2	1	0.11	0.0608	0.10	0.0422	0.48	0.0654	-0.097	0.0045
Residual	10								
Lack of fit	5		0.3054		0.8060		0.1548		0.8788
Pure error	5								
Total	19								
R^2		0.9824		0.9844		0.9857		0.9785	
R^2_{adj}		0.9666		0.9703		0.9729		0.9592	
Predicted R^2		0.9012		0.9462		0.9155		0.9342	
Press		2.35		0.87		46.63		0.31	
Adeq. precision		28.828		29.035		32.471		25.576	
CV%		5.76		4.24		1.30		1.40	

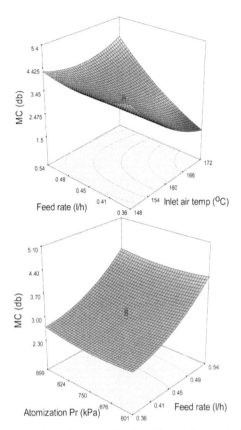

FIGURE 8.1   Response surface plots showing the effects of spray drying process variables on moisture content.

8.3.1.3   ACETALDEHYDE RETENTION (%AR)

Acetaldehyde is considered as the major flavoring compound in yoghurt that establishes a balanced flavor profile along with acetone (Cheng, 2010). The acetaldehyde content of sweetened yoghurt was 24.56 ± 0.12 mg/kg. The acetaldehyde retention in the present study varied from 15.33 to 19.73 mg/kg, that is, 62.42–80.36% of the initial amount. The highest retention of acetaldehyde was found at the lowest temperature of 140°C. The inlet air temperature ($p < 0.001$) and feed rate ($p < 0.01$) had significant effect on acetaldehyde retention (Table 8.1). The regression coefficients of first-order term indicated that the acetaldehyde retention decreased with increase in inlet air temperature and it increased with increase in feed rate. It is reported that inlet air temperature is more detrimental to flavor and sensory properties

(Bhandari et al., 1992). Higher feed rate minimizes the evaporation rate; so, it may not pose drastic effect on acetaldehyde. Interaction effect of inlet air temperature and atomization pressure evinced a negative correlation ($p <$ 0.01) (Fig. 8.2a).

8.3.1.4   OVERALL ACCEPTABILITY

The overall sensory acceptability score of reconstituted SYP varied from 6.25 to 8.0 based on a 9-point hedonic scale. The highest and lowest scores were obtained at inlet air temperatures of 148 and 180°C, respectively. The interaction effects of all the independent variables on overall acceptability can be observed in Figure 8.2b. The results of the analysis of variance demonstrated that the overall acceptability of reconstituted SYP is significantly affected by the inlet air temperature in linear, quadratic, and interaction terms and the acceptability decreased with increase in inlet air temperature ($p < 0.01$) (Table 8.1). Our results are in line with the findings of Koc et al. (2010) where they reported a negative effect of overall acceptability with outlet air temperature.

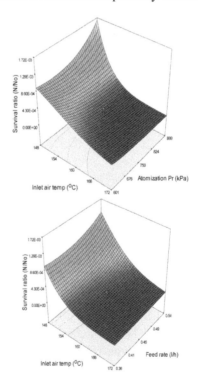

FIGURE 8.2   Response surface plots showing the effects of spray drying process variables on survival ratio.

Since outlet air temperature is controlled by inlet air temperature, it would be justifiable to show the effect of inlet temperature in similar manner as shown by outlet temperature. It was obvious that the overall acceptability increased with increase in feed rate ($p < 0.01$).

8.3.1.5 MORPHOLOGY

The scanning electron microscopy images of yoghurt powder produced at the optimum spray-drying conditions are illustrated in Figure 8.3 with a magnification level of 200× and 300×, and particles were mostly spherical and smooth in surface. Similar micrographs of yoghurt powder has also been reported by Koc et al. (2010). A continuous protein matrix can be observed due to the gel structure of casein protein (Rascon-Diaz et al., 2012). Sugar present in SYP could have some contribution to the matrix structure. The cracks on the surface indicate the distortion of protein matrix. At the same magnification (50 μm size), very much similar micrographs were presented by Yousefi et al. (2011) and attributed to carrier agents like maltodextrin and can markedly influence the morphology.

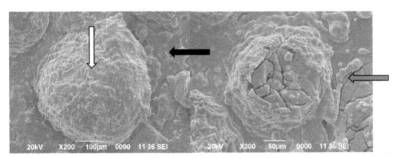

FIGURE 8.3 Scanning electron microscopy images of sweetened yoghurt powder produced at the optimum spray-drying conditions; black arrow indicating protein matric, gray arrow indicating the distortion of matric, and white arrow indicating the smooth spherical particle.

8.3.1.6 PARTICLE SIZE

The reconstitution properties of powder depend on the particle size of powder to a greater extent. Nath and Satpathy (1998) found that the particle size of powder increases with feed pumping rate and decreases with compressed air flow rate. Similar trend was observed in our findings (Fig. 8.4a). The particle size of SYP varied from 2.13 to 5.15 μm.

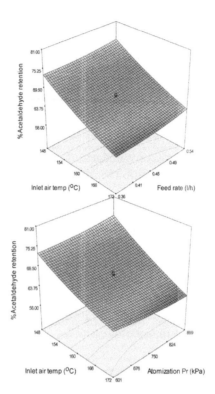

FIGURE 8.4   Response surface plots showing the effects of spray drying process variables on acetaldehyde retention.

The atomization pressure demonstrated a significant effect on particle size. Although the effect of inlet air temperature on particle size was not significant either in linear or quadratic terms, the interaction effect with feed rate and atomization pressure showed increase in particle size. The particle size distribution pattern is given in Figure 8.4b.

8.3.1.7   BULK, TAPPED, AND PARTICLE DENSITIES

The results demonstrated that the variation in bulk density of SYP was from 344.80 to 475.41 kg/m³. Figure 8.5a indicates that the bulk density increased with increase in inlet air temperature to certain extent, then decreased. Mastes (1991) explained the positive effect of temperature on bulk density by correlating slurry density as a function of temperature. Lower the density, better atomization we can achieve with a result of smaller particle size.

Tonon et al. (2008) reported similar findings while drying acai juice powder. However, Reddy et al. (2014) found a negative effect of air temperature on bulk density. The decreasing bulk density with increased air temperature could be due to faster evaporation rate and more porous and fragmented structure of the product (Goula and Adamopoulos, 2003). Increased atomization pressure demonstrated a positive effect on bulk density. This could again be explained by smaller diameter particle formation at higher pressure. A negative relation of bulk density with feed rate was established in the current work. Tapped density of SYP varied from 551.72 to 782.50 kg/m³.

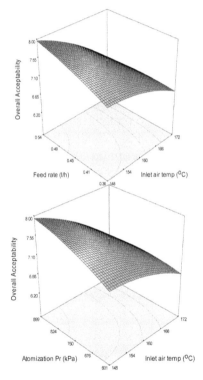

FIGURE 8.5 Response surface plots showing the effects of spray drying process variables on overall acceptability.

The results of tapped density demonstrated similar trend as bulk density and the effect of drying parameters found to be significant on tapped density (Fig. 8.3b). The range of particle density of SYP was between 1187.50 and 1666.67 kg/m³. An increase in inlet air temperature resulted in a decreased particle density (Fig. 8.3c). The findings were in support of the work carried out by Walton (2000).

8.3.1.8 FLOWABILITY AND COHESIVENESS

The flowability and cohesion in terms of CI and HR of SYP were in the range of 25.74–47.60% and 1.39–1.90, respectively. As per the categorization of flowability by Lebrun et al. (2012), powder with CI greater than 38 and HR greater than 1.6 is treated as very poor. Hence, SYP falls in the category of lower flowability. This could be due to presence of sucrose and maltodextrin in SYP. Increasing inlet air temperature resulted in a dramatic increase in both CI and HR values.

8.3.2 RECONSTITUTION PROPERTIES OF SYP

8.3.2.1 SOLUBILITY

Solubility represents the complete dissolution of the soluble components like lactose, undenatured whey protein, salts, as well as the dispersion components like casein (Fang et al., 2008). The rate of insoluble material formation in dairy products depends on the temperature treatment before and during drying. The solubility of SYP varied from 72% to 88%. Koc et al. (2014), however, found the solubility in the range 65.0–72.5%. Higher solubility value in our work could be the presence of sucrose in SYP. The analysis of variance indicated that solubility of SYP was affected by all the process variables. The results demonstrated that the solubility of SYP decreased with increase in inlet air temperature and atomization pressure, and increased with feed rate.

TABLE 8.2 Analysis of Variance of the Observed Results of Reconstitution Properties of Sweetened Yoghurt Powder.

Source	df	Solubility index		Wettability		Dispersibility	
		Sum of square	p value	Sum of square	p value	Sum of square	p value
Model	9	330.98	<0.0001	72,727.84	<0.0001	357.20	<0.0001
x_1	1	10.62	0.0127	64,216.40	<0.0001	0.98	0.4556
x_2	1	7.86	0.0262	387.77	0.3143	7.22	0.0610
x_3	1	39.02	0.0002	4.34	0.9130	218.34	<0.0001
x_{12}	1	55.13	<0.0001	496.13	0.2584	12.45	0.0198
x_{13}	1	28.12	0.0006	190.13	0.4752	8.78	0.0423

TABLE 8.2   *(Continued)*

Source	df	Solubility index		Wettability		Dispersibility	
		Sum of square	*p* value	Sum of square	*p* value	Sum of square	*p* value
x_{23}	1	6.13	0.0442	36.13	0.7531	25.99	0.0025
x_1^2	1	77.90	<0.0001	811.67	0.1563	60.35	0.0001
x_2^2	1	87.42	<0.0001	1380.52	0.0735	0.044	0.8719
x_3^2	1	1.65	0.2606	4517.07	0.0047	16.59	0.0095

8.3.2.2   WETTABILITY

The wettability has direct relation with the dissolution of powder components. Powder with increased solubility takes less time to wet their surface,
thus lower wetting time or wettability. The wetting time of SYP varied from
132 to 378 s. The analysis of variance demonstrated a positive effect of
inlet air temperature with wettability (Table 8.2). It is evident that at higher
temperature, powder components are susceptible to structural damage and
thus take long time to wet. Reddy et al. (2014) reported similar findings
in their research and they explained the lower wetting in terms of reduced
residual moisture in goat milk powder at higher inlet air temperature. Kim et
al. (2002) explained the wettability and dispersibility as functions of particle
size, density, porosity, surface charge, surface area, and the presence of
amphipathic substances and surface activity of the particles.

8.3.2.3   DISPERSIBILITY

The results of dispersibility of SYP indicates a variation in the range of
70.62–88.74%. The analysis of variance demonstrated a negative effect
of atomization pressure and inlet air temperature on dispersibility (Table
8.2). Lower dispersibility at higher temperature could be a cause of higher
denaturation rate of protein (Fang et al., 2008). Koc et al. (2014), however,
did not find any specific trend of dispersibility in yoghurt powder.

8.4   CONCLUSION

Spray-drying process parameters have shown significant effect on flavor
retention, starter bacteria mortality, and functional and reconstitution

properties of yoghurt powder. Air inlet temperature was found to be critical process parameter affecting the functional properties. Results indicate that reconstitution properties are mostly affected by atomization pressure and inlet air temperature. Evaluating reconstitution behavior by correlating with process conditions could permit us to determine optimum spray-drying conditions for better consumer acceptability. The optimized conditions were inlet air temperature 148°C, feed rate 0.54 L/h, and atomization pressure 898 kPa.

The study gives a wider spectrum of spray-drying sugar-rich fermented foods inclusive of better survival rate and flavor retention against extreme processing conditions. Further, understanding the effect of sugar on reconstitution and functional properties of SYP will help us to improve the quality.

KEYWORDS

- misti dahi
- sweetened yoghurt powder
- reconstitution properties
- bacteria survival ratio
- response surface methodology

REFERENCES

Adhikari, B.; Howes, T.; Bhandari, B. R.; Truong, V. Stickiness in Foods: A Review of Mechanisms and Test Methods. *Int. J. Food Prop.* **2001,** *4* (1), 1–33.

Adhikari, B.; Howes, T.; Wood, B. J.; Bhandari, B. R. The Effect of Low Molecular Weight Surfactants and Proteins on Surface Stickiness of Sucrose during Powder Formation through Spray Drying. *J. Food Eng.* **2009,** *94* (2), 135–143.

Alkahtani, H. A.; Hassan, B. H. Spray Drying of Roselle (*Hibiscus sabdariffa* L.) Extract. *J. Food Sci.* **1990,** *55* (4), 1073–1076.

Aneja, R. P.; Mathur, B. N.; Chandan, R. C.; Banerjee, A. K. *Technology of Indian Milk Products*; Dairy India Publication: New Delhi, India, 2002.

Baisya, R. K.; Bose, A. N. Studies on the Dehydration of Dahi (Milk Curd). *J. Food Sci. Technol.—Mysore* **1974,** *11* (3), 128–131.

Barak, S.; Mudgil, D. Locust Bean Gum: Processing, Properties and Food Applications—A Review. *Int. J. Biol. Macromol.* **2014,** *66,* 74–80.

Bhandari, B. R.; Datta, N.; Howes, T. Problems Associated with Spray Drying of Sugar-Rich Foods. *Dry. Technol.* **1997,** *15* (2), 671–684.

Bhandari, B. R.; Dumoulin, E. D.; Richard, H. M. J.; Noleau, I.; Lebert, A. M. Flavor Encapsulation by Spray Drying—Application to Citral and Linalyl Acetate. *J. Food Sci.* **1992,** *57* (1), 217–221.

Bielecka, M.; Majkowska, A. Effect of Spray Drying Temperature of Yoghurt on the Survival of Starter Cultures, Moisture Content and Sensoric Properties of Yoghurt Powder. *Nahrung-Food* **2000,** *44* (4), 257–260.

Cheng, H. F. Volatile Flavor Compounds in Yogurt: A Review. *Crit. Rev. Food Sci. Nutr.* **2010,** *50* (10), 938–950.

Dave, R. I.; Shah, N. P. Ingredient Supplementation Effects on Viability of Probiotic Bacteria in Yogurt. *J. Dairy Sci.* **1998,** *81* (11), 2804–2816.

De, S. *Outlines of Dairy Technology*; Oxford University Press: New Delhi, 1991.

Downtown, G. E.; Flores-Luna, J. L.; King, C. J. Mechanisms of Stickiness in Hygroscopic, Amorphous Powders. *Indian Eng. Chem. Fundam.* **1982,** *21* (4), 447–451.

Fang, Y.; Selomulya, C.; Chen, X. D. On Measurement of Food Powder Reconstitution Properties. *Dry. Technol.* **2008,** *26* (1), 3–14.

FSSAI (Food Safety and Standards Authority of India). Vol. III; Govt. of India, 2011; p 292.

Ghosh, J. *Production, Packaging and Preservation of Misti Dahi*; Kurukshetra University: Kurukshetra, Haryana, India, 1986.

Ghosh, J.; Rajorhia, G. S. Technology for Production of Misti Dahi—A Traditional Fermented Milk Product. *Indian J. Dairy Sci.* **1990,** *43*, 239–246.

Gong, Z. Q.; Zhang, M.; Mujumdar, A. S.; Sun, J. C. Spray Drying and Agglomeration of Instant Bayberry Powder. *Dry. Technol.* **2008,** *26* (1), 116–121.

Goula, A. M.; Adamopoulos, K. G. Spray Drying Performance of a Laboratory Spray Dryer for Tomato Powder Preparation. *Dry. Technol.* **2003,** *21* (7), 1273–1289.

ICMR-DBT. ICMR-DBT Guidelines for Evaluation of Probiotics in Food. *Indian J. Med. Res.* **2011,** *134*, 22–25.

Izadi, M.; Eskandari, M. H.; Niakousari, M.; Shekarforoush, S.; Hanifpour, M. A.; Izadi, Z. Optimisation of a Pilot-Scale Spray Drying Process for Probiotic Yoghurt, Using Response Surface Methodology. *Int. J. Dairy Technol.* **2014,** *67* (2), 211–219.

Jaya, S. Microstructure Analysis of Dried Yogurt: Effect of Different Drying Methods. *Int. J. Food Prop.* **2009,** *12* (3), 469–481.

Jinapong, N.; Suphantharika, M.; Jamnong, P. Production of Instant Soymilk Powders by Ultrafiltration, Spray Drying and Fluidized Bed Agglomeration. *J. Food Eng.* **2008,** *84* (2), 194–205.

Kim, S. S.; Bhowmik, S. R. Survival of Lactic Acid Bacteria during Spray Drying of Plain Yogurt. *J. Food Sci.* **1990,** *55* (4), 1008–1010.

Kim, S. S.; Shin, S. G.; Chang, K. S.; Kim, S. Y.; Noh, B. S.; Bhowmik, S. R. Survival of Lactic Acid Bacteria during Microwave Vacuum-Drying of Plain Yoghurt. *Food Sci. Technol.—Lebensmittel-Wissensch. Technol.* **1997,** *30* (6), 573–577.

Kim, E. H. J.; Chen, X. D.; Pearce, D. Surface Characterization of Four Industrial Spray-Dried Dairy Powders in Relation to Chemical Composition, Structure and Wetting Property. *Colloids Surf.; B—Biointerfaces* **2002,** *26* (3), 197–212.

Koc, B.; Yilmazer, M. S.; Balkır, P.; Ertekin, F. K. Spray Drying of Yogurt: Optimization of Process Conditions for Improving Viability and Other Quality Attributes. *Dry. Technol.* **2010,** *28* (4), 495–507.

Koc, B.; Sakin-Yilmazer, M.; Kaymak-Ertekin, F.; Balkir, P. Physical Properties of Yoghurt Powder Produced by Spray Drying. *J. Food Sci. Technol.—Mysore* **2014,** *51* (7), 1377–1383.

Lebrun, P.; Krier, F.; Mantanus, J.; Grohganz, H.; Yang, M. S.; Rozet, E.; Boulanger, B.; Evrard, B.; Rantanen, J.; Hubert, P. Design Space Approach in the Optimization of the Spray-Drying Process. *Eur. J. Pharm. Biopharm.* **2012,** *80* (1), 226–234.

Lloyd, R. J.; Chen, X. D.; Hargreaves, J. B. Glass Transition and Caking of Spray Dried Lactose. *Int. J. Food Sci. Technol.* **1996,** *31*, 305–311.

Mastes, K. *Spray Drying Handbook*, 5th ed.; Longman Scientific and Technical: Harlow, 1991.

Nath, S.; Satpathy, G. R. A Systematic Approach for Investigation of Spray Drying Processes. *Dry. Technol.* **1998,** *16* (6), 1173–1193.

Rao, K. N.; Solanki, D. C. Fermented Dairy Products—Lab to Industry: Research Creativity and Experiences. In *3rd International Conference on Fermented Foods Health Status and Social Well-being*; Anand: Gujarat, India, 2007; pp 70–73.

Rascon-Diaz, M. P.; Tejero, J. M.; Mendoza-Garcia, P. G.; Garcia, H. S.; Salgado-Cervantes, M. A. Spray Drying Yogurt Incorporating Hydrocolloids: Structural Analysis, Acetaldehyde Content, Viable Bacteria, and Rheological Properties. *Food Bioprocess Technol.* **2012,** *5* (2), 560–567.

Rathi, S. D.; Deshmukh, D. K.; Ingle, V. M.; Syed, H. H. Studies on Physicochemical Properties of Freeze Dried Dahi. *Indian J. Dairy Sci.* **1990,** *43* (2), 249–251.

Rattes, A. L. R.; Oliveira, W. P. Spray Drying Conditions and Encapsulating Composition Effects on Formation and Properties of Sodium Diclofenac Microparticles. *Powder Technol.* **2007,** *171* (1), 7–14.

Reddy, R. S.; Ramachandra, C. T.; Hiregoudar, S.; Nidoni, U.; Ram, J.; Kammar, M. Influence of Processing Conditions on Functional and Reconstitution Properties of Milk Powder Made from Osmanabadi Goat Milk by Spray Drying. *Small Rumin. Res.* **2014,** *119* (1–3), 130–137.

Rennie, P. R. A Study of the Cohesion of Dairy Powders. *J. Food Eng.* **1999,** *39*, 277–284.

Sarkar, S. P.; Dave, J. M.; Sannabhadti, S. S. Effect of Thermization of Misti Dahi on Shelf Life and Beta-d-Galactosidase Activity. *Indian J. Dairy Sci.* **1992,** *45* (3), 135–139.

Seth, D.; Mishra, H. N.; Deka, S. C. Effect of Spray Drying Process Conditions on Bacteria Survival and Acetaldehyde Retention in Sweetened Yoghurt Powder: An Optimization Study. *J. Food Process Eng.* **2016.** DOI:10.1111/jfpe.12487.

Silva, A. P.; Cervantes, M. A. S.; Galindo, H. S. G. Acetaldehyde Retention during Spray Drying of Yoghurt. *Milchwissensch.—Milk Sci. Int.* **1997,** *52* (2), 89–93.

Silva, J.; Freixo, R.; Gibbs, P.; Teixeira, P. Spray-Drying for the Production of Dried Cultures. *Int. J. Dairy Technol.* **2011,** *64* (3), 321–335.

Teixeira, P. C.; Castro, M. H.; Malcata, F. X.; Kirby, R. M. Survival of *Lactobacillus-Delbrueckii* ssp. *bulgaricus* Following Spray-Drying. *J. Dairy Sci.* **1995,** *78* (5), 1025–1031.

Telang, A. M.; Thorat, B. N. Optimization of Process Parameters for Spray Drying of Fermented Soy Milk. *Dry. Technol.* **2010,** *28* (12), 1445–1456.

Tonon, R. V.; Brabet, C.; Hubinger, M. D. Influence of Process Conditions on the Physicochemical Properties of Açai (*Euterpe oleraceae* Mart.) Powder Produced by Spray Drying. *J. Food Eng.* **2008,** *88* (3), 411–418.

Walton, D. E. The Morphology of Spray-Dried Particles a Qualitative View. *Dry. Technol.* **2000,** *18* (9), 1943–1986.

Yousefi, S.; Emam-Djomeh, Z.; Mousavi, M. S. Effect of Carrier Type and Spray Drying on the Physicochemical Properties of Powdered and Reconstituted Pomegranate Juice (*Punica granatum* L.). *J. Food Sci. Technol.—Mysore* **2011,** *48*, 677–684.

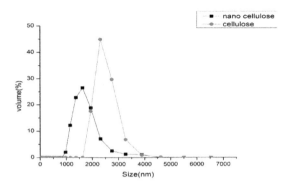

FIGURE 3.3   Particle size analysis of cellulose and nanocellulose.

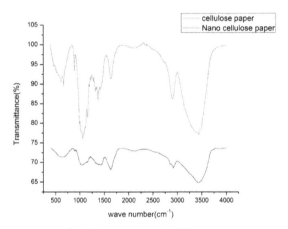

FIGURE 3.5   FT-IR curve of cellulose paper and cellulose nanopaper.

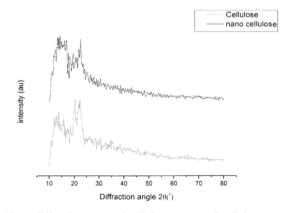

FIGURE 3.6   X-ray diffraction curve of cellulose paper and cellulose nanopaper.

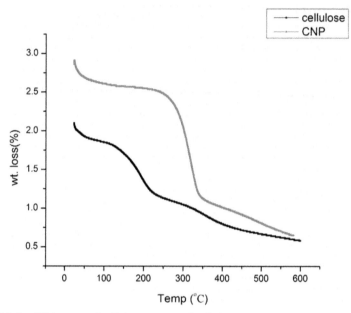

FIGURE 3.8   TGA curve of cellulose paper and cellulose nanopaper.

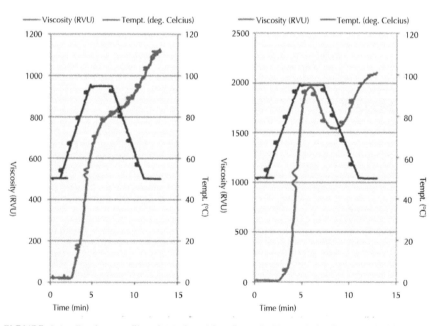

FIGURE 6.1   Pasting profile of (a) dry chips flour (DCF) and (b) fermented chips flour (FCF) of cassava.

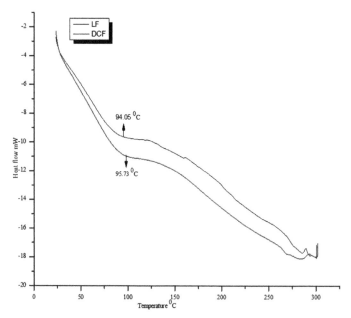

FIGURE 6.4   DSC graph of dry chips flour (DCF) and fermented chips flour (FCF).

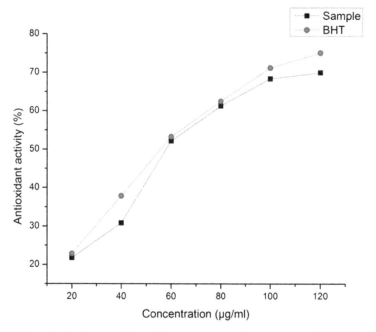

FIGURE 7.3   Antioxidant activity of the *A. myriophylla* extract by DPPH free radical scavenging activity assay.

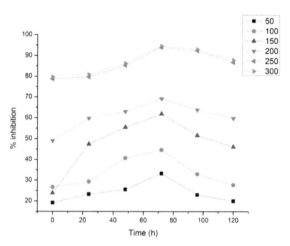

FIGURE 7.4 Inhibition of lipid peroxidation (%) of the extract by thiobarbituric acid (TBA) assay.

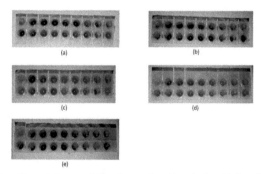

FIGURE 7.5 Microtiter plates used for the study of antimicrobial activity of the extract against different microbes: (a) *S. aureus*, (b) *E. coli*, (c) *L. plantarum*, (d) *L. sakei*, and (e) *S. cerevisae.*

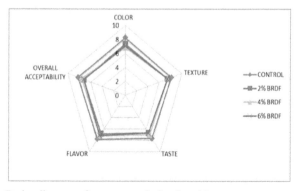

FIGURE 9.6 Radar diagram of sensory analysis of cookies.

CHAPTER 9

CULINARY BANANA RHIZOME: A SOURCE OF DIETARY FIBER AND ITS APPLICATION

GARIMA SHARMA, YESMIN ARA BEGUM, and
SANKAR CHANDRA DEKA*

Department of Food Engineering and Technology, Tezpur University, Napaam 784028, Sonitpur, Assam, India

Corresponding author. E-mail: sankar@tezu.ernet.in

ABSTRACT

The objective of the present study was on extraction of dietary fiber (DF) from culinary banana rhizome, its various physicochemical and functional properties, and incorporation into cookies. 2,2-Diphenyl-1-picrylhydrazyl (DPPH) and total polyphenol content of the rhizome powder were found to be 68.94% and 40 mg gallic acid equivalent/g rhizome extract, respectively. Functional properties, scanning electron microscopy (SEM), and X-ray diffraction (XRD) of soluble, insoluble, and total dietary fiber (TDF) concentrates were evaluated. Banana rhizome showed better functional properties when compared to corn fiber. SEM analysis revealed that soluble fiber particles were more amorphous in structure and less fibrous than insoluble fiber, and XRD evinced A-type crystalline diffraction patterns. Cookies were incorporated with 0%, 2%, 4%, and 6% of extracted DF concentrates. Physical properties, textural analysis, color measurement, antioxidant activity, and sensory analysis of the DF-enriched cookies were performed. Cookies incorporated with 4% banana rhizome DF concentrate showed 7.86% TDF content and 44.17% antioxidant activity and were found more acceptable than the other incorporated cookies based on sensory analysis.

9.1   INTRODUCTION

Dietary fiber (DF) is defined as the edible parts of the plants or analogous carbohydrates that are resistant to digestion and absorption in the human small intestine, with complete or partial fermentation in the large intestine, as adopted by the AACC (2000). Many people confuse DF with crude fiber, which actually is one of its parts. It remains as residue after the acid–alkali test that dissolves all the soluble (SF) and some part of insoluble fibers (IFs). Crude fiber composed of cellulose and lignin, while DFs include polysaccharides, oligosaccharides, lignin, and associated plant substances. Based on their simulated intestinal solubility, DFs are classified as insoluble dietary fiber (IDF) and soluble dietary fiber (SDF). IFs include lignin, cellulose, and hemicelluloses, while SFs include pectins, β-glucans, galactomanan, gums, and a large range of nondigestible oligosaccharides including inulin. DF helps in reducing cardiovascular diseases, diverticulosis, constipation, irritable colon, colon cancer, and diabetes (Tavaini and La Vecchi, 1995; Pietinen, 2001). The insoluble fraction of the fiber (IF) seems to be related to the intestinal regulation, whereas the SF is associated to the decrease of cholesterol levels and the adsorption of intestinal glucose as reported by Scheneeman (1987).

In recent years, there becomes a trend to search new sources of DF such as agronomic by-products that have traditionally been undervalued and can be used in the food industry. DF has developed, in last few years, an additional importance related to its use as functional ingredients. The importance of food fibers has led to the development of a large and potential market for fiber-rich products and ingredients (Chau and Huang, 2003). Hesser (1994) reported that supplementation has been used to enhance fiber content of foods, like in cookies, crackers, and other cereal-based products; enhancement of fiber content in snack foods, beverages, spices, imitation cheeses, sauces, frozen foods, canned meats, meat analogues, etc.

Banana plant is the world's largest herbaceous perennial plant which belongs to the family Musaceae and is one of the most widely grown tropical fruits along the tropics and subtropics of Capricorn (Mohapatra et al., 2010). India is the major producer of banana contributing 27% of the total banana production in the world; accumulation of the massive biomass is also a perennial unresolved problem. After harvesting, rhizome of banana plant contributes about 12.67% of the total plant biomass waste (Saravanan and Aradhya, 2011). A very negligible percent of these biowaste is used, though it has many medicinal and nutraceutical benefits (Saravanan and Aradhya, 2014). Research by Pushpangadan et al. (1989) showed that banana rhizome was used by villagers as a medicine for treating pyorrhea, piles, diabetes, food poisoning, intestinal worms, mental diseases, acidity, burns, and wounds. In this context, use of banana rhizome with rich nutrient and medicinal properties felt worth pursuing. In the present study, DF was extracted from culinary banana rhizome; its various physicochemical and functional properties were evaluated as affected by fiber and was further incorporated into cookies and its properties were assessed.

9.2   MATERIALS AND METHODS

9.2.1   PREPARATION OF SAMPLE

Banana rhizome (Fig. 9.1) was collected from Tezpur Central University campus, Assam (India), and then washed, shredded, dried, and grounded into powder and was stored in plastic container for further analysis.

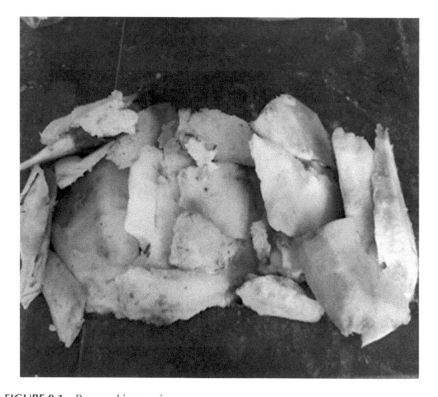

FIGURE 9.1   Banana rhizome pieces.

9.2.2 CHEMICAL ANALYSIS

Moisture content, ash, and fat content of rhizome powder were determined according to AOAC (1965) methods. Protein content was estimated according to Lowry et al. (1951), crude fiber content was determined according to Maynard (1970), and total carbohydrate was determined by difference. SDF, IDF, and total dietary fiber (TDF) was estimated according to AOAC enzymatic gravimetric method of Prosky et al. (1988). All analyses were carried out in triplicates and expressed as mean value and standard deviation was calculated.

9.2.3 TOTAL POLYPHENOL CONTENT AND ANTIOXIDANT ACTIVITY

The total polyphenol content (TPC) of banana rhizome powder was determined by the Folin–Ciocalteau colorimetric method. 2,2-Diphenyl-1-picrylhydrazyl (DPPH) assay of the rhizome powder was determined by the methods of Brand-Williams et al. (1995). The scavenging activity of the extract was calculated as follows:

$$\text{Scavenging activity (\%)} = \left\{ 1 - \left(\frac{AS}{AC} \right) \right\} \times 100$$

where AS is the absorbance of DPPH and extract and AC is the absorbance of the control and all samples were tested in triplicate.

9.2.4 EXTRACTION OF DIETARY FIBER FROM Musa ABB RHIZOME

DF from *Musa* ABB rhizome was extracted by Bouaziz et al. (2014) with some modifications and Figure 9.2 illustrates the extraction process from powder to produce DF concentrates. Hot water was used to extract DF from ground rhizome powder in a jar, homogenized using a mechanical stirrer 2021 (Heidolph rzr, Metrohm, USA) and maintained in a thermostatic bath (Raypa, Spain) and operating conditions used for the extraction were powder:water ratio 0.05, temperature 80°C, pH 4, ionic strength 2, time 30 min, agitation speed 400 rpm. After solubilization of free sugars and fructans, insoluble DFs were recuperated by centrifugation (5000 rpm, 20 min). Its concentration was carried out by a succession of five rinsing (at

40°C) and of five centrifugations until the residue was free of sugars. On the other hand, soluble DFs (fructans) were settled overnight in ethanol 95% (extract:ethanol ratio 0.25, precipitation temperature 5°C) and stirred gently for 1 h at 5°C and then recovered by centrifugation at 5000 rpm for 20 min (Masmoudi et al., 2008). Its concentration was carried out by rinsing three times with 50%, 70%, and 100% ethanol. The residues obtained were dried in a hot air oven (50°C) and then ground in laboratory grain mill and passed through 0.5 mm mesh to give the DF concentrates and stored in an airtight container for subsequent analyses.

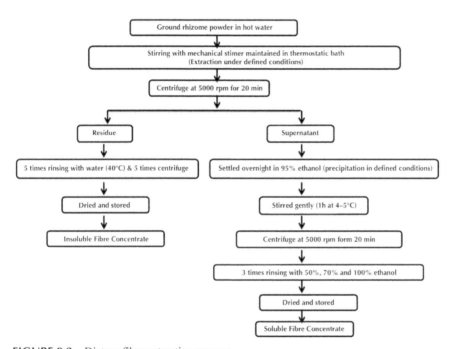

FIGURE 9.2   Dietary fiber extraction process.

9.2.5   PHYSICOCHEMICAL PROPERTIES OF DIETARY FIBER CONCENTRATE

To determine water (WHC) and oil-holding capacity (OHC) of the extracted DF, 20 mL of distilled water or commercial olive oil were added to 1 g of dry sample, stirred, and incubated at room temperature for 1 h. Tubes were centrifuged at 3000g for 20 min and the supernatant was decanted and the tubes were allowed to drain for 10 min to ensure the proper removal of water or oil but not the residue. The residue was weighed and WHC and OHC were calculated as g water or oil per g dry sample, respectively. The method used was slight modification of the method by Rodriguez-Ambriz et al. (2008).

Swelling power and solubility were determined according to the method as described by Leach et al. (1959) with some modifications. In a preweighed centrifuge tube, 1 g of powder was placed and 20 mL of distilled water was added and stirred at room temperature for 30 min and then centrifuged at 4000 rpm for 20 min. The supernatant was poured into a preweighed petri plate and placed in the oven at 105°C to evaporate. The solid residue in the petri plate was weighed again and the difference in weight calculated as percentage solubility. The residue in the tube was then weighed and the swelling power was determined by the following equation:

$$\% \text{ Solubility} = \frac{\text{weight of dried sediment}}{\text{weight of original sample}} \times 100$$

$$\% \text{ Swelling power} = \frac{\text{weight of residue in tube}}{\text{weight of original sample} \times (100 - \text{percent solubility})} \times 100$$

Bulk density of the sample was determined by the method of Wang and Kinsella (1976) and 3 g of the powdered sample was placed in a 50-mL graduated cylinder and packed by tapping the cylinder on a rubber sheet until a constant volume was obtained. The bulk density was expressed as g of sample/mL. For determination of pH, suspension of powder [8% (w/v)] was stirred for 5 min, allowed to stand for 30 min, filtered, and the pH of the filtrate was measured (Suntharalingam and Ravindran, 1993).

9.2.6   ANTIOXIDANT ACTIVITY

Antioxidant activity (AOA) was measured using the method described by of Brand-Williams et al. (1995) as described above and sample was tested in triplicate.

9.2.7   SCANNING ELECTRON MICROSCOPY

Particle structures of SDF and IDF extracted from *Musa* ABB rhizome were evaluated by scanning electron microscopy (SEM) (JSM-6390 LV, Japan, PN junction type, semiconducting detector). On a double-sided adhesive tape, rhizome powder was attached and kept on a metal stud, which was coated with silver under vacuum and was examined at 20 kV and magnification of 1000× (scale bar 10 μm).

9.2.8   X-RAY DIFFRACTION

X-ray diffraction (XRD) was performed (Rigaku Miniflex) to examine the phase composition and crystallinity of the rhizome DF powder. XRD was operated at 30 kV and 15 mA, and the assay was recorded in room temperature over the 2θ range of 10°–70°. The degree of crystallinity (Cleven et al., 1978) was calculated from the peak area under the curve using the following equation:

$$D_c(\%) = A_c \times \frac{(100)}{(A_c + A_a)}$$

where D_c is the degree of crystallinity, A_c is the crystallized area on the X-ray diffractogram, and A_a is the amorphous area on the X-ray diffractogram.

9.2.9   COOKIE FORMULATION AND PREPARATION

Blends of 2%, 4%, and 6% for cookies were prepared by substituting wheat flour with extracted DF powder from banana rhizome (BRDF). The formula used was according to Ajila et al. (2008) with some modifications and the various ingredients were as follows: 200 g wheat flour, 60 g sugar, 90 g vegetable fat, 2 g sodium chloride, 2 g sodium bicarbonate, and 30 mL water. Vegetable fat and ground sugar were mixed to a cream than the mixture of flour, BRDF powder, sodium bicarbonate, and sodium chloride was added and mixed thoroughly to form dough and was kept in refrigerator for 20 min. The dough was kneaded and divided into small equal portions, rolled into small balls, and pressed slightly to give shape of small cookies. Baking was carried out at 190°C for 20 min. Cookie samples were cooled and stored in airtight containers and cookies made from wheat were used as a control.

9.2.10   PHYSICAL AND TEXTURAL ANALYSIS OF COOKIES

Diameter (W) of cookies was measured by laying six cookies edge to edge with the help of a scale. The same set of cookies was rotated 90°C and the diameter was remeasured. Average values of these cookies were taken. Thickness (T) of cookies was measured by stacking six of them on top of one another and taking the average of six readings. Spread ratio was calculated by dividing diameter (W) by thickness (T) (Ajila et al., 2008). A texture analyzer (TAXT2i/50, Stable Microsystems, USA) equipped with a 50-kg load cell was used for cookie texture evaluation. Cookies were evaluated for hardness within 24 h by measuring the peak force during penetration using a 5-mm cylinder probe. The analyzer was set at a "return to start" cycle, a pretest, test, and posttest speed of 1.00, 0.5, and 10.00 mm/s, respectively, and a penetration distance of 5 mm. A force/penetration plot was made for every test (Handa et al., 2012) and measurements were conducted three times and results are expressed as mean ± SD values.

9.2.11 PROXIMATE ANALYSIS OF COOKIES

Overall nutritional composition of cookies, moisture, total ash, crude fat, protein, and total carbohydrate were determined by previously described methods.

9.2.12 TOTAL DIETARY FIBER AND ANTIOXIDANT ACTIVITY

TDF content was determined according to the AOAC enzymatic gravimetric method of Prosky et al. (1988). AOA was measured using the method described by Brand-Williams et al. (1995) as described above and all samples were tested in triplicate.

9.2.13 COLOR MEASUREMENT

L^*, a^*, and b^* color values of the cookies were measured using Hunter color measuring spectrophotometer (UltraScan VIS, Hunter Lab), where L^* indicates degree of lightness or darkness [$L^* = 0$ (perfect black) and $L^* = 100$ (perfect white)]; a^* indicates degree of redness (+) and greenness (−), whereas b^* indicates degree of yellowness (+) and blueness (−).

9.2.14 SENSORY EVALUATION

Cookies made from DF from *Musa* ABB rhizome was subjected to sensory evaluation using 20 semitrained panelists drawn within the University community. Cookies were evaluated for color, texture, taste, flavor, and overall acceptability. The ratings were carried on a 9-point hedonic scale ranging from 9 (like extremely) to 1 (dislike extremely).

9.2.15 STATISTICAL ANALYSIS

The data were analyzed by the analysis of variance and were expressed as mean values from three replicates with standard deviations.

9.3 RESULTS AND DISCUSSIONS

9.3.1 CHEMICAL ANALYSIS

The proximate composition of *Musa* ABB rhizome (Table 9.1) revealed moisture content 93.25%, ash 0.72%, protein 0.1%, fat 0.2%, crude fiber 0.78%, and total carbohydrate 4.63%, and all results were in wet basis. Crude fiber was higher than ash, protein, fat, and the results are in line with Gopalan et al. (1989) for banana rhizome.

TDF content of *Musa* ABB rhizome powder was found to be 62.11%, of which SDF and IDF contents were 9.33% and 52.78%, respectively. DF content of dried apple pomace as reported by Sudha et al. (2007b) was 51.10% TDF, 36.50% IDF and is lower than BRDF. TDF content of *Musa* ABB BRDF was also higher than apple DF, pear DF, orange DF, peach DF, artichoke DF, asparagus DF, wheat bran, and oat bran which were 60.1%, 36.1%, 37.8%, 35.8%, 58.8%, 49.05%, 44.0%, and 23.8%, respectively (Grigelmo-Miguel and Martin-Belloso, 1999a). The relationship between IDF and SDF fractions in banana rhizome was found to be 5.65, which is higher than apple, pear, orange, peach, artichoke, and asparagus DF but lower than wheat bran and almost equal to oat bran as reported by Grigelmo-Miguel and Martin-Belloso (1999b). This ratio should be in the range of 1.0–2.3 to obtain the high-quality DF (Spiller, 1985).

TABLE 9.1 Chemical Composition of Banana Rhizome.

Chemical composition	Amount (%)
Moisture	93.25 ± 0.29
Ash	0.72 ± 0.03
Protein	0.1 ± 0.002
Fat	0.2 ± 0.18
Crude fiber	1.1 ± 0.16
Soluble dietary fiber	9.33 ± 0.64
Insoluble dietary fiber	52.78 ± 0.58
Total dietary fiber	62.11 ± 1.02

All data are mean ± SD of three replicates.

9.3.2　TPC AND ANTIOXIDANT ACTIVITY

The scavenging percentage of rhizome powder methanolic extract was 68.94% (Table 9.2) and found higher than the methanol extract of rhizome of different banana cultivars, namely, Ney Mannan (ABB), Safed Velchi (AB), Red Banana (AAA), Giant Cavendish (AAA), Poovan (AAB), Nendran (AAB), Monthan (ABB), and Nanjanagudu Rasabale (AAB) as reported by Kandasamy and Aradhya (2014).

TABLE 9.2 DPPH and TPC Content of Banana Rhizome.

% Scavenging	68.94 ± 0.10
TPC (mg GAE/g rhizome extract)	40.05 ± 0.02

All data are mean ± SD of three replicates.
DPPH, 2,2-Diphenyl-1-picrylhydrazyl; TPC, total polyphenol content; GAE, gallic acid equivalent.

TPC content of rhizome powder (Table 9.2) was found to be 40 mg GAE/g rhizome extract, which was higher than the TPC of hexane extract of different banana cultivars, namely, Ney Mannan (ABB), Safed Velchi (AB), Red Banana (AAA), Giant Cavendish (AAA), Poovan (AAB), Nendran (AAB), Monthan (ABB), and Nanjanagudu Rasabale (AAB). The TPC content of ethyl acetate extract of Poovan (AAB), Nendran (AAB), Monthan (ABB) found lower than *Musa* ABB rhizome but higher in Ney Mannan

(ABB), Safed Velchi (AB), Red Banana (AAA), Giant Cavendish (AAA), and Nanjanagudu Rasabale (AAB) (Kandasamy and Aradhya, 2014).

9.3.3 SOLUBLE AND INSOLUBLE DIETARY FIBER EXTRACTION YIELD

The results of the extraction process from banana rhizome are presented in Table 9.3. Extraction yield of SDF, IDF, and TDF of these two fibers were 9.93%, 60.08%, and 70.01%, respectively, and is higher than the SDF extraction yield of date seeds (1.53%) and lower than IDF (81.97%) and TDF yield (83.50%) (Al-Farsi and Lee, 2008). Under selected optimal conditions, Bouaziz et al. (2014) reported the extraction yield of SDFs and IDFs from *Agave americana* L. and was 81.53% and 92.56%, respectively, and much higher than our results.

TABLE 9.3 Extracted Soluble, Insoluble, and Total Dietary Fiber Yield.

Dietary fiber concentrate	Extraction yield (%)
SDF	9.93 ± 0.45
IDF	60.08 ± 0.49
TDF	70.01 ± 0.27

All data are mean ± SD of three replicates. SDF, soluble dietary fiber; IDF, insoluble dietary fiber; TDF, total dietary fiber.

9.3.4 PHYSICOCHEMICAL PROPERTIES OF DIETARY FIBER CONCENTRATE

Functional properties of the extracted DF powder are shown in Table 9.4. Extracted soluble and insoluble concentrates were mixed in 1:2.3 proportions, and for high-quality DF, insoluble and soluble ratio should be in the range of 1.0–2.3 (Grigelmo-Miguel and Martin-Belloso, 1999b). So to obtain the physiological effects of both insoluble and SFs, it was mixed in 2:3 ratio and properties were evaluated. WHC, OHC, and swelling power of the TDF concentrate were found to be 4.72 g water/g dry sample, 1.53 g oil/g dry sample, and 5.20%, respectively, which were compared with corn (Mora et al., 2013) and gave better functional properties. DFs with high OHC evince stabilization of high-fat food product and emulsions and DF with high WHC could be used as functional ingredients to avoid synersis and can modify the viscosity and texture of formulated foods (Grigelmo-Miguel and Martin-Belloso, 1999a). Swelling capacity depends on the IDF/SDF ratio, particle size, extraction condition, and source (Jamie et al., 2002) and this property for TDF concentrate was found higher than corn. Swelling capacity is observed to be dependent on characteristic of individual component and physical structure of fiber matrix like porosity and crystallinity (Auffret et al., 1994).

TABLE 9.4 Functional Properties of Extracted Dietary Fiber and Comparison with Potato Peel Fiber, Wheat Bran, and Corn.

Sample	WHC (g water/g dry sample)	OHC (g oil/g dry sample)	Swelling power (%)	Solubility (%)	Bulk density (g/mL)	pH
TDF concentrate	4.72 ± 0.05	1.53 ± 0.05	5.20 ± 0.02	5.33 ± 0.05	6.47 ± 0.06	5.49 ± 0.04
SDF concentrate	4.49 ± 0.34	0.61 ± 0.08	9.35 ± 0.06	18.36 ± 0.05	–	–
IDF concentrate	3.35 ± 0.21	1.32 ± 0.08	6.66 ± 0.17	6.01 ± 0.12	–	–
Corn (for comparison)	2.05 ± 0.09	0.87 ± 0.05	0.79 ± 0.0	–	–	–

All data are mean ± SD of three replicates. WHC, water holding capacity; OHC, oil holding capacity; SDF, soluble dietary fiber; IDF, insoluble dietary fiber; TDF, total dietary fiber.

9.3.5 ANTIOXIDANT ACTIVITY OF DIETARY FIBER CONCENTRATES

DF, in recent years, has been recognized as a carrier of antioxidant properties (Saura-Calixto, 2011). The results in Table 9.5 show the antioxidant properties of soluble, insoluble, and TDF concentrates as 4.82%, 35.62%, and 28.94%, respectively. This shows that DF also has the scavenging property which makes it more beneficial for health.

TABLE 9.5 Radical Scavenging Activity of Fiber Concentrates.

Dietary fiber concentrate	% Scavenging activity
Soluble dietary fiber	4.82 ± 0.23
Insoluble dietary fiber	35.62 ± 0.09
Total dietary fiber	28.94 ± 0.16

All data are mean ± SD of three replicates.

9.3.6 SCANNING ELECTRON MICROSCOPY

SEM micrographs of extracted SDFs and IDFs are illustrated in Figure 9.3. The SEM analysis revealed the surface structure of SF particles and was amorphous in structure and less fibrous than IF. The insoluble particle showed more fibrous structure which could affect the final product making harder and stronger because it can absorb less water and form stronger and compact bonds of fibers. SEM analysis of IDF obtained from *A. americana* L. was cylindrical-shaped structures with lot of vacuum spaces and unlike our results, which showed more fibrous and no vacuum spaces which justify the less water and OHC (Bouaziz et al., 2014). The SDF of *A. americana* L. showed heterogeneous structure with pronounced morphological and dimensional variability which is similar in structure with SF of our present investigation.

FIGURE 9.3   SEM micrographs of (a) soluble DF and (b) insoluble DF.

9.3.7 X-RAY DIFFRACTION

XRD patterns for SDF and IDF are shown in Figure 9.4a and b. It has been reported that DF consists of 70% crystalline and 30% amorphous region (Meng-Mei and Tai-Hua, 2016). Figure 9.4b and a depicted that IDF has sharp peaks at 18°–24° and for SDF at 15°–23° which indicated the presence of A-type crystalline diffraction patterns (Aravind et al., 2012). The crystallinity percent was 4% and 3% for IDF and SDF, respectively, and our results are in line with the crystallinity of inulin and uncooked pasta sample (Aravind et al., 2012).

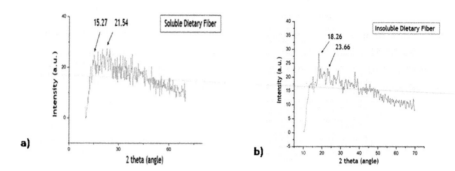

FIGURE 9.4 XRD patterns of (a) soluble DF and (b) insoluble DF.

9.3.8 PHYSICAL AND TEXTURAL ANALYSIS OF COOKIES

Physical characteristics of the cookies, namely, diameter, thickness, and spread ratio, were affected with increase in level of DF (Table 9.6; Fig. 9.5). Thickness slightly decreased and diameter slightly increased with the increase in level of BRDF, while spread ratio increased. This might be attributed to the presence of high amount of IF which formed network and has low WHC. Spread ratio of oat bran and barley-bran-enriched biscuits also reduced with increase in the incorporation without much change in the thickness (Sudha et al., 2007a). Hardness of the cookies decreased with increase in level of DF and this decrease in strength might be ascribed to the amount of IDF present in BRDF.

TABLE 9.6 Physical Characteristics of Cookies.

Parameters	Control	2% BRDF	4% BRDF	6% BRDF
Diameter (D, mm)	36.28 ± 0.73^a	36.34 ± 0.61^b	37.26 ± 0.31^c	37.87 ± 0.11^d
Thickness (T, mm)	12.71 ± 0.25^d	12.09 ± 0.32^c	11.32 ± 0.09^b	11.14 ± 0.10^a
Spread ratio (D/T)	2.85 ± 0.01^a	3.00 ± 0.13^b	3.29 ± 0.01^c	3.39 ± 0.03^d
Weight (g)	9.70 ± 0.05^d	7.68 ± 0.17^a	9.18 ± 0.17^b	9.19 ± 0.08^c
Hardness (kg force)	25.95 ± 0.83^d	20.79 ± 1.14^c	18.74 ± 1.08^b	16.23 ± 0.66^a

All data are mean ± SD of three replicates. Mean followed by different letter in same column differs significantly ($p \leq 0.05$). BRDF, banana rhizome dietary fiber.

9.3.9 PROXIMATE COMPOSITION OF COOKIES

Proximate composition of cookies was evaluated and is shown in Table 9.7. Moisture content of control, 2%, 4%, and 6% BRDF-added cookies were 4.05%, 4.12%, 4.34%, and 4.45%, respectively. Significant increase in moisture content is found which is may be due to presence of DF. Protein, fat, ash, and carbohydrates ranged from 8.23% to 7.23%, 26.18% to 25.62%, 1.18% to 1.28%, and 60.36% to 61.42% for control, 2%, 4%, and 6% BRDF, respectively. Slight decrease in protein content was observed which may be due to substitution of BRDF as it has very low protein content (Table 9.2). Fat content decreased slightly and could be due to low OHC of the DF. The overall nutritional compositions of the cookies prepared are comparable with the commercially available Britannia Nutrichoice biscuits, one of India's leading brands which has protein, fat, carbohydrates, and DF at 7.5%, 19.6%, 62%, and 9%, respectively.

TABLE 9.7 Proximate Composition of Cookies.

Parameters	Control	2% BRDF	4% BRDF	6% BRDF
Moisture	4.05 ± 0.02^a	4.12 ± 0.05^b	4.34 ± 0.05^c	4.45 ± 0.03^d
Ash	1.18 ± 0.04^a	1.19 ± 0.03^b	1.21 ± 0.05^c	1.28 ± 0.02^d
Protein	8.23 ± 0.01^d	7.81 ± 0.01^c	7.62 ± 0.01^b	7.23 ± 0.01^a
Fat	26.18 ± 0.04^d	25.82 ± 0.03^c	25.71 ± 0.02^b	25.62 ± 0.06^a
Total carbohydrate	60.36 ± 0.05^a	61.06 ± 0.04^b	61.12 ± 0.06^c	61.42 ± 0.68^d

All data are mean ± SD of three replicates. Mean followed by different letter in same column differs significantly ($p \leq 0.05$). BRDF, banana rhizome dietary fiber.

9.3.10 TOTAL DIETARY FIBER AND ANTIOXIDANT ACTIVITY

The TDF content and radical scavenging activity of the ethanol extract of the cookies are presented in Table 9.8. The cookies formulated with BRDF powder revealed increase in TDF content from 3.78% (control) to 9.93% (6% BRDF). With increase in the level of BRDF incorporation, the DPPH radical scavenging activity also increased from 18.37% (control) to 52.26% (6% BRDF). The increase in AOA might be associated with the formation of dark color pigments during baking process and these pigments impart the antioxidant properties (Xu and Chang, 2008).

TABLE 9.8   TDF and Antioxidant Assay of Cookies.

Parameters	Control	2% BRDF	4% BRDF	6% BRDF
TDF	3.78 ± 0.23^a	5.62 ± 0.19^b	7.86 ± 0.45^c	9.93 ± 0.31^d
% Scavenging	18.37 ± 0.05^a	30.09 ± 0.94^b	44.17 ± 0.45^c	52.26 ± 0.06^d

All data are mean ± SD of three replicates. Mean followed by different letter in same column differs significantly ($p \leq 0.05$). BRDF, banana rhizome dietary fiber.

9.3.11 COLOR MEASUREMENT

Color of the cookies were measured by Hunter system using $L*$, $a*$, and $b*$ values (Table 9.9) and prepared cookies became darker with increasing level of BRDF. $L*$ value decreased with increasing level of BRDF. Control cookies showed the highest brightness compared to the DF-enriched cookies, and on the other hand, $a*$ value which indicates redness decreased significantly. Brightness may be decreased due to the Maillard reaction during baking process and also because of the BRDF powder which itself is dark in color. Similar changes in color values were also reported by Sudha et al. (2007a) in biscuits incorporated with different brans.

TABLE 9.9   Color of Cookies.

Cookies	$L*$	$a*$	$b*$
Control	61.41 ± 0.51^d	11.89 ± 0.12^d	11.84 ± 0.07^d
2% BRDF	55.78 ± 0.61^b	10.20 ± 0.44^c	8.90 ± 0.05^c
4% BRDF	55.71 ± 0.60^b	8.85 ± 0.66^b	7.91 ± 0.59^b
6% BRDF	55.21 ± 0.35^a	8.23 ± 0.20^a	7.65 ± 0.15^a

All data are mean ± SD of three replicates. Mean followed by different letter in same column differs significantly ($p \leq 0.05$). BRDF, banana rhizome dietary fiber.

9.3.12 SENSORY EVALUATION

Sensory analysis data (Table 9.10) unveiled five sensory attributes, namely, color, texture, taste, flavor, and overall acceptability; Figure 9.6 illustrates the radar diagram of the sensory analysis of cookies. Surface color of 2% and 6% BRDF-added cookies was slightly less acceptable compared to the control. Texture acceptability was similar in all three DF-incorporated cookies, but slightly lower than the control. Acceptability of taste and flavor was little less in 4% than control, but higher than 2% and 6%. Except the control, the overall acceptability was higher in 4% than 2% and 6%.

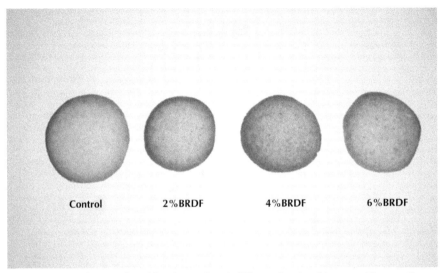

FIGURE 9.5   Picture of cookies (control and different levels of banana rhizome dietary fiber). BRDF, banana rhizome dietary fiber.

Results revealed (Table 9.8) that in 4% and 6% incorporated BRDF cookies, the TDF increased, and can be considered "high-fiber cookies." But from sensory analysis perspective, in reference to the color, texture, taste, flavor, and overall acceptability 4% BRDF-added cookies were more acceptable over 6% BRDF-added cookies. So, the cookies incorporated with 4% of BRDF concentrate can be claimed as "high-fiber cookies with antioxidant property."

TABLE 9.10   Sensory Analysis of Cookies.

Parameter	Control	2% BRDF	4% BRDF	6% BRDF
Color	8.23 ± 0.40	7.33 ± 0.58	6.83 ± 0.76	7.08 ± 0.14
Texture	8.17 ± 0.28	7.50 ± 0.50	7.50 ± 0.50	7.50 ± 0.50
Taste	7.67 ± 0.76	6.67 ± 0.58	7.00 ± 0.87	6.67 ± 0.76
Flavor	7.83 ± 0.29	6.83 ± 0.76	7.17 ± 0.29	7.17 ± 0.29
Overall acceptability	8.17 ± 0.29	7.17 ± 0.29	7.50 ± 0.50	7.17 ± 0.29

All data are mean ± SD of three replicates. BRDF, banana rhizome dietary fiber.

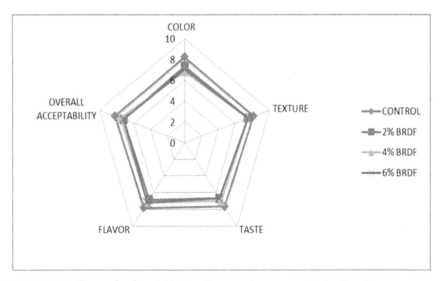

FIGURE 9.6   **(See color insert.)** Radar diagram of sensory analysis of cookies.

9.4   CONCLUSIONS

The chemical composition of the *Musa* ABB banana rhizome powder (BR) evidenced that it is an extremely good source of ash, crude fiber, and DFs and has a good free radical scavenging activity. Results of various physico-chemical properties and functional properties of DF of banana rhizome have the credibility to support that the fiber is an excellent source for developing various functional foods. Cookies enriched with BRDF showed significantly higher TDF and AOA and can be regarded as "high-fiber cookies" with high AOA and has potential for industrial exploitation.

ACKNOWLEDGMENT

The financial help received from DRDO, Ministry of Defence, Govt. of India, is duly acknowledged.

KEYWORDS

- banana rhizome
- dietary fiber
- physicochemical properties
- functional properties
- XRD
- antioxidant activity

REFERENCES

AACC (American Association of Cereal Chemists). *Approved Methods of the AACC*, 10th ed.; American Association of Cereal Chemists: St. Paul, MN, 2000; pp 22–85.

Ajila, C. M.; Leelavathi, K.; Prasada Rao, U. J. S. Improvement of Dietary Fibre Content and Antioxidant Properties in Soft Dough Biscuits with the Incorporation of Mango Peel Powder. *J. Cer. Sci.* **2008**, *48*, 319–326.

Al-Farsi, M. A.; Lee, C. Y. Optimization of Phenolics and Dietary Fibre Extraction from Date Seeds. *Food Chem.* **2008**, *108*, 977–985.

AOAC. *Official Methods of Analysis of the Association of Official Agricultural Chemists*, 10th ed.; AOAC: Washington, DC, 1965.

Aravind, N.; Sissons, M. J.; Fellows, C. M.; Blazek, J.; Gilbert, E. P. Effect of Inulin Soluble Dietary Fibre Addition on Technological, Sensory, and Structural Properties of Durum Wheat Spaghetti. *Food Chem.* **2012**, *132*, 993–1002.

Auffret, A.; Ralet, M. C.; Guillon, F.; Barry, J. L.; Thibault, J. F. Effect of Grinding and Experimental Conditions on the Measurement of Hydration Properties of Dietary Fiber. *LWT—Food Sci. Technol.* **1994**, *27*, 166–172.

Bouaziz, A.; Masmoudi, M.; Kamoun, A.; Besbes, S. Optimization of Insoluble and Soluble Fibres Extraction from *Agave americana* L. Using Response Surface Methodology. *J. Chem.* **2014**. DOI:10.1155/2014/627103.

Brand-Williams, W.; Cuvelier, M. E.; Berset, C. Use of a Free Radical Method to Evaluate Antioxidant Activity. *LWT—Food Sci. Technol.* **1995**, *28* (1), 25–30.

Chau, C. F.; Huang, Y. L. Comparison of the Chemical Composition and Physicochemical Properties of Different Fibers Prepared from Peel of *Citrus sinensis* L. cv. Liucheng. *J. Agric. Food Chem.* **2003**, *51*, 2615–2618.

Cleven, R.; Van den, B. C.; Van der, P. L. Crystal Structure of Hydrated Potato Starch. *Starch—Stärke* **1978**, *30* (7), 223–228.

Gopalan, C.; Rama Sastri, B. V.; Balasubramanian, S. C. In *Nutritive Value of Indian Foods*; Narasinga Rao, B. S.; Deosthale, Y. G.; Pant, K. C. National Institute of Nutrition: Hyderabad, India, 1989; pp 47–58.

Grigelmo-Miguel, N.; Martin-Belloso, O. Characterization of Dietary Fibre from Orange Juice Extraction. *Food Res. Int.* **1999a**, *131*, 355–361.

Grigelmo-Miguel, N.; Martin-Belloso, O. Comparison of Dietary Fibre from By-Products of Processing Fruits and Greens and from Cereals. *Lebens.—Wissensch. Technol., Food Sci. Technol.* **1999b**, *32*, 503–508.

Handa, C.; Goomer, S.; Siddhu, A. Physicochemical Properties and Sensory Evaluation of Fructoligosaccharide Enriched Cookies. *J. Food Sci. Technol.* **2012**, *49* (2), 192–199.

Hesser, J. M. Applications and Usage of Dietary Fibre in the USA. *Int. Food Ingred.* **1994**, *2*, 50–52.

Jamie, L.; Molla, E.; Fernandez, A.; Martin-Cabrejas, M.; Lopez, A. F.; Esteban, R. Structural Carbohydrates Differences and Potential Source of Dietary Fibre of Onion (*Allium cepa* L.) Tissues. *J. Agric. Food Chem.* **2002**, *50*, 122–128.

Kandasamy, S.; Aradhya, S. M. Polyphenolic Profile and Antioxidant Properties of Rhizome of Commercial Banana Cultivars Grown in India. *Food Biosci.* **2014**, *8*, 22–32.

Leach, H. W.; Mc Cowen, L. D.; Schoch, T. J. Structure of the Starch Granule. I. Swelling and Solubility Patterns of Various Starches. *Cer. Chem.* **1959**, *36*, 534–544.

Lowry, O. H.; N. J. Rosebrouugh, N. J.; Farr, A. L.; Randall, R. J. Protein Measurement with the Folin Phenol Reagent. *J. Biol. Chem.* **1951**, *193*, 265–275.

Masmoudi, M.; Besbes, S.; Chaabouni, M. Optimization of Pectin Extraction from Lemon By-Product with Acidified Date Juice Using Response Surface Methodology. *Carbohydr. Polym.* **2008**, *74* (2), 185–192.

Maynard, A. J. *Methods in Food Analysis*; Academic Press: New York, 1970; p 176.

Mohapatra, D.; Mishra, S.; Sutar, N. Banana and Its By-Product Utilization: An Overview. *J. Sci. Ind. Res.* **2010**, *69*, 323–329.

Meng-Mei, M.; Tai-Hua, M. Effects of Extraction Methods and Particle Size Distribution on the Structural, Physicochemical, and Functional Properties of Dietary Fibre from Deoiled Cumin. *Food Chem.* **2016**, *194*, 237–246.

Mora, Y. N.; Contreras, J. C.; Aguilar, C. N.; Meléndez, P. De la Garza, I.; Rodríguez, R. Chemical Composition and Functional Properties from Different Sources of Dietary Fibre. *Am. J. Food Nutr.* **2013**, *1* (3), 27–33.

Pietinen, P. Dietary Fibre and Coronary Heart Disease: Epidemiology. *COST Actions Bioactive Micronutrients in Mediterranean Diet and Health*; European Commission, 2001; pp 23–25.

Prosky, L.; Asp, N. G.; Schweizer, T. F.; Devries, J. W.; Furda, I. Determination of Insoluble, Soluble, and Total Dietary Fibre in Foods and Food Products: Interlaboratory Study. *J. Assoc. Off. Anal. Chem.* **1988**, *71* (5), 1017–1023.

Pushpangadan, P.; Kaur, J.; Sharma, J. Plantain or Edible Banana (*Musa × paradisiac* Var. *sapiemtum*) Some Lesser Known Folk Uses in India. *Anc. Sci. Life* **1989**, *9* (1), 20–24.

Rodriguez-Ambriz, S. L.; Islas-Hernandez, J. J.; Agama-Acevedo, E.; Tovar, J.; Bello-Perez, L. A. Characterization of a Fibre-Rich Powder Prepared by Liquefaction of Unripe Banana Flour. *Food Chem.* **2008**, *107*, 1515–1521.

Saravanan, K.; Aradhya, S. M. Potential Nutraceutical Food Beverage with Antioxidant Properties from Banana Plant Bio-Waste (Pseudostem and Rhizome). *Food Funct.* **2011**, *2*, 603–610.

Saravanan, K.; Aradhya, S. M. Polyphenolic Profile and Antioxidant Properties of Rhizome of Commercial Banana Cultivars Grown in India. *Food Biosci.* **2014**, *8*, 22–32.

Saura-Calixto, F. Dietary Fibre as Carrier of Dietary Antioxidants: An Essential Physiological Function. *J. Agric. Food Chem.* **2011**, *59*, 43–49.

Scheneeman, B. O. Soluble vs. Insoluble Fibre—Different Physiological Responses. *Food Technol.* **1987,** *41*, 81–82.

Spiller, G. A. Suggestions for a Basis on Which to Determine a Desirable Intake of Dietary Fibre. In *CRC Handbook of Dietary Fibre in Human Nutrition*; Spiller, G. A., Ed.; CRC Press: Boca Raton, FL, 1986; pp 281–283.

Sudha, M. L.; Vetrimani, R.; Leelavathi, K. Influence of Fibre from Different Cereals on the Rheological Characteristics of Wheat Flour Dough and on Biscuit Quality. *Food Chem.* **2007a,** *100*, 1365–1370.

Sudha, M. L.; Bhaskaran, V.; Leelavathi, K. Apple Pomace as a Source of Dietary Fibre and Polyphenols and Its Effect on the Rheological Characteristics and Cake Making. *Food Chem.* **2007b,** *104*, 686–692.

Suntharalingam, S.; Ravindran, G. Physical and Biochemical Properties of Green Banana Flour. *Plant Foods Hum. Nutr.* **1993,** *43* (1), 19–27.

Tavaini, A.; La Vecchia, C. Fruit and Vegetable Consumption and Cancer Risk in a Mediterranean Population. *Am. J. Clin. Nutr.* **1995,** *61*, 1374–1377.

Wang, J. C.; Kinsella, J. E. Functional Properties of Novel Proteins: Alfalfa Leaf Protein. *J. Food Sci.* **1976,** *41* (2), 286–292.

Xu, B.; Chang, S. K. C. Total Phenolics, Phenolic Acids, Isoflavones, and Anthocyanins and Antioxidant Properties of Yellow and Black Soybeans as Affected by Thermal Processing. *J. Agric. Food Chem.* **2008,** *56*, 7165–7175.

BIOCARBON-STORAGE POTENTIAL OF TEA PLANTATIONS OF NILGIRIS, INDIA, IN RELATION TO LEAF CHLOROPHYLL AND SOIL PARAMETERS

L. ARUL PRAGASAN[*] and R. SARATH

Department of Environmental Sciences, Bharathiar University, Coimbatore 641046, Tamil Nadu, India

[*]*Corresponding author. E-mail: arulpragasan@yahoo.co.in*

ABSTRACT

Biocarbon storage is the sinking of atmospheric carbon in plants. It helps to reduce the alarming level of increase in greenhouse gases, particularly carbon dioxide, in the atmosphere. The concentration of atmospheric CO_2 was 270 ppm before the industrial revolution period, and now, it has crossed 500 ppm. This is one of the main reasons for global warming and climate change. Currently, there are lots of measures taken for mitigation of this global threat, and one such measure is biocarbon storage. This chapter aims to provide the biocarbon-storage potential of tea plantation in Gudalur located in Nilgiris district of Tamil Nadu state in India.

10.1 INTRODUCTION

Biocarbon storage or sequestration refers to the storage of carbon in a stable system, and that occurs through direct fixation of atmospheric carbon dioxide (CO_2) by means of plant photosynthesis. In this process, plants absorb CO_2 from atmosphere and store as biomass, and thereby, it reduces the air pollution caused by the greenhouse gas. It is well known that the increasing concentration of greenhouse gases particularly CO_2 is the reason behind global warming and climate change. Reduction of such harmful gas is vital for mitigation of the drastic change in climate. Thus, biocarbon storage is a handy technique which helps in the reduction of atmospheric carbon concentration.

Although biocarbon storage is important, so far only a few number of studies particularly related to the effect of crop plantations (such as coffee, cocoa, and rubber) on carbon fixation in tropical regions were carried out (Steffan-Dewenter et al., 2007; Li et al., 2008). Besides the growing awareness, information to support the sustainable management of crop plantations, conditions of biocarbon storage, and dynamics in the systems are still poorly understood (Li et al., 2011).

Tea (*Camellia sinensis* L.) is a perennial evergreen broad-leaved cash crop cultivated for continuous growth of young shoots. Tea leaves contain more than 700 chemicals; among those, useful compounds for human health include flavonoids, amino acids, vitamins (C, E, and K), caffeine, and polysaccharides. Recently, it is proven that tea drinking has been associated with cell-mediated immune function in human; tea improves beneficial intestinal microflora, provides immunity against intestinal disorders, and also protects cell membrane from oxidative damage (Bhagat et al., 2010). In India, tea was best known as a medical plant and not as a drink for pleasure, until 19th century during when tea plantations were established by the British. Tea is one of the three common beverages (coffee, tea, and cocoa) in the world, consumed by a large number of people. Due to high growing demand, it is considered as one of the major components of world beverage market. Nevertheless, tea cultivation is restricted to certain regions of the world due to specific climate and soil conditions (Li et al., 2011).

Worldwide, about 36.9 lakh hectares of land area is used for tea cultivation. With increasing global tea trade, area of tea cultivation as well as tea production have increased significantly. As of 2013, the world annual tea production has crossed 5.34 million tons which is 66% higher than the value for the year 2003. Globally, China (1.94 million tons) and India (1.21 million tons) are the top tea-producing nations, and they account for 59% of the total tea production of the world (FAO, 2016). India is the world's largest tea-drinking nation; however, the per capita of tea consumption is average 750 g/person/year. In India, tea plantations are concentrated in Assam, Himachal Pradesh, West Bengal, and parts of Karnataka, Kerala, and Tamil Nadu. The total tea harvest area as of 2003 was 516,000 ha and it had increased to 563,980 by 2013, and the tea production for 2013 was 1208,780 t which was 846,000 t for the year 2003 (FAO, 2016). Tea industries in India provide job opportunity for about one million people (Kalita et al., 2015).

Although, today, tea production for beverage is the main reason for the establishment of tea plantations, its long rotational lifecycle (from 40 to 90 years) may represent a high potential for biocarbon storage. Unlike the intensively studied biocarbon storage dynamics of forest systems (Ren et al., 2011; Fonseca et al., 2011; Hendri et al., 2012; Sheikh et al., 2012; Pragasan and Karthick, 2013; Sundarapandian et al., 2013; Pragasan, 2014, 2015a, 2015b, 2015c, 2016; Shah et al., 2014), tea plantations are poorly understood in terms of its biocarbon storage. To understand the biocarbon storage in tea crop, the present work was carried out in a tea plantation in India. The main objectives of the present study were (1) to determine biocarbon storage of tea plantation, and (2) to find the relationship between biocarbon storage of tea plant with leaf chlorophyll content and soil parameters, such as pH and conductivity, nitrogen (N), phosphorus (P), and potassium (K).

10.2 MATERIALS AND METHODS

10.2.1 STUDY AREA

The present study was carried out in Gudalur (11.50°N, 76.50°E) located at Nilgiris district of Tamil Nadu state in India (Fig. 10.1). The human population of Gudalur (survey 2011) was 49,535 individuals with a sex ratio of 1032 females for every 1000 males, and the density of population was 200/ km^2. Gudalur is located at an altitude ca. 1100 m asl. This region has hills with varied altitudinal range, rich in diverse flora and fauna.

Besides tea (Fig. 10.2) and coffee plantation, vegetables such as potato, cabbage, and cauliflower, spices such as cardamom, pepper, and ginger are cultivated in this area. Tea cultivation was introduced by British during the pre-independence period. Many small and large private tea estates exist in the region. In recent years, potato and other root growers in the region have switched to tea cultivation.

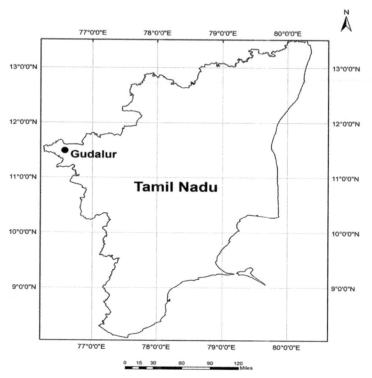

FIGURE 10.1   Map showing the location of the study area.

FIGURE 10.2   Tea plantation of the study area.

10.2.2   CLIMATE

Gudalur falls in tropical climatic zone. This region experiences four seasons, namely, southwest monsoon (SWM), northeast monsoon (NEM), winter (WS) and summer (SS) seasons (Fig. 10.3). The mean annual rainfall for a 10-year period (1999–2008) available for the district was 1652 ± 353 mm ($\pm$SD). Rainfall was maximum during SWM (836 ± 214 mm) followed by NEM (494 ± 129 mm), SS (277 ± 97 mm), and WS (46 ± 53 mm). The mean monthly maximum temperature (for 2008–09) was 21.4 ± 5.6°C and the mean monthly minimum temperature for the same period was 15.0 ± 3.8°C.

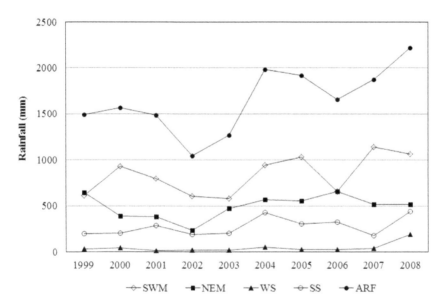

FIGURE 10.3   Rainfall pattern for a 10-year period available for the study area.

10.2.3   SOIL

The soil type of this region can be categorized into five major types, namely, alluvial and colluvial soil, black soil, lateritic soil, red loam, and red sandy soil. The predominant type being lateritic soil, in contrast the red loam and red sandy soils are seen in small patches. In the valleys, where the water logging is common during the monsoons are dominated by the black soil. While, alluvial, and colluvial soils are common along the valleys and river courses.

10.2.4 METHODS

Biocarbon storage of tea plantation was calculated based on field survey and laboratory analysis. To determine the total biocarbon storage (tC/ha) for tea plantation, density (individuals/ha), biomass (kg/individual), and carbon content (%) data were used. Density of tea crop was determined by quadrat method. Ten quadrats of size 10 m × 10 m were randomly placed and all the tea crops were counted from each quadrat to determine the density. Biomass of a single tea crop (13.6 kg/individual; leaf—5%, stem—84%, root—11%) was adopted from a 43-year-old tea plantation study (Kamau et al., 2008) as destruction sampling was not possible. From each quadrat, leaf and stem samples were collected, to determine the carbon content (%) of leaf, stem, and root components for calculation of biocarbon storage. Chlorophyll content of tea crop was detected using chlorophyll meter (SPAD-502 Plus). Soil samples were collected to determine the soil pH, conductivity, and soil nutrients such as nitrogen (N, g/kg), phosphorus (P, %), potassium (K, %).

In the laboratory, the collected leaf and stem samples were air dried for 2 days. Then, the samples were dried in hot air oven for 24 h at 105°C to get dry weights. Then, 1 g of oven-dried grind samples was taken separately in preweighted crucibles. The crucibles were placed in the furnace at 550 ± 5°C for 2 h. Then, the crucibles were cooled slowly inside the furnace. After cooling, the crucibles with ash were weighted for determination of carbon content using the following equations adopted earlier (Pragasan and Karthick, 2013):

$$C_c = (100 - A_{sh}) \times 0.58$$

$$A_{sh} = \frac{(W_3 - W_1)}{(W_2 - W_1)} \times 100$$

where C_c is carbon content (%); A_{sh} is ash percentage (%); W_1 is the weight of crucible (g); W_2 is the sum of the weight of oven dried grind sample and crucible (g); and W_3 is the sum of the weight of ash and crucible (g).

Soil pH and conductivity were estimated using the following procedure. One gram of soil sample was mixed with 50 mL of water and equilibrated for 1 h using a mechanical shaking incubator (Wisecube, Daihan Scientific, Korea) at 200 rpm at 35°C. The supernatant was analyzed for pH using digital pH meter (Elico-LI 127) and conductivity using digital conductivity meter (Elico-CM 180). Estimation of soil N, P, and K was carried out using standard methods such as Kjeldahl method, Bray's method, and flame photometry method, respectively.

Further, carbon dioxide sequestration was determined from biocarbon storage data using conversion ratio of 1 kg of carbon = 3.68 kg of carbon dioxide, as this ratio was calculated by dividing the atomic weight of CO_2 by the atomic weight of carbon.

10.2.5 STATISTICAL ANALYSIS

Statistical analysis of variance (ANOVA) was carried out to check for significance in variation of carbon content of tea plant components such as leaf, root, and stem samples. Linear regression was adopted to find the relationship between biocarbon storage of tea plantation with tealeaf chlorophyll content and soil parameters such as pH, conductivity, N, P, and K.

10.3 RESULTS

10.3.1 DENSITY

Density of plants calculated for the tea plantation was 18,540 ± 250 individuals/ha. Tea crop in the plantation had approximately 60–75 cm interdistance between two individuals. The abundance of tea plant for the ten 10 m × 10 m sample plots varied from 179 individuals to 187 individuals. The mean abundance per sample was 185.4 ± 2.5 (±SD).

10.3.2 CARBON CONTENT

The carbon content for both the plant components of tea crop was determined through loss on ignition method. The biocarbon storage assessed for the leaf samples ($n = 30$) ranged from 53.94% to 54.52%, and the average value was 54.15 ± 0.28%. The carbon content determined for the stem samples ($n = 20$) ranged from 56.26% to 57.42%; the average carbon storage for the stem samples was 56.78 ± 0.26%. And, the carbon content calculated for root ($n = 20$) varied from 55.1% to 55.68%, and the average value was 55.49 ± 0.19%. One-way ANOVA revealed that there existed no significant variation in carbon content values among the leaf ($F(9, 20) = 1.095$, $p > 0.05$), stem

($F_{(9, 10)} = 1.000$, $p > 0.05$) as well as for root ($F_{(9, 10)} = 1.603$, $p > 0.05$) samples.

10.3.3 CHLOROPHYLL CONTENT

The chlorophyll content observed for the 10-sample plots ranged from 61.71 ± 23.3 to 83.55 ± 28.9. The average chlorophyll content for the total leaf samples ($n = 100$) of tea plantation was **69.31 ± 20.54%**. One-way ANOVA performed to check for variation in chlorophyll content showed no significant variation in chlorophyll content among the leaf samples ($F_{(9,90)} = 1.188$, $p > 0.05$).

10.3.4 SOIL PARAMETERS

The soil parameters (pH, conductivity, N, P, and K) were determined using standard methods. Soil pH is considered a master variable in soils as it controls many chemical processes that take place. It specifically affects plant nutrient availability by controlling the chemical forms of the nutrient. The optimum pH range for most plants is between 5.5 and 7.0. The soil pH value for the tea plantation ($n = 5$) varied from 4.9 to 5.73, and the mean pH value was 5.4 ± 0.4. The soil conductivity determined for the tea plantation ($n = 5$) varied from 13.7 to 17.6 µs, and the average value was 16.31 ± 1.84 µs. Soil N, P, and K values ranged from 1.68 to 3.43 g/kg, 0.05–0.29%, and 0.02–0.04%, respectively, for the total samples ($n = 5$ for each parameters) collected from the tea plantation. The average value for the soil N, P, and K was 2.42 ± 0.65 g/kg, 0.13 ± 0.09%, and 5.38 ± 0.37%, respectively.

10.3.5 BIOCARBON STORAGE

The total biocarbon storage (tC/ha) determined for the tea plantation of Gudalur was 142.47 ± 1.92 tC/ha, and the biocarbon value varied from 137.56 to 143.71 tC/ha among the 10-sample plots. The total biocarbon storage of tea plantation was maximum shared by stem (84%, 120.26 ± 1.62 tC/ha), followed by root (11%, 15.39 ± 0.21) and leaf (5%, 6.83 ± 0.09) (Fig. 10.4). One-way ANOVA revealed a strong significant variation in biocarbon stock among the three plant components of tea plant ($F_{(2, 27)} = 44{,}521.42$, $p < 0.00001$).

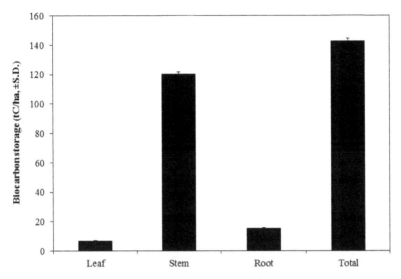

FIGURE 10.4   Biocarbon storage of tea plantation by different plant components.

10.3.6   RELATIONSHIP OF BIOCARBON STORAGE WITH LEAF CHLOROPHYLL AND SOIL PARAMETERS

Regression analysis done between the biocarbon storage (tC/ha) values of tea plantations with leaf chlorophyll and soil parameters such as pH, conductivity, N, P, and K of the tea plantation under the study revealed that biocarbon storage (tC/ha) of tea plantation had positive relation with only soil N (Table 10.1). This reveals that N is a necessary component for sequestering more atmospheric carbon as biocarbon in tea plantations.

TABLE 10.1   Relationship of Biocarbon Storage (tC/ha) with Other Parameters.

Parameters	Leaf carbon	
	Equation	R^2
Leaf chlorophyll	$y = 0.0327x + 140.57$	0.0300
Soil pH	$y = -0.7892x + 147.02$	0.0867
Soil conductivity	$y = 0.1499x + 140.34$	0.0754
Soil N	$y = 1.2947x + 139.64$	0.7091
Soil P	$y = 3.7166x + 142.31$	0.1228
Soil K	$y = 18.592x + 142.33$	0.0213

10.3.7   CARBON DIOXIDE SEQUESTRATION

The estimated value of CO_2 sequestration by the tea plantation varied between 506.22 and 528.84 t/ha, and the average value was 524.32 ± 7.1 t/ha. For different components of tea plant, the CO_2 sequestration value was 25.12 ± 0.3, 442.56 ± 6.0, and 56.64 ± 0.8 t/ha for leaf, stem, and root part, respectively.

10.4   DISCUSSIONS

Quantifying biocarbon storage is a necessary task when assessing the particular carbon budget of an ecosystem. Tea is an important cash crop, and tea plantations take large amounts of cultivable land worldwide. Plants trap atmospheric carbon dioxide through photosynthesis and store in leaves, stems, and roots. The tea plantations absorb more carbon and serve as a good source for mitigation for global warming. The present study revealed that the biocarbon-storage potential of tea plantation in Gudalur, Nilgiris district of Tamil Nadu state, India, was high 142.47 ± 1.92 tC/ha when compared to a few other studies available in literature (Table 10.2). The biocarbon storage (tC/ha) of the tea plantation of Gudalur falls within the range reported for tropical forest of West Papua, Indonesia (Hendri et al., 2012), higher than the value reported for tea plantation in China (44.99–60.64 tC/ha, Li et al., 2011), Kenya (43.0–72.0 tC/ha, Kamau et al., 2008), tropical forest in Costa Rica (13.2–65.4 tC/ha, Fonseca et al., 2011), natural forest and plantations in Puducherry (19.5–131.8 tC/ha, Sundarapandian et al., 2013), noncommercial plantation in China (8.72–8.80 tC/ha, Chen et al., 2009), agroforestry in India (31.95–83.07 tC/ha, Rizvi et al., 2011), Scots pine plantation in Finland (13.87–22.58 tC/ha, Cao et al., 2010), subtropical Pinus forest in Himachal Pradesh, India (47.3 tC/ha, Shah et al., 2014), and lesser than the value reported for subtropical forest in Garhwal Himalayas, India (203.02–230.84 tC/ha, Sheikh et al., 2012). The difference in biocarbon values (tC/ha) may be influenced by different factors including ecosystem type, plant density, rainfall, temperature, and soil parameters.

TABLE 10.2 Comparison of Biocarbon Storage of Tea Plantation of Gudalur with Other Available Studies.

Location	Ecosystem	tC/ha	Annual rainfall (mm)	Temperature (°C)	Reference
Gudalur, India	Tea plantation	142.47	1652	15.0–21.4	Present study
Mali, West Africa	Agroforestry	0.70–54.00	300–700	29.0	Takimoto et al. (2008)
Guanacaste, Costa Rica	Silvopastoral	3.50–12.50	1725	28.0	Andrade et al. (2008)
West Papua, Indonesia	Tropical forest	54.90–153.40	2394	26.9	Hendri et al. (2012)
Xiamen, China	Forest	4.56–13.52	1100	21.0	Ren et al. (2011)
Costa Rica	Tropical forest	13.20–65.40	3420–6840	25.0–27.0	Fonseca et al. (2011)
Bodamalai hills, India	Tropical forest	10.94	1058	28.3	Pragasan (2015a)
Kalrayan hills, India	Tropical forest	38.88	1058	28.3	Pragasan (2015b)
Shervarayan hills, India	Tropical forest	56.55	1058	28.3	Pragasan (2015c)
Chitteri hills, India	Tropical forest	58.55	1058	28.3	Pragasan (2014)
Kolli hills, India	Tropical forest	73.70	1058	28.3	Pragasan (2016)
Coimbatore, India	Eucalyptus plantation	27.72	646	26.3–38.8	Pragasan and Karthick (2013)
Coimbatore, India	Mixed species plantation	22.25	646	26.3–38.8	Pragasan and Karthick (2013)
Puthupet, Puducherry	*Anacardium occidentale* plantation	19.50	1311	28.5	Sundarapandian et al. (2013)
Puthupet, Puducherry	*Casuarina equisetifolia* plantation	23.90	1311	28.5	Sundarapandian et al. (2013)
Puthupet, Puducherry	*Cocos nucifera* plantation	83.30	1311	28.5	Sundarapandian et al. (2013)

TABLE 10.2 *(Continued)*

Location	Ecosystem	tC/ha	Annual rainfall (mm)	Temperature (°C)	Reference
Puthupet, Puducherry	*Mangifera indica* plantation	70.50	1311	28.5	Sundarapandian et al. (2013)
Puthupet, Puducherry	Natural forest	131.80	1311	28.5	Sundarapandian et al. (2013)
Saharanpur (Uttar Pradesh), India	Agroforestry	31.95	–	23.3	Rizvi et al. (2011)
Haryana, India	Agroforestry	83.07	970	–	Rizvi et al. (2011)
Garhwal Himalayas, India	Subtropical Pinus forest	203.02–230.84	960	10.0–23.0	Sheikh et al. (2012)
Himachal Pradesh, India	Subtropical Pinus forest	47.30	1000	15.0–36.0	Shah et al. (2014)
Kericho, Kenya	Tea plantation	43–72	2150	15.5–18.0	Kamau et al. (2008)
China	Tea plantation	50.90	700–2600	12.0–22.0	Li et al. (2011)
Yunnan Province, China	Noncommercial plantation	8.72–8.80	–	–	Chen et al. (2009)
Finland	Scots pine plantation	13.87–22.58	–	–	Cao et al. (2010)

From a few available literature (Table 10.2), linear regression analysis was performed to check for any relation between biocarbon storage (tC/ha) and annual rainfall (mm) and temperature (°C). The analysis revealed that there existed no significant relationship of biocarbon storage (tC/ha) with annual rainfall (Fig. 10.5) as well as temperature (Fig. 10.6). In the present study, the plant density was 18,540 ± 250 individuals/ha which is almost threefolds higher than a 43-year old tea plantation in China (6730 individuals/ha, **Kamau et al., 2008**), and the biocarbon storage (tC/ha) of Gudalur (142.47 tC/a) is also threefolds higher than the value (43.0 tC/ha) reported for the 43-year old tea plantation in China (**Kamau et al., 2008**); this supports a positive relation between tea plant density (individuals/ha) and biocarbon storage (tC/ha). However, more in-depth studies are warranted to draw a clear conclusion on the relationship between biocarbon storage and other influencing factors.

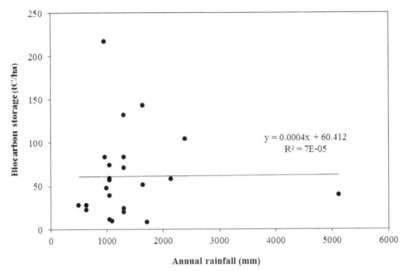

FIGURE 10.5   Impact of rainfall on biocarbon storage.

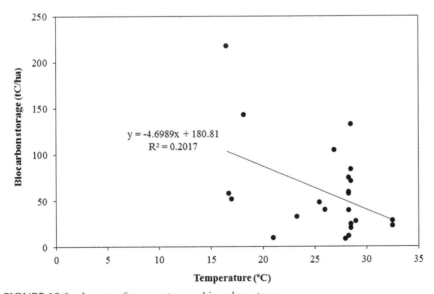

FIGURE 10.6   Impact of temperature on biocarbon storage.

Among the three plant components, the total biocarbon storage of tea plantation in Gudalur was maximum shared by stem (120.26 ± 1.62 tC/ha), followed by root (15.39 ± 0.21) and leaf (6.83 ± 0.09), and a strong significant ($p < 0.00001$) variation in the biocarbon value was observed among them. The amount of biocarbon stock is directly proportional to the total biomass stock. It is widely applied that 50% of the total biomass is considered as the total carbon stock of tropical tree species (Atjay et al., 1979; Brown and Lugo, 1982; Dixon et al., 1994; Takimoto et al., 2008; Pragasan, 2014; Timilsina et al., 2014). In the present study, the carbon content (%) of the three plant components was 54.15%, 56.78%, and 55.49% for leaf, stem, and root, respectively. While the value recorded respectively for the three components was 51.97%, 49.73%, and 46.62% for the tea plantation in China (Li et al., 2011).

The results of regression analysis carried out between the biocarbon storage (tC/ha) and leaf chlorophyll content (%) and soil parameters, such as pH, conductivity, N, P, and K, for the present study revealed that biocarbon storage (tC/ha) of tea plantation at Gudalur had positive relation with only soil N ($R^2 = 0.71$, see Table 10.1); this suggests that soil N is an influencing factor for biocarbon storage in tea plantation. Studies have proved earlier that nutrients, especially nitrogen, is known to be key determinant of biocarbon storage in forest ecosystems (Kimmins, 1996; Kamau et al., 2008).

In terms of CO_2 sequestration, the value determined for the tea plantation at Gudalur was 524.32 ± 7.1 tCO_2/ha. This suggests that the tea plantation in Gudalur traps 524 t of atmospheric carbon dioxide as biocarbon storage.

10.5 CONCLUSIONS

The present study signifies the potential role of tea plantation in biocarbon storage in the present scenario of global warming and climate change. The results of the present study conclude that the biocarbon storage in the tea plantation of Gudalur was 142.47 tC/ha, which is high when compared to other available studies. The biocarbon storage (tC/ha) potential of tea plantation in Gudalur was positively influenced by soil N which is in accordance with Kimmins (1996) and Kamau et al. (2008). The global review of tea in terms of land cover, production, yield, exports, and imports indicated overall increase in the quantity of tea in the world market in recent years. If the existing tea plantations are managed sustainably, definitely tea plantations can work wonders in reduction of the alarming increase in the concentration of the atmospheric carbon dioxide gas that leads to global warming one of the grave environmental problems at present. Hence, tea plantations are not only an economic resource but also provide a long-term cost-effective means of biocarbon storage and thereby play a pivotal role in mitigation of global warming and climate change. Hence, creating awareness on biocarbon storage and encouragement of tea plantation and sustained harvest of tea leaf is required for improved biocarbon storage to mitigate global warming and climate change.

KEYWORDS

- biocarbon storage
- atmospheric carbon
- carbon dioxide
- plant photosynthesis
- global warming

REFERENCES

Andrade, H. J.; Brook, R.; Ibrahim, M. Growth, Production and Carbon Sequestration of Silvopastoral Systems with Native Timber Species in the Dry Lowlands of Costa Rica. *Plant Soil* **2008**. DOI:10.1007/s11104-008-9600-x.

Atjay, G. L.; Ketner, P.; Duvignead, P. Terrestrial Primary Production and Phytomass. In *The Global Carbon Cycle*; Bolin, B., Degens, E. T., Kempe, S., Eds.; Wiley and Sons: New York, 1979; pp 129–182.

Bhagat, R. M.; Baruah, R. D.; Safique, S. Climate and Tea (*Camellia sinensis* (L.) O. Kuntze) Production with Special Reference to North Eastern India: A Review. *J. Environ. Res. Develop.* **2010**, *4*, 1017–1028.

Brown, S.; Lugo, A. E. The Storage and Production of Organic Matter in Tropical Forests and Their Role in the Global Carbon Cycle. *Biotropica* **1982**, *14*, 161–187.

Cao, T.; Valsta, L.; Makela, A. A Comparison of Carbon Assessment Methods for Optimizing Timber Production and Carbon Sequestration in Scots Pine Stands. *For. Ecol. Manage.* **2010**, *260*, 1726–1734.

Chen, X.; Zhang, X.; Zhang, Y.; Wan, C. Carbon Sequestration Potential of the Stands under the Grain for Green Program in Yunnan Province, China. *For. Ecol. Manage.* **2009**, *258*, 199–206.

Dixon, R. K.; Brown, S.; Solomon, R. A.; Trexler, M. C.; Wisniewski, J. Carbon Pools and Flux of Global Forest Ecosystems. *Science* **1994**, *263*, 185–190.

FAO. Food and Agriculture Association of the United Nations (FAO): Rome, 2016. www.faostat-fao.org.

Fonseca, W.; Benayas, J. M. R.; Alice, F. E. Carbon Accumulation in the Biomass and Soil of Different Aged Secondary Forests in the Humid Tropics of Costa Rica. *For. Ecol. Manage.* **2011**, *262*, 1400–1408.

Hendri; Yamashita, T.; Kuntoro, A. A.; Lee, H. S. Carbon Stock Measurements of a Degraded Tropical Logged-Over Secondary Forest in Manokwari Regency, West Papua, Indonesia. *For. Stud. China* **2012**, *14* (1), 8–19.

Kalita, R. M.; Das, A. K.; Nath, A. J. Allometric Equations for Estimating Above- and Belowground Biomass in Tea (*Camellia sinensis* (L.) O. Kuntze) Agroforestry System of Barak Valley, Assam, Northeast India. *Biomass Bioenerg.* **2015**, *83*, 42–49.

Kamau, D. M.; Spiertz, J. H. J.; Oenema, O. Carbon and Nutrient Stocks of Tea Plantations Differing in Age, Genotype and Plant Population Density. *Plant Soil* **2008**, *307*, 29–39.

Kimmins, J. P. Importance of Soil and Role of Ecosystem Disturbance for Sustainable Productivity of Cool Temperature and Boreal Forest. *Soil Sci. Soc. Am. J.* **1996**, *60*, 1643–1654.

Li, H. M.; Ma, Y. X.; Aide, T. M.; Liu, W. J. Past, Present and Future Land-Use in Xishuangbanna China and the Implications for Carbon Dynamics. *For. Ecol. Manage.* **2008**, *255*, 16–24.

Li, L.; Wu, X.; Xue, H.; et al. Quantifying Carbon Storage for Tea Plantations in China. *Agric. Ecosyst. Environ.* **2011**, *141*, 390–398.

Pragasan, L. A.; Karthick, A. Carbon Stock Sequestration by Tree Plantations in University Campus at Coimbatore, India. *Int. J. Environ. Sci.* **2013**, *3*, 1700–1710.

Pragasan, L. A. Carbon Stock Assessment in the Vegetation of the Chitteri Reserve Forest of the Eastern Ghats in India Based on Non-Destructive Method Using Tree Inventory Data. *J. Earth Sci. Clim. Change* **2014**, *S11*, 001.

Pragasan, L. A. Tree Carbon Stock Assessment from the Tropical Forests of Bodamalai Hills Located in India. *J. Earth Sci. Clim. Change* **2015a**, *6*, 314.

Pragasan, L. A. Assessment of Tree Carbon Stock in the Kalrayan Hills of the Eastern Ghats, India. *Walailak J. Sci. Technol.* **2015b**, *12* (8), 659–670.

Pragasan, L. A. Total Carbon Stock of Tree Vegetation and Its Relationship with Altitudinal Gradient from the Shervarayan Hills Located in India. *J. Earth Sci. Clim. Change* **2015c**, *6*, 273.

Pragasan, L. A. Assessment of Carbon Stock of Tree Vegetation in the Kolli Hill Forest Located in India. *Appl. Ecol. Environ. Res.* **2016**, *14* (2), 169–183.

Ren, Y.; Wei, X.; Wei, X.; et al. Relationship between Vegetation Carbon Storage and Urbanization: A Case Study of Xiamen, China. *For. Ecol. Manage.* **2011**, *261*, 1214–1223.

Rizvi, R. H.; Dhyani, S. K.; Yadav, R. S.; Singh, R. Biomass Production and Carbon Stock of Popular Agroforestry Systems in Yamunanagar and Saharanpur Districts of Northwestern India. *Curr. Sci.* **2011**, *100* (5), 736–742.

Shah, S.; Sharma, D. P.; Pala, N. A.; et al. Temporal Variations in Carbon Stock of *Pinusroxburghii* Sargent Forests of Himachal Pradesh, India. *J. Mt. Sci.* **2014**, *11* (4), 959–966.

Sheikh, M. A.; Kumar, S.; Kumar, M. Above and Below Ground Organic Carbon Stocks in a Sub-Tropical *Pinusroxburghii* Sargent Forest of the Garhwal Himalayas. *For. Stud. China* **2012**, *14* (3), 205–209.

Steffan-Dewenter, I.; Kessler, M.; Barkmann, J.; et al. Tradeoffs between Income, Bio-Diversity, and Ecosystem Functioning during Tropical Rainforest Conversion and Agroforestry Intensification. *PNAS* **2007**, *104*, 4973–4978.

Sundarapandian, S. M.; Amritha, S.; Gowsalya, L.; et al. Estimation of Biomass and Carbon Stock of Woody Plants in Different Land-Uses. *For. Res.* **2013,** *3,* 115.

Takimoto, A.; Nair, P. K. R.; Nair, V. D. Carbon Stock and Sequestration Potential of Traditional and Improved Agroforestry Systems in the West African Sahel. *Agric. Ecosyst. Environ.* **2008,** *125,* 159–166.

Timilsina, N.; Escobedo, F. J.; Staudhammer, C. L.; Brandeis, T. Analyzing the Causal Factors of Carbon Stores in a Subtropical Urban Forest. *Ecol. Complex.* **2014,** *20,* 23–32.

CHAPTER 11

HORCHATA DE CHUFA: HEALTHY, DIGESTIVE-DISEASE-PREVENTION PROPERTIES, TREATMENT, AND CAUTIONS

FRANCISCO TORRENS[1*] and GLORIA CASTELLANO[2]

[1]Institut Universitari de Ciència Molecular, Universitat de València, Edifici d'Instituts de Paterna, PO Box 22085, E-46071 València, Spain

[2]Departamento de Ciencias Experimentales y Matemáticas, Facultad de Veterinaria y Ciencias Experimentales, Universidad Catylica de Valencia San Vicente Mártir, Guillem de Castro-94, E-46001 València, Spain

*Corresponding author. E-mail: torrens@uv.es

ABSTRACT

Horchata de chufa is a tigernut-based drink. Most nonalcoholic soft drinks are not nutrients. However, nutritional advantages exist in horchata without industrial processing. Natural horchata is rich in *phospholipids*, *arginine*, and *biotin* than industrial one. It presents healthy properties and properties of prevention of digestive diseases. The nutritional content of natural horchata turns it into an interesting vegetal *milk*. It is ideal in human nutrition, as it not only presents nutritional components but also has features of *functional food*, because of its additional values such as digestive *eupeptic*, *hypolipemiant*, and *immunomodulator* , and should be considered as a component of *Mediterranean diet*. It has a loss of nutritional value when it is subjected to the thermal processes that prolong its expiration date.

11.1 INTRODUCTION

Cyperus esculentus L. is a crop of the sedge family widespread across many places of the world. In València (Spain), *C. esculentus* L. var. *sativus* Boeck. is cultivated for its edible tubers (yams), called earth almonds or tigernuts and for the preparation of *horchata de chufa*, a sweet, milk-like beverage (cf. Fig. 11.1) (Soriano del Castillo, 2014).

FIGURE 11.1 *Cyperus esculentus* L. var. *sativus* Boeck. tigernuts and *horchata de chufa*. *Source*: Casas (2007).

Nowadays, population lifestyle reveals a rise in pathologies related to food by different factors, for example, getting of a job of family members in charge of home-food organization, less time devoted to food care, great food offer, and low knowledge for the interpretation of products nutritional labeling, greater food processing, rise in prepared food consumption, low general fruits and vegetables consumption, and great proteic food consumption, etc. Moisturizing and soft drinks can be divided into two groups: those containing alcohol (ethanol, EtOH, e.g., beer, cider, white wine, champagne, sangria) and the ones not containing EtOH (nonalcoholic, NA). Two types of NA soft drinks are available in shops: those bringing caffeine (Caff, e.g., energy, cola carbonated drinks) and the ones derived from fruits, carbonated or not, with a maximum of 8–10% fruit juice. Main nutritional problems of fruit-derived soft drinks are listed in Table 11.1 (Fundación Valenciana de Estudios Avanzados, 2003).

TABLE 11.1   Main Nutritional Problems That Usually Consumed Soft Drinks Present.

Type of drink	Main nutritional problems[a]
Beer	Contains EtOH
	High caloric content (EtOH + sucrose)
Cola drinks	Stimulating effect
	High caloric content (except *lights*)
	Too much Na^+ and K^+
	Contain H_3PO_4
	Cause gastric acid hypersecretion
Soft drinks derived from fruits	A lot contain carbonic gas CO_2
	Too much Na^+
	Poor content of fruit juice

[a]In addition, none of these drinks is nutrient, because:
- the CHOs are added.
- they contain neither proteins nor lipids.
- their mineral content is unbalanced.
- poor or null content of vitamins; sometimes, vitamin C is added.

Source: Fundación Valenciana de Estudios Avanzados (2003).

One of the most traditional drinks in Spain, especially in Mediterranean area, is horchata de chufa, classically reputed as a natural, wholesome, and energy NA soft drink, with beneficial and healthy digestive properties empirically shown since hundreds of years. It does not contain EtOH and is nutritive because it contains carbohydrates (CHOs), lipids, proteins, a number of minerals, and vitamins (Vits) but not cholesterol (CHOL). It presents global advantages over NA soft drinks: (1) it lacks of Caff and other stimulants, which authorizes its consumption in children, pregnant women, and old people; (2) in its composition, it does not bring phosphoric acid H_3PO_4, with which it does neither *steal* Ca^{2+} from bones nor alter teething; (3) if no sugar is added, it presents lower caloric density than usual sweetened soft drinks, and its calories are not *empty* as it is accompanied by other nutrients of which these features are lacking; (4) it presents digestive properties derived from its content of amino acids (AAs) and starch, which is *astringent* (tending to shrink or constrict body tissues), for example, *rice water*, and functions as *soluble fiber* at the colon level because 20%

is not absorbed by the small intestine, and enzymes (e.g., amylase, lipase), which support normal gastrointestinal digestion, so that it can be used as *prebiotic*, that is, substance that favors the action of *probiotics* or beneficial intestinal acidophilic bacteria [e.g., in prevention, cure of diarrheas because of viral infection (e.g., summer ones), *traveler's* and post-antibiotherapy diarrhea]. Horchata was, since hundreds of years, a habitual component of *Mediterranean diet*. The prestige position, which Mediterranean diet presents in modern human food, comes from their properties such as anti-arteriosclerosis and reduction of risk of suffering certain cancers (not only digestive), and fact that Mediterranean populations live longer than those of North Europe. In earlier publication, the molecular categorization of yams by principal component and cluster analyses (Torrens Zaragozá, 2013), beer, all a science, alcohol, analysis, and quality testing at multiple brewery stages (Torrens and Castellano, in press (a)); quantitative structure–property relationship (QSPR) prediction of retention times of methylxanthines and cotinine by bioplastic evolution (Torrens and Castellano, 2018); molecular classification of Caff, its metabolites, and nicotine metabolite (Torrens and Castellano, 2016); and QSPR prediction of retention times of chlorogenic acids in coffee by bioplastic evolution was reported (Torrens and Castellano, in press (b)).

11.2 CHEMICAL–NUTRITIONAL COMPOSITION OF TIGERNUT AND HORCHATA

Tigernut proximates [in g/100 g dry matter (DM)] are CHOs (51 g: *starch*, 31 g; *sucrose*, etc., 20 g), lipids [23 g; *unsaturated fatty acids* (UFAs), 83%, mainly *oleic acid* (C18:1)], and *proteins* (8–9 g, standing out *albumin*). It is rich in *Vits C* and *E*, and mineral salts (e.g., Mg, P, and K). Between AAs, *arginine* (Arg, R) stands out. Because of its great content of starch, it results as a good *fiber* (8–10 g) source.

Tigernut oil is richer in UFAs (83.8%) than *saturated fatty acids*, standing out oleic acid (C18:1) as the major (cf. Table 11.2), followed by palmitic (C16:0), linoleic (C18:2), stearic (C18:0), myristic (C14:0), and palmitoleic (C16:1) ones.

TABLE 11.2   Fatty Acid (FA) Profile (%) of València Tigernut Oil.

FA	Percentage (%)
Oleic (C18:1)	68.0
Palmitic (C16:0)	19.1
Linoleic (C18:2)	9.3
Stearic (C18:0)	2.5
Myristic (C14:0)	0.7
Palmitoleic (C16:1)	0.2

Source: Fundación Valenciana de Estudios Avanzados (2003).

There are nutritional advantages in horchata de chufa without industrial processing. Table 11.3 lists the energetic and nutritional content of components in València tigernut and horchata, and their diet contributions.

TABLE 11.3   Nutritional Content of Components in València Tigernut and Horchata.

Component	Tigernut contribution by portion[a]	Horchata contribution by portion[b]
Macronutrients		
Energy	351.0 kJ	817.6 kJ
CHOs	8.5 g	31.0 g
Sugars	1.6 g	24.7 g
Polysaccharides	6.6 g	6.1 g
Total fiber	3.5 g	0.2 g
Water	1.6 g	167.0 g
Proteins	1.5 g	1.7 g
AAs		
Ala (A)	58.1 mg	67.0 mg
Arg (R)	280.8 mg	536.5 mg
Asp (D)	150.2 mg	119.2 mg
Cys (C)	179.9 mg	48.7 mg
Glu (E)	166.5 mg	135.2 mg
Gly (G)	41.4 mg	52.6 mg

TABLE 11.3　*(Continued)*

Component	Tigernut contribution by portion[a]	Horchata contribution by portion[b]
His (H)	35.6 mg	19.9 mg
Ile (I)	32.8 mg	37.9 mg
Leu (L)	32.8 mg	64.9 mg
Lys (K)	60.5 mg	70.6 mg
Met (M)	32.7 mg	17.9 mg
Phe (F)	43.3 mg	35.5 mg
Pro (P)	44.8 mg	27.9 mg
Ser (S)	42.3 mg	40.7 mg
Thr (T)	44.1 mg	50.1 mg
Trp (W)	72.1 mg	13.8 mg
Tyr (Y)	23.3 mg	16.3 mg
Val (V)	45.2 mg	49.1 mg
Lipids/fats		
Total lipids/fats	4.7 g	7.2 g
SFAs	0.8 g	1.1 g
MUFAs	3.3 g	5.3 g
PUFAs	0.4 g	0.7 g
Oleic acid (C18:1)	3.2 g	5.3 g
Linoleic acid (C18:2)	0.4 g	0.7 g
α-Linolenic acid (C18:3)	–	20.9 mg
CHOL	–	–
Liposoluble vits		
Vit A	–	–
Vit D	–	–
Vit E	2.0 mg	–
Vit K	–	–

TABLE 11.3   *(Continued)*

Thiamine (Vit B$_1$)	46.0 mg	41.0 mg
Riboflavin (Vit B$_2$)	20.0 mg	–
Niacin (Vit B$_3$)	0.4 mg	0.3 mg
Vit B$_6$	66.1 mg	–
Folic acid (Vit B$_9$)	28.2 mg	4.1 mg
Vit B$_{12}$	–	–
Biotin (Vit B$_7$)	–	–
Pantothenic acid (Vit B$_5$)	–	–
Vit C	1.2 mg	–
Minerals		
Ca	14.0 mg	59.5 mg
Mg	17.4 mg	38.9 mg
Na	7.6 mg	18.5 mg
K	11.4 mg	188.8 mg
P	5.6 mg	102.6 mg
Fe	0.7 mg	2.5 mg
Zn	0.8 mg	2.0 mg
Cu	36.2 mg	1.4 mg
Mn	4.1 mg	–
Other components of interest for health		
Total polyphenols	34.8 mg[c]	37.1 mg[d]

[a]It is considered 1 portion = 25 tigernuts (0.80 g × 25 units = 20 g of whole València tigernut).
[b]It is considered 1 portion = 200 mL.
[c]Total polyphenols are expressed in mg of gallic acid/portion of València tigernut.
[d]Total polyphenols are expressed in mg of gallic acid/portion of València horchata de chufa.
CHOs, carbohydrates; AAs, amino acids; Ala, alanine; Arg, arginine; Asp, aspartic acid; Cys, cysteine; Glu, glutamic acid; Gly, glycine; His, histidine; Ile, isoleucune; Leu, leucine; Lys, lysine; Met, methionine; Phe, phenylalanine; Pro, proline; Ser, serine; Thr, threonine; Trp, tryptophan; Tyr, tyrosine; Val, valine; SFAs, saturated fatty acids; MUFAs, monounsaturated fatty acids; PUFAs, polyunsaturated fatty acids; CHOL, cholesterol; vits, vitamins.
Source: Soriano del Castillo (2014).

In horchata, the proximate is water; so, tigernut percentages perceptibly decay and sugar quality decreases because of the low starch solubility in water and sugars addition in its manufacturing. However, *oligosaccharides* (containing a small number of *monosaccharides*) are also present in minimum quantity, which have *prebiotic* activity and health beneficial properties. In diet, horchata is, in tea or as a snack, a good milk, chocolate milkshakes, or cola soft-drinks replacement.

Natural (artisanal, not thermally treated) horchata is richer in *phospholipids* (PLs), *Arg,* and *biotin* (Vit $B_{7/8}$/H) than industrial (enzymatical and thermally treated) one (cf. Fig. 11.2) (Casas, 2007; Rubert Bassedas et al., personal communication). Phospholipids are a group of lipids found in the cell membranes of plants and animals, emphasizing phosphatidic acid. In fresh horchata, its concentration is significantly higher than industrial horchata. Phosphatidic acid helps the smooth functioning of cell membranes, ensuring their permeability, which is fundamental for the normal activity of membrane proteins, for example, their receptors.

FIGURE 11.2 *Món Orxata* València *horchata de chufa.*
Source: Casas (2007).

Arginine is an important AA that takes part in the formation of nitric oxide •NO in the body, giving rise to important physiological properties, for example, ability to produce vasodilation, which would contribute to improve circulation and blood pressure. It is in a concentration five times greater in artisanal horchata than industrial one.

11.3 COMPARATIVE COMPOSITION OF YAM TIGERNUT AND HORCHATA DE CHUFA

Horchata de chufa presents digestive properties. (1) It is *eupeptic* (having good digestion) by its content of amylase and lipase, facilitating CHOs and fats digestion, respectively. It eases the discomfort of *flatulent dyspepsia* (slow digestion) and avoids *meteorism* (*strain* by gas). (2) It is diuretic because of its abundant content of water and scarce in Na^+ (so it is not retained). (3) It is *antidiarrheal* because of its content of starch, which presents properties thickener of feces, and, as *prebiotic*, it favors the growth of *beneficial fermentation gut biota* (*lactic acid-forming bacilli*); *rise* and *tigernut waters* were known cures of child diarrheas in the 1950s. (4) By its content of *peroxidase* and *catalase*, it would be useful in oral sores as oropharyngeal *disinfectant*, although it is not habitual. In comparison with milk, properties are emphasized: it presents no lactose, less Na/K^+, and contains a far-from-negligible amount of soluble fiber, which is lacking in milk. However, its content of AAs and monounsaturated fatty acids (MUFAs) offers healthy properties, which make it incomparable with other soft drinks or nutrients. *Arginine takes part in ˙NO formation, which* presents important functions in digestive pathophysiology (cf. Table 11.4).

TABLE 11.4 Functions of Nitric Oxide ˙NO in the Digestive System.

Conditions	Functions of ˙NO
In the normal functionalism of the digestive tract (synthesis of *constitutional* ˙NO)	Neural inhibition of smooth muscle (e.g., in the relaxation of LOS)
	Maintenance of the blood flow of the GIM
	Maintenance of the normal gastric chlorhydropeptic secretion
	Contributing to the protection of GIM vs. exogenous (e.g., NSAIDs) or endogenous (bacterial endotoxins) stimuli
	Maintenance of the hepatocellular functions of depuration and excretion
In certain pathological conditions (synthesis of ˙NO *inducible* by inflammatory mechanisms)	Possible role in the gut inflammatory mechanisms
	Possible role in development of toxic megacolon
	Possible role in portal hypertension

LOS, lower esophageal sphincter; GIM, gastrointestinal mucosa; NSAIDs, nonsteroidal anti-inflammatory drugs.
Source: Fundación Valenciana de Estudios Avanzados (2003).

11.4 HEALTHY DIGESTIVE PROPERTIES OF HORCHATA DE CHUFA

Healthy digestive properties of horchata de chufa are as follows: (1) As *eupeptic* and *make digestion easier* by its enzyme content. It avoids *meteorism* (*strain* by gas) and *flatulency* (excess of winds). (2) As provider of readily assimilable energy in collectives (e.g., children, old people) by its content of CHOs but without lactose and fructose, sugars for which 1/3–1/2 of Spanish population is intolerant. (3) By its content of starch, *antidiarrheal* properties in those owing to viral infections and summer ones, in which no pharmacological treatment is required. (4) By its content of Arg (semi-essential AA, $\cdot NO$ donor), it exhibits a specific effect on *immunocompetence*. Arginine supplementation increases lymphocyte cellularity, which is important in *renal insufficiency*, which alters Arg endogenous release. Arginine makes *wound healing* easier and reduces the energy cost in *hypercatabolic* states. Arginine is an essential component of the oral samples used in *enteral nutrition* (tube feeding) in human clinic, and horchata is a good source of it. (5) It is a modest, but not inconsiderable, source of Fe. It can be a supplement in children and pregnant women. (6) It is an important source of P, Mg, and Zn. (7) Tigernut mineral contribution is similar to dry oleaginous fruits, so it has *hypolipemiant* [reducing CHOL and *triglycerides* (TGs) concentrations in blood] properties on both CHOL/TGs, as its main component is oleic MUFA. It is an ideal complement for *hypolipemiant* diets.

11.5 USE OF HORCHATA AS A TREATMENT OF COMMON DIGESTIVE DISEASES

With regard to its composition, horchata de chufa can be used as a natural treatment in certain common digestive diseases: (1) *functional* and *flatulent dyspepsias*; (2) *irritable bowel syndrome*, with predominance of *chronic diarrhea* and *abdominal strain* by *meteorism*; and (3) *acute diarrhea* by drugs, viruses, or summer one.

11.6 CAUTIONS IN THE CONSUMPTION OF HORCHATA DE CHUFA

Cautions in the consumption of horchata de chufa are as follows: (1) they are not necessary in *high blood pressure* patients, provided almost null content of Na^+ in the drink and the absence of stimulating components (Caff, tyramine). By its content of Arg, $\cdot NO$ precursor, it can have *vasodilating effects*. (2) Not in patients with *chronic hepatopathy, minor* or *moderate renal insufficiency*, neither in patients with *nephritic syndrome* nor pregnant women, all by its low content of Na^+, which makes it proportionally like a tomato, pepper, cucumber, and carrot. (3) If there is no addition of extra sucrose, horchata presents lesser caloric and simple-sugar content than many commercial fruit juices; so, if there is no abusive consumption, there would be necessity of special cautions in neither *compensated diabetics* nor *excess-weight* patients. (4) It does not contain *gluten*; so, that is no necessity of preventing *celiac* (permanent intolerant of gluten) patients from its consumption. (5) In spite of the fact that horchata is widely consumed in Spain, especially in summer, only two cases of *food allergy* with *cutaneous* and *bronchial hypersensitivity* reactions were described in the specialized medical journals. Tigernut allergy is much less common than soya ones.

11.7 DISCUSSION

Most NA soft drinks are not nutrients as they hardly contain proteins and lipids; CHOs that they contain are added as sweeteners, and their mineral content is unbalanced because they contain a Na^+ and K^+ excess; they are not natural but artificial, and bring preservers, colors, and flavors.

However, nutritional advantages exist in horchata de chufa without industrial processing. Natural horchata is richer in PLs, *Arg,* and Vit B_7 than industrial one. The nutritional content of natural horchata turns it into an interesting vegetal *milk*, especially in the no-sugar-added format. Horchata is ideal for human nutrition, as it not only presents nutritional components but also features as *functional food*, because of its additional value such as digestive *eupeptic, hypolipemiant,* and *immunomodulator* and should be considered as a component of *Mediterranean diet*.

Horchata has a loss of nutritional value when it is subjected to the thermal processes that prolong its expiration date.

11.8 CONCLUSION

From the preceding results and discussion, the following conclusions can be drawn:

(1) Nutritional advantages exist in horchata de chufa without industrial processing.
(2) Natural horchata is richer in PLs, arginine, and biotin than industrial one.
(3) Horchata is a *functional food* because of its additional values such as digestive *eupeptic*, *hypolipemiant*, and *immunomodulator* and should be considered a component of *Mediterranean diet*.
(4) Horchata has a loss of nutritional value when it is subjected to the thermal processes that prolong its expiration date.

ACKNOWLEDGMENTS

The authors thank support from Generalitat Valenciana (Project No. PROMETEO/2016/094) and Universidad Catolica de Valencia San Vicente Martir (Project No. UCV.PRO.17-18.AIV.03).

KEYWORDS

- sedge
- *horchata de chufa*
- edible tubers
- milk-like beverage
- pathologies

REFERENCES

Casas, A. Descobrint L'orxata: Un Estudi Revela els Avantatges Nutricionals de la Beguda Sense Processament Industrial. *Mètode* **2007,** *2007* (93), 1.

Fundación Valenciana de Estudios Avanzados (FVEA). *Jornada Chufa y Horchata: Tradición y Salud*; FVEA–Generalitat Valenciana: València, 2003.

Rubert Bassedas, J.; Pérez, G.; Navarro, J. L.; Soriano, J. M.; Blesa, J.; Monforte, A.; Martí, C.; Balaguer, C., personal communication.

Soriano del Castillo, J. M., Ed. *El Gran Libro de la Horchata y la Chufa de Valencia*; Universitat de València–Fundació Lluís Alcanyís: València, 2014.

Torrens, F.; Castellano, G. Beer, All a Science, Alcohol, Analysis and Quality Testing at Multiple Brewery Stages. In *Innovations in Physical Chemistry*; Haghi, A. K., Ed.; Apple Academic–CRC: Waretown, NJ, in press (a).

Torrens, F.; Castellano, G. QSPR Prediction of Retention Times of Chlorogenic Acids in Coffee by Bioplastic Evolution. In *Quantitative Structure–Activity Relationship*; Kandemirli, F., Ed.; InTechOpen: Vienna, in press (b).

Torrens, F.; Castellano, G. QSPR Prediction of Retention Times of Methylxanthines and Cotinine by Bioplastic Evolution. *Int. J. Quant. Struct.—Prop. Relat.* **2018**, *3*, 74–78.

Torrens, F.; Castellano, G. Molecular Classification of Caffeine, Its Metabolites and Nicotine Metabolite. In *Frontiers in Computational Chemistry*; Ul-Haq, Z., Madura, J. D., Eds.; Bentham: Hilversum, Holland, 2016; Vol. 4, pp 3–51.

Torrens Zaragozá, F. Molecular Categorization of Yams by Principal Component and Cluster Analyses. *Nereis* **2013**, *2013* (5), 41–51.

CHAPTER 12

WATER TREATMENT: APPLICATIONS OF ADSORPTION AND ION-EXCHANGE CHROMATOGRAPHY IN THE CHEMICAL, PHARMACEUTICAL, AND FOOD INDUSTRIES

SAJAD AHMAD GANAI[1*] and AMJAD MUMTAZ KHAN[2]

[1]*Department of Chemistry, National Institute of Technology, Srinagar, Jammu and Kashmir, India*

[2]*Department of Chemistry, AMU, Aligarh, Uttar Pradesh 202001, India*

Corresponding author. E-mail: sajadali16@gmail.com

ABSTRACT

Water, nature's most wonderful, abundant, and useful compound, which is unique to Earth, covers near about 75% of the Earth's surface. It is the only known substance that can naturally exist as a gas, solid, and liquid within the relatively small range of air temperatures and pressures found on Earth's surface. In many ways, water is a miracle liquid. It is essential for all living things on this planet. Water is considered to be of greatest importance for the existence of animal and plant life. Without food, an animal can survive for a number of days, but without it one cannot survive.

Water is not only important for living things but also occupies a unique position in the industrial world. It is used in steam generation and as a coolant in power and chemical plants. It is also used in the other fields such as production of steel, rayon, paper, textiles, etc.

234

Applied Food Science and Engineering with Industrial Applications

12.1 SOURCES OF WATER

There are two important sources of water: (a) surface waters and (b) underground waters.

(a) Surface waters: Surface water is categorized as follows: (1) rain water, (2) river water, (3) sea water, and (4) lake water.

(1) *Rain water*: It is considered as the purest form of natural water, because it is obtained as a result of evaporation from the surface water. But, when it comes toward earth through the atmosphere, it dissolves a considerable amount of industrial gases (e.g., CO_2, SO_2, NO_2, etc.) and suspended particles, both of organic and inorganic nature.

(2) *River water*: Rivers are fed by rain and spring waters. Water from these sources flow over the surface of land, dissolves the soluble minerals of the soil, and ultimately finds its way into the river. River water dissolves minerals of the soil like chlorides; sulfates; and bicarbonates of calcium, magnesium, and iron. This water also contains organic matters derived from the decomposition of plant.

(3) *Lake water*: This water has more constant chemical composition. It usually contains lesser amounts of dissolved minerals than even well water but contains higher content of organic matter.

(4) *Sea water*: This is the highly impure form of natural water. This water contains the highest content of dissolved impurities. Sea water contains about 3.5% of dissolved salts, out of which about 2.6% is sodium chloride.

Surface water, generally contains suspended matter, which often contains the disease-producing bacteria and thus not fit for human consumption.

(b) Underground waters: As the rain water percolates into the earth, it comes in contact with a number of mineral salts present in the soil and dissolves some of them. Water continues its downward journey, till it meets a hard rock, when it retreats upward, it may come out in the form of a spring. The underground water, in general, is clearer in appearance due to filtering action of the soil but contains more of the dissolved salts. Thus, water from springs and wells contains more hardness. Generally, underground water is of high organic purity.

12.2 IMPURITIES IN WATER

The various types of impurities found in water have been put into the following categories:

(1) Suspended impurities
(2) Colloidal impurities
(3) Dissolved impurities
(4) Biological impurities

(1) *Suspended impurities*: These include the organic matter in the shape of branches, leaves, dead waste, etc. Organic matter imparts color, odor to water, and alters its taste. Water has to undergo screening, filtration, sedimentation, and disinfection for the removal of organic matter.
Inorganic suspended impurities may also be present in water, which include sediments, clay, and sand. They impart turbidity.

(2) *Colloidal impurities*: These are solid impurities that may be present in water in the finely divided and suspended state. Such impurities are not easy to settle and impart color and turbidity to water.

(3) *Dissolved impurities*: These include dissolved gases and dissolved inorganic salts. Organic impurities are generally not water soluble and do not exist in the dissolved state. Dissolved gases like ammonia and hydrogen sulfide impart a bad odor to water. Dissolved oxygen, nitrogen oxides, and carbon dioxide lead to boiler corrosion. These gases can be removed by de-aeration and using specific chemical treatments. Dissolved inorganic salts like bicarbonates, chlorides, and sulfates of calcium and magnesium impart hardness to water. These impurities can be removed by using ion-exchange process, lime soda treatment, and reverse treatment.

12.3 HARDNESS OF WATER

Hardness was originally defined as the soap-consuming capacity of water. This soap-consuming capacity is mainly due to the presence of calcium and magnesium ions in water; these react with the sodium salts of long-chain fatty acids present in soap to form insoluble scums of Ca^- and Mg^- soaps having no detergent value.

$$2C_{17}H_{35}COONa + CaCl_2 \rightarrow (C_{17}H_{35}COO)_2 Ca + 2NaCl$$
Sodium stearate   Calcium stearate

Other metal ions such as those of Fe, Mn, Al, Ba, and Sr react in the same way and so could also be referred to as hardness, but natural waters only contain traces of these ions. In addition, acids such as carbonic acid can cause free fatty acid to separate from a soap solution and thus contribute to the hardness. In practice, however, the hardness of water is taken as a measure of its Ca(II) and Mg(II) content.

Thus, hard water is water that contains an appreciable quantity of dissolved minerals like calcium and magnesium and does not form lather with soap. In other words, hardness of water is the soap-consuming capacity of water. It can also be defined as the capacity of precipitation of soaps or as a characteristic property of water that prevents the lathering of soap (Khan et al., 2005; Zhang et al., 2005).

12.3.1 TYPES OF HARDNESS

12.3.1.1 TEMPORARY AND PERMANENT HARDNESS

Temporary hardness is caused by the presence of dissolved bicarbonates of calcium and magnesium. It can be removed by boiling. The bicarbonates decompose on boiling giving rise to carbonates and hydroxides that are insoluble and deposit at the bottom of the vessel and thus can be removed by simple physical methods.

$$Ca(HCO_3)_2 \rightarrow CaCO_3 \downarrow + H_2O + CO_2$$

$$Mg(HCO_3)_2 \rightarrow Mg(OH)_2 \downarrow + 2CO_2$$

Permanent hardness is due to the presence of chlorides and sulfates of calcium and magnesium in water. Hence, it can be caused mainly by the presence of $MgCl_2$, $MgSO_4$, $CaCl_2$, and $CaSO_4$ and also salts of heavier elements such as iron and aluminum. This type of hardness cannot be removed by boiling. These terms, that is, temporary and permanent hardness, are getting gradually out of use and are being replaced by the more preferred terms, alkaline, and nonalkaline hardness.

12.3.1.2 ALKALINE AND NONALKALINE HARDNESS

Alkaline hardness is defined as the hardness due to bicarbonates, carbonates, and hydroxides of the hardness causing metals. In raw water, the alkaline hardness almost is always associated with the bicarbonates, but a treated or boiler water may also contain hardness due to carbonates and hydroxides in solution. The alkalinity is equal to the sum of the concentrations of the bicarbonate, carbonate, and hydroxide, expressed in equivalents. If this alkalinity is less than the total hardness, the alkaline hardness is equal to alkalinity. But, when the alkalinity to methyl orange is equal or greater than the total hardness, the alkaline hardness is equal to the total hardness.

12.3.2 DEGREE OF HARDNESS

The degree of hardness is defined as the number of parts by weight of calcium carbonate hardness per particular number of parts of water, depending upon the unit employed. Though the hardness does not always arise due to the presence of $CaCO_3$ in water but is expressed in terms of equivalents of $CaCO_3$ because of the following reasons:

(1) Its molecular weight is exactly hundred, which makes calculations convenient.
(2) It is the most insoluble salt that can be precipitated in water treatment.

12.3.3 UNITS OF HARDNESS

(1) *Parts per million (ppm)*: One part per million is a unit weight of solute per million weight units of solution. In dilute solutions, density is nearly equal to unity, 1 ppm = 1 mg/L. While describing hardness, all the hardness-causing impurities are expressed in terms of their respective weights equivalent to $CaCO_3$. Thus, 1 ppm means that 1 part of $CaCO_3$ equivalent is present in 106 parts of water.
(2) *Equivalents per million (epm)*: One equivalent per million is a unit chemical equivalent weight of solute per million weight unit of solution. In dilute solutions, density does not differ very much from unity, 1 epm = 1 milligram equivalent per liter, and in titrimetry, 1 epm is conventionally taken as equal to 1 mL of 1 N solution per liter.
(3) *Grains per imperial gallon (gpg)*: In English system, hardness is expressed in terms of grains (1 grain = 1/7000 Ib) per gallon (10 Ib),

that is, parts per 70,000 parts. 1 grain per gallon is also called as a degree Clark. Thus, 7 degree Clark means that 7 grains in terms of $CaCO_3$ are present per gallon of water, or 7 parts are present per 70,000 parts of water. On ppm scale, it means that $7 \times 1,000,000/70,000 = 100$ ppm of hardness (as $CaCO_3$) is present in the water sample.

Relationship between various units of hardness:

$$1 \text{ ppm} = 1 \text{ mg/L} = 0.1 \text{ °Fr} = 0.07 \text{ °Cl}$$

$$1 \text{ mg/L} = 1 \text{ ppm} = 0.1 \text{ °Fr} = 0.07 \text{ °Cl}$$

$$1 \text{ °Fr} = 10 \text{ ppm} = 10 \text{ mg/L} = 0.7 \text{ °Cl}$$

$$1 \text{ °Cl} = 1.43 \text{ °Fr} = 14.3 \text{ ppm} = 14.3 \text{ mg/L}$$

12.3.4 DISADVANTAGES OF HARD WATER

Water is extensively used in industry and for domestic purposes. The quality of water required for both purposes differs, but the presence of hardness in water has many disadvantages both in industrial and domestic purposes.

12.3.5 DOMESTIC PURPOSES

Presence of hardness affects the domestic uses of water as described below:

(1) **Washing**: When hard water is used for washing purposes, does not lather freely with soap but produces sticky precipitates. The formation of such precipitates continues till all calcium and magnesium salts present in water are precipitated. After that, the soap gives lather with water. This causes wastage of lot of soap being used. Moreover, the sticky precipitate adheres on the fabric giving spots and streaks. Also, presence of iron salts may cause staining of cloth.

(2) **Bathing**: Soap does not form a soft lather with hard water. Since lather is not formed, the dirt does not get trapped in soap bubbles but remains on the skin and is difficult to remove. Thus, the cleansing quality of soap is depressed and a lot of soap is wasted. The curdy precipitate also sticks to the hair and makes it greasy and lifeless.

(3) **Dishwashing**: The cleaning action of soaps is reduced even further when it comes in contact with hot hard water, as it releases more

minerals on contact. Thus, the dishes, instead of being washed, become more spotted and this reduces the quality of the crockery over a few washes.

(4) **Drinking**: Hard water causes bad effect on our digestive system. Moreover, the possibility of forming calcium oxalate crystals in urinary tracks is increased.

(5) **Cooking**: Due to the presence of dissolved hardness-producing salts, the boiling point of water is elevated. Thus, more time and fuel are required for cooking. Certain foods like pulses, beans, and peas do not cook soft in hard water. Also tea or coffee, prepared in hard water, has an unpleasant taste and muddy-looking extract.

12.3.6 INDUSTRIAL PURPOSES

In industry, water is mainly used to generate steam for cooling and sometimes directly as solvent. The disadvantages of hard water used for industrial purposes are described below:

(1) **Sugar industry**: Water used in the sugar industry must be free from sulfates, nitrates, carbonates, and bacteria. If these impurities are present in water that has to be used in the sugar refining causes difficulties in the crystallization of sugar. Moreover, the sugar so produced may be deliquescent.

(2) **Dyeing industry**: The water used for dyeing, if contains cations can result in a change in color from what is desired. Hence, water for dying industry needs to be free from color-altering cations.

(3) **Textile industry**: Hard water causes much of soap to go as waste during washing yarn, fabric, etc. Also, precipitates of calcium and magnesium soaps adhere to the fabrics. These fabrics, when dyed later on, do not produce exact shades of color. Iron and manganese salts containing water may cause colored spots on fabrics, thereby spoiling their beauty.

(4) **Concrete industry**: Water free from chloride and sulfate ions is mandatory for the concrete industry, as the presence of these ions compromises the hardenability of concrete.

(5) **Paper industry**: Calcium and magnesium salts tend to react with chemicals and other materials employed to provide a smooth and shining finish to paper. Moreover, iron salts may even affect the color of the paper being produced.

(6) **Pharmaceutical industry**: If hard water is used for preparing pharmaceutical products, certain undesirable products in them may be produced.

(7) **Water boilers and pipes**: Hard water contains dissolved salts of calcium and magnesium. When this water comes in contact with heating appliances, these salts precipitate out and form limescale deposits on the appliances. Due to these scales, there is wastage of energy for heating purposes and the appliances may also fail after regular usage. Pipes carrying hard water often get clogged with the limescale deposits and thus have to be replaced after only a few years.

12.3.7 ESTIMATION OF HARDNESS

The estimation of hardness of water is of great importance. Determination of hardness of water can be done by using the following methods:

(1) O. Hehner's method
(2) Soap titration method
(3) Complexometric titration method using ethylene diamine tetra acetic acid (EDTA).

Out of these, complexometric titration or EDTA method is definitely preferable because of the greater accuracy, convenience, and more rapid procedure.

12.3.7.1 COMPLEXOMETRIC TITRATION METHOD

EDTA is a well-known complexing agent, which is widely used in analytical work, on account of its powerful complexing action and commercial availability. This is also available under trade names as Versene or Tritriplex III.

EDTA in the form of its sodium salt yields the anion, which forms complex ions with Ca(II) and Mg(II).

EDTA is generally used in the form of disodium salt or tetrasodium salt on account of their greater stability.

Lowering of the pH will decrease the stability of the metal–EDTA complex. It has been observed that by large, the EDTA complexes of divalent metals are stable in alkaline or slightly acid solution, while those of trivalent and tetravalent metal ions may exist in solutions of much higher acidity.

TABLE 12.1   Stability of Some Metal–EDTA Complexes with Respect to pH.

Metal ions	Minimum pH at which their respective EDTA complexes exist
Ca(II), Mg(II), Ba(II), Sr(II)	8–10
Cu(II), Pb(II), Zn(II), Co(II), Ni(II), Mn(II), Fe(II), Al(III), Cd(II), Sn(II)	4–6
Zr(IV), Hf(IV), Th(IV), Bi(III), Fe(III)	1–3

The completion of the reaction is identified using Eriochromr black-T (EBT) indicator. This is an organic dye, blue in color. It also forms relatively less stable complexes with bivalent metal ions of Ca, Mg, etc., which are wine red in color. Therefore, on addition of the indicator to hard water, it first reacts with free metal ions to give a wine red color. When EDTA is added, it attacks the metal–indicator complex and sets the indicator free. The reaction can be represented as

$$M^{2+} \text{ indicator complex} + EDTA \rightarrow M^{2+} \text{ EDTA complex} + \text{Free indicator}$$

So at the end point, a change from wine red to blue color is observed. Since the reaction involves the liberation of H^+ ions, and the indicator is sensitive to the concentration of H^+ ions (pH) of the solution, a constant pH of around 10 has to be maintained. For this purpose, ammonia–ammonium chloride buffer is used.

12.3.7.1.1   Notes

(1) It may be noted that the presence of Mg(II) is must for a sharp color change at the end point with EBT indicator.

(2) Ca(II) forms a relatively weak, soluble, colored complex with EBT, but this complex is not sufficiently stable to be useful as indicator in the titration of Ca(II) with EDTA.

(3) The magnesium–indicator complex is more stable than the calcium–indicator complex but is less stable than the magnesium–EDTA complex. This is the reason why during the titration of a solution containing Mg(II) and Ca(II) using EBT, the EDTA first reacts with the free Ca(II), then with the free Mg(II), and finally with the Mg–indicator complex at the end point.

(4) If magnesium ions are not present in the solution containing Ca(II), they must be added, because they are important for the color change the at end point.

(5) pH is rather critical in this titration. A pH of 10 is most satisfactory. At higher pH, Mg(II) may be precipitated as $Mg(OH)_2$. In strongly acidic solutions, the indicator tends to polymerize to a red-brown product.

12.3.7.1.2 Procedure

A standard solution of EDTA is prepared by dissolving a known amount of EDTA (4–6 g) in 100–150 mL deionized water and 4 ml NaOH, and then making up the volume in a 250-mL volumetric flask. About 25 mL of the given sample of hard water is taken in a clean titration flask and about 3 mL of NH_3–NH_4Cl buffer followed by 2–3 drops of EBT indicator is added to it. This solution is titrated against standard EDTA solution to get concordance values (V mL).

12.3.7.1.3 Calculations

The strength of the EDTA solution may be calculated as

Weight of EDTA dissolved in 250 mL of water = ...g
Weight of EDTA dissolved per liter = ...g × 4
Molarity of EDTA solution = ...g × 4/372.4 = a M (say)
We know that 1000 mL of 1 M EDTA = 100 g of $CaCO_3$. So,
V mL of a M EDTA = 100 × V × a/1000 g of $CaCO_3$ = y g of $CaCO_3$.
25 mL of hard water contains y g of $CaCO_3$
Therefore, 106 mL of hard-water sample contains
4 × 106/25 g of $CaCO_3$ = y/25 ppm
So, total hardness of water = ...ppm.

Permanent hardness of water can be determined by titrating the water sample against standard EDTA solution after removal of temporary hardness by boiling the water sample. The difference between total and permanent hardness gives the temporary hardness.

12.4   TECHNIQUES FOR WATER SOFTENING

Water softening is the process of reducing hardness producing salts from water, therefore reducing hardness. Water that is used in the chemical industry should be free from calcium and magnesium salts. These salts tend to precipitate out as hard deposits on the pipes and heat exchanger surfaces. The resulting buildup of scale can restrict water flow in pipes. In boilers, the deposits act as an insulation that impairs the flow of heat into water, reducing the heating efficiency and in some cases leading to failure of the boiler. In industrial boilers, most of the troubles, for example, scale and sludge, priming, foaming, corrosion, arise due to hardness of boiler feed water. It is important to remove hardness-causing salts from boiler feed water. For this, different treatments are used, for example, lime soda process, zeolite process, ion-exchange process. But out of these processes, ion-exchange process is preferred because of the following reasons:

(1) This process can be used to soften highly acidic or alkaline waters.
(2) It produces water of very low hardness.
(3) The ion-exchange apparatus, once set up, is easy to operate and control.
(4) It works well for variations in hardness of water which do not have to be uniform at all times.
(5) The entire unit takes up less space, produces better quality water and is more economical.

12.5   ION-EXCHANGE PROCESS OR DEIONIZATION PROCESS

Ion exchange is an adsorption phenomenon where the mechanism of adsorption is electrostatic. Electrostatic forces hold ions to charged functional groups on the surface of the ion-exchange resin. The absorbed ions replace ions that are on the resin surface on a 1:1 charge basis. Ion exchange is a process of softening of water by exchanging harmful ions with harmless ions from an ion-exchange resin. Ion-exchange resins have high molecular weight and are cross-linked polymers with a porous structure. The ion-exchange properties arise due to the presence of acidic (sulfonic or carbonyl) or basic (substituted amino group) functional group attached to the polymeric chains. The acidic groups exchange their H^+ ions for cations like $Ca(II)$, $Mg(II)$ present in water are known as cation-exchange resins. The basic groups exchange OH^- for anions present in water are known as anion-exchange resins.

In the process, hard water is first passed through cation-exchange column, which removes all the cations like Ca(II), Mg(II), etc. from it and an equivalent amount of H^+ ions are released from this column to water under treatment. Thus,

$$2RH^+ + Ca^{2+} \rightarrow R_2Ca^{2+} + 2H^+$$
$$2RH^+ + Mg^{2+} \rightarrow R_2Mg^{2+} + 2H^+$$

After this, the water is passed through anion-exchange resin, which removes all the anions present in the water and also an equivalent amount of OH^- ions are released from this column to water. Thus,

$$ROH^- + Cl^- \rightarrow RCl^- + OH^-$$

The H^+ and OH^- ions released combine to form demineralized water (DMW).

$$H^+ + OH^- \rightarrow H_2O$$

Thus, the water coming out from the exchanger is free from cations and anions. When the resins are exhausted and lose their capacity to exchange H^+ and OH^- ions, they are regenerated which is the beauty of this type of materials.

The exhausted cation-exchange column is regenerated by passing a solution of dil. HCl or dil. H_2SO_4 and then washing with DMW.

$$R^2Ca^{2+} + 2H^+ \rightarrow 2RH^+ + Ca^{2+}$$

Similarly, anion-exchange resin is regenerated by passing a dil. NaOH and then washing with DMW.

$$R_2SO_4^{2-} + 2OH^- \rightarrow 2ROH^- + SO_4^{2-}$$

The regenerated ion-exchange resins are used again and again.

KEY WORDS

- water
- earth's surface
- food
- miracle liquid
- coolant
- importance

REFERENCES

Khan, A. A.; Alam, M. M.; Inamuddin. Preparation, Characterization and Analytical Applications of a New and Novel Electrically Conducting Fibrous Type Polymeric–Inorganic Composite Material: Polypyrrole Th(IV) Phosphate Used as a Cation-Exchanger and Pb(II) Ion-Selective Membrane Electrode. *Mater. Res. Bull.* **2005,** *40,* 289.

Zhang, H.; Pang, J. H.; Wang, D.; Li, A.; Li, X.; Jiang, Z. Sulfonated Poly(Arylene Ether Nitrile Ketone) and Its Composite with Phosphotungstic Acid as Materials for Proton Exchange Membranes. *J. Membr. Sci.* **2005,** *264,* 56–64.

CHAPTER 13

META-ANALYSIS OF UNDERUTILIZED BEANS WITH NUTRIENTS PROFILE

FRANCISCO TORRENS[1*] and GLORIA CASTELLANO[2]

[1]*Institut Universitari de Ciència Molecular, Universitat de València, Edifici d'Instituts de Paterna, PO Box 22085, E-46071 València, Spain*

[2]*Departamento de Ciencias Experimentales y Matemáticas, Facultad de Veterinaria y Ciencias Experimentales, Universidad Catylica de Valencia San Vicente Mártir, Guillem de Castro-94, E-46001 València, Spain*

Corresponding author. E-mail: torrens@uv.es

ABSTRACT

The effect of seed coat color on the nutraceutical compounds content is distinguished, where black bean stands out because its content of anthocyanins, polyphenols, and flavonoids, for example, quercetin, which confers black bean with an elevated antioxidant capacity. Conversely, the analyzed evidence shows that more studies are needed to expand the knowledge of the nutraceutical quality of bean genotypes, grown or wild type, and their impact on health, to be used in genetic improvement programs or as a strategy to encourage their consumption, which is based on the potential it presents for health preservation and disease prevention.

13.1 INTRODUCTION

Beans are a source of essential nutrients in diet, providing *protein*, complex *carbohydrate* (CHO), vitamins, minerals, and dietary *fiber*, while being low in *fat*, Na and cholesterol (CHOL)-free (Edem et al., 1990; Broughton et al., 2003; Khalil and Khan, 1995; Mazur et al., 1998). Beans are the most important legume staple for human consumption (López-Pedrouso et al., 2012). The high *protein* content makes them cheap alternative *protein* source compared to priced animal or meat-*protein* sources (Achinewhu, 1982). In Nigeria, beans are cultivated [e.g., bean *Phaseolus vulgaris*; African yam bean (AYB) *Sphenostylis stenocarpa*; mung bean *Vigna radiata*; soybean *Glycine max* L.; pigeon pea *Cajanus cajan*; African oil bean *Pentaclethra macrophylla*; black-eyed pea (BEP, cowpea) *V. unguiculata*]. The seeds are processed into flours and used as food or feed. Their uses include the forti-fication of indigenous starchy meals as composites, production of flavoring fermented food condiments employed in traditional dishes, beverage (soya milk), and cake (soya cake, tofu) preparation. Bioactive compounds from bean *P. vulgaris* varieties with implications for health were reviewed (Chávez-Mendoza and Sánchez, 2017). Soyasaponin I (cf. Fig. 13.1) is predominant saponin in bean cotyledons and seed coats.

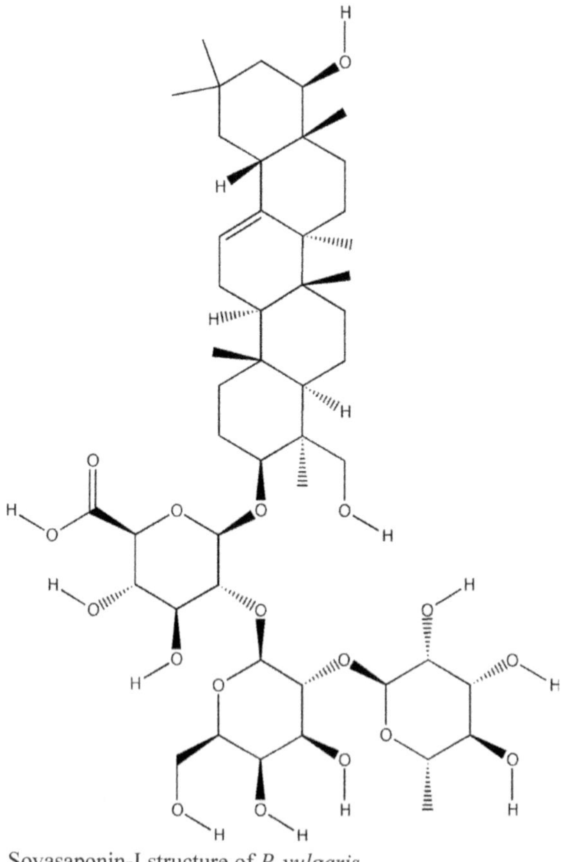

FIGURE 13.1 Soyasaponin-I structure of *P. vulgaris*.

The main flavonoids in bean are catechin (C), kaempferol, quercetin, myricetin, and procyanidin (cf. Fig. 13.2).

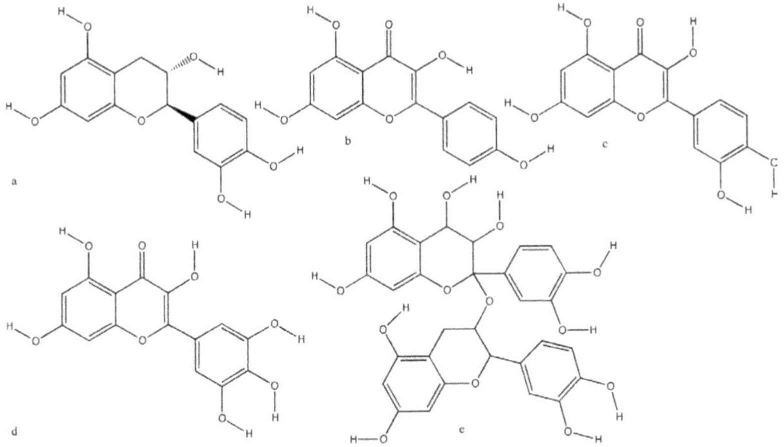

FIGURE 13.2 Structures of major flavonoids: (a) C, (b) kaempferol, (c) quercetin, (d) myricetin, and (e) procyanidin.

Gallic, vanillic, coumaric, sinapic, ferulic, and chlorogenic acids (CGAs, cf. Fig. 13.3) are phenolic acids mainly found in bean.

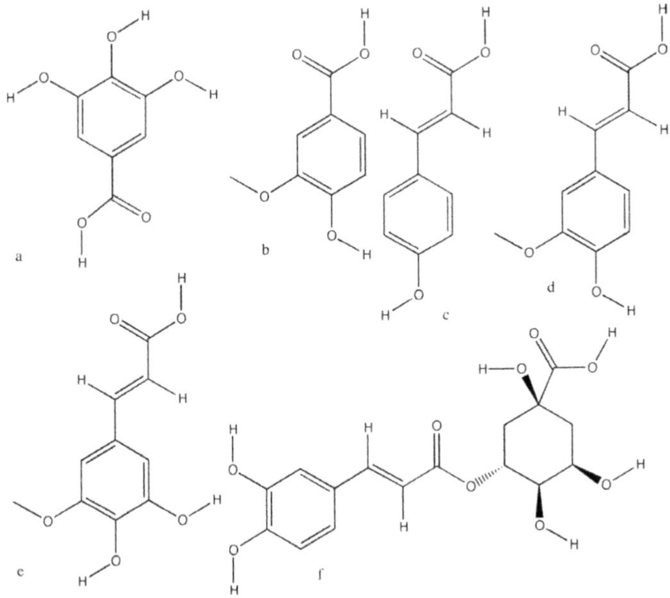

FIGURE 13.3 Structures of major phenolic acids: (a) gallic, (b) vanillic, (c) coumaric, (d) ferulic, (e) sinapic, and (f) CGAs.

Bean is one of the main sources of vitamin B (Vit B) complex in the diet of Mexican people (especially thiamine, riboflavin, niacin, folic acid, cf. Fig. 13.4).

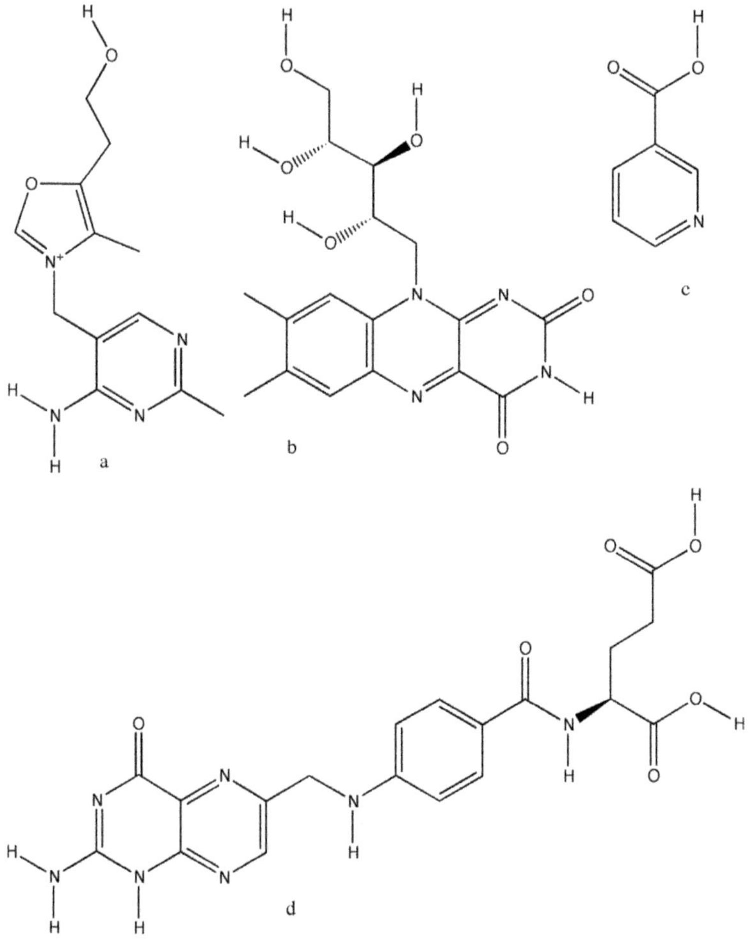

FIGURE 13.4 Structures of major vitamins of *P. vulgaris*: (a) thiamine, (b) riboflavin, (c) niacin, and (d) folic acid.

Adamu et al. (2015) reported nutrients profile of underutilized beans in southwestern Nigeria. Proximates of common and underutilized beans (cf. Table 13.1) showed that *protein*, *ash* (mineral), and reducing sugars were highest in AYB. Soybean and BEP CHO was the highest. Crude *fiber* and *fat* were highest in mung bean.

TABLE 13.1 Results of Proximate Composition of Some Underutilized Beans in Southwestern Nigeria.

Species	Moisture[a]	Ash	Fat	Protein	CHO	Fiber	Reducing sugar	Energy
1. Common bean *Phaseolus vulgaris*	7.09	3.26	2.52	21.07	60.36	5.71	37.90	348.38
2. AYB *Sphenostylis stenocarpa*	1.93	4.30	2.70	24.96	59.85	6.26	38.37	363.54
3. Mung bean *Vigna radiata*	4.74	3.82	6.60	22.07	47.47	15.24	27.53	338.10
4. Pigeon pea (red) *Cajanus cajan*	5.81	3.55	2.57	18.67	60.12	9.28	28.46	338.29
5. Pigeon pea (gray)	1.24	3.48	0.79	17.98	61.52	15.00	18.58	325.07
6. BEP (cowpea) *Vigna unguiculata*	7.48	3.41	1.89	16.59	62.81	7.82	21.92	334.61
7. Soybean *Glycine max* L.	3.58	4.01	5.89	15.45	62.81	8.26	23.36	366.05

[a]Compositions: proximates (i_1, moisture in %; i_2, ash; i_3, fat; i_4, crude protein; i_5, carbohydrate; i_6, crude fiber; i_7, reducing sugar; i_8, energy in kcal/100 g), minerals in mg/100 g [macrominerals (i_9, K; i_{10}, Ca; i_{11}, Mg; i_{12}, Na; i_{13}, Fe; i_{14}, Zn; i_{15}, P), microminerals (i_{16}, Co; i_{17}, Cu; i_{18}, Mn), and i_{19}, relative], and phaseolin protein fraction (i_{20}, albumin in %/mg protein; i_{21}, globulin; i_{22}, prolamin; i_{23}, total protein).

CHO, carbohydrate; AYB, African yam bean; BEP, black-eyed pea.

In the dendrogram of some underutilized beans in southwestern Nigeria according to proximate contents (cf. Fig. 13.5), mung bean groups with bean and AYB; pigeon, BEP, and soybean class together. Beans turn out to be in separate groupings, while peas cluster together. The greatest similarity is detected between pigeon (red variety) and BEP.

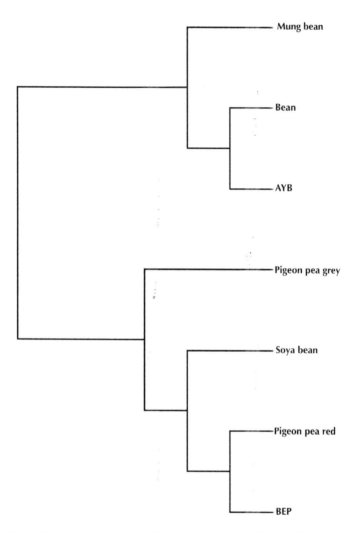

FIGURE 13.5 Dendrogram of underutilized beans in South Western Nigeria according to proximates.

In dendrogram of some underutilized beans in southwestern Nigeria with regard to mineral nutrients (cf. Fig. 13.6), mung and soybeans class together; bean, AYB, and BEP belong to the same grouping. Maximum resemblance results between bean and AYB.

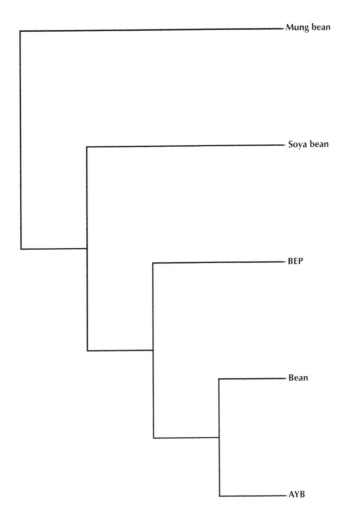

FIGURE 13.6 Dendrogram of underutilized beans in South Western Nigeria according to mineral nutrients.

Soybean contents are proximates, minerals, amino acids, isoflavones [daidzein (cf. Fig. 13.7a), genistein (Fig. 13.7b), glycitein (Fig. 13.7c)], and lecithin (Fig. 13.7e) (Yang et al., 2011; Chang et al., 2014; Maria John et al., 2013; Maria John et al., 2015). Isoflavones are similar to 17β-estradiol [estrogen (OES), Fig. 13.7d]; 100 g of soybean contain 130–500 mg of isoflavones (Kaufman et al., 1997).

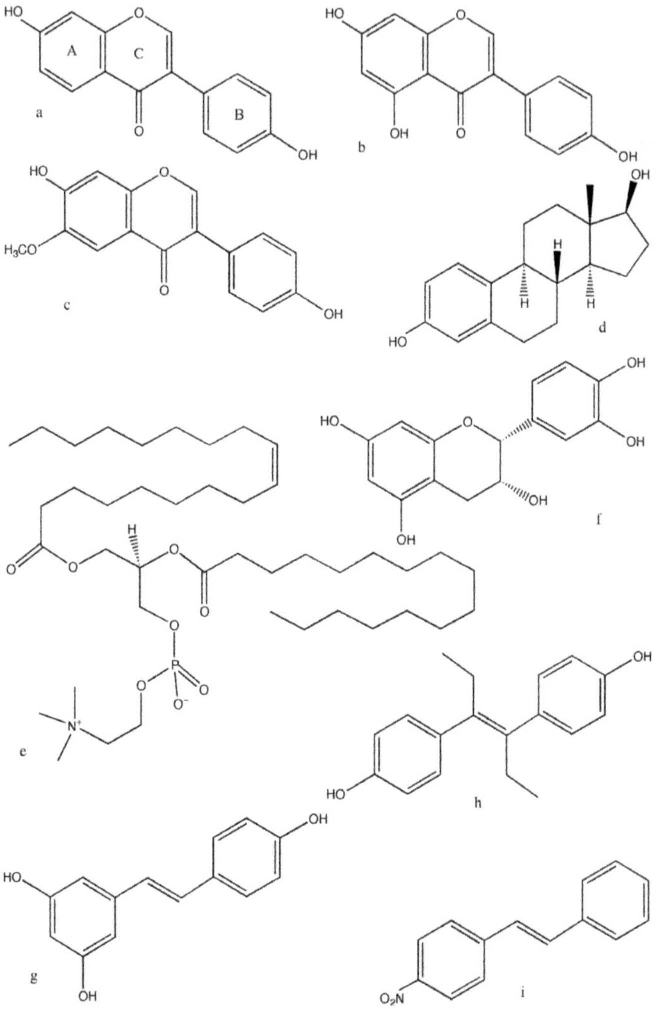

FIGURE 13.7 (a) Daidzein, (b) genistein, (c) glycitein, (d) 17β-oestradiol, (e) lecithin, (f) EC, (g) RES, (h) DES, and (i) 4-NO$_2$TS.

Earlier publications in *Nereis* classified yams (Torrens-Zaragozá, 2013), lactic acid bacteria (Torrens-Zaragozá, 2014), fruits (Torrens Zaragozá, 2015), food spices (Torrens-Zaragozá, 2016), and legumes (Torrens and Castellano, 2014b; Torrens and Castellano, 2018) by principal component (PCA), cluster (CA), and meta-analyses. The molecular classifications of 33 phenolic compounds derived from the cinnamic and benzoic acids from *Posidonia oceanica* (Castellano et al., 2012), 74 flavonoids (Castellano et al., 2013), 66 stilbenoids (Castellano et al., 2014), 71 triterpenoids and steroids from *Ganoderma* (Castellano and Torrens, 2015a) and 17 isoflavonoids from *Dalbergia parviflora* (Castellano and Torrens, 2015b), artemisinins (Torrens et al., 2017, 2018) were reported. A tool for interrogation of macromolecular structure was published (Torrens and Castellano, 2014a). Mucoadhesive polymer hyaluronan favors transdermal penetration absorption of caffeine (Torrens and Castellano, 2014c, 2015). Endocrine disruptor diethylstilbestrol (DES, Fig. 13.7h), bisphenol-A (similar to DES), polycarbonate, and epoxy-silica nanocomposites were reported (Torrens-Zaragozá, 2011; Torrens and Castellano, 2013). The main aim of the present report is to develop code-learning potentialities and, since legumes nutrients are more naturally described via varying size-structured representation, find approaches to structured information processing. In view of beans nutritional benefits, the objective was to categorize them with PCA/CA to differentiate legumes groups and identify characteristic compounds of various vegetables.

13.2   COMPUTATIONAL METHOD

PCA is a dimension-reduction technique (Hotelling, 1933; Kramer, 1998; Patra et al., 1999; Jolliffe, 2002; Xu and Hagler, 2002; Shaw, 2003). From starting variables X_j, PCA builds orthogonal variables $\tilde{P}_j$, linear combinations of mean-centerd ones $\tilde{X}_j = X_j - \bar{X}_j$, corresponding to eigenvectors of sample covariance matrix $S = 1/(n-1)\sum_{i=1}^{n}(x_i - \bar{x})(x_i - \bar{x})'$. For every loading vector $\tilde{P}_j$, matching eigenvalue $\tilde{l}_j$ of S tells how much data variability is explained: $\tilde{l}_j = \mathrm{Var}\left(\tilde{P}_j\right)$. Loading vectors are sorted in decaying eigenvalues. First k PCs explain most variability. After selecting k, one projects p-dimensional data on to subspace spanned by k loading vectors and computes coordinates versus $\tilde{P}_j$, yielding scores:

$$\tilde{t}_i = \tilde{P}'\left(x_i - \bar{x}\right) \tag{13.1}$$

for every $i = 1, \ldots, n$ having trivially zero mean. With regard to original coordinate system, projected data point is computed fitting:

$$\hat{x}_i = \bar{x} + \tilde{P}\tilde{t}_i \tag{13.2}$$

Loading matrix $\tilde{p}$ ($p \times k$) contains loadings column-wise and diagonal one $\tilde{L}=\left(\tilde{l}_j\right)_j$ ($k \times k$), eigenvalues. Loadings k explain variation:

$$\frac{\left(\sum_{j=1}^{k} \tilde{l}_j\right)}{\left(\sum_{j=1}^{p} \tilde{l}_j\right)^3} \geq 80 \tag{13.3}$$

CA starting point is $n \times p$ data matrix $\mathbf{X}$ containing p components measured in n samples (IMSL, 1989; Tryon, 1939). One assumes data were preprocessed to remove artifacts, and missing values, imputed. The CA organizes samples into a small number of clusters such that examples within a bunch are similar. Distances l_q between samples $x, x' \in \Re^p$ result:

$$\left\| x - x' \right\|_q = \left(\sum_{i=1}^{p} \left| x_i - x_i' \right|^q \right)^{1/q} \tag{13.4}$$

(e.g., Euclidean l_2, Manhattan l_1 distances). *Pearson's correlation coefficient* (PCC) compares samples:

$$r\left(x - x'\right) = \frac{\sum_{i=1}^{p}\left(x_i - \overline{x}\right)\left(x_i' - \overline{x'}\right)}{\left[\sum_{i=1}^{p}\left(x_i - \overline{x}\right)^2 \sum_{i=1}^{p}\left(x_i' - \overline{x'}\right)^2\right]^{1/2}} \tag{13.5}$$

where $\overline{x} = \left(\sum_{i=1}^{p} x_i\right)/p$ is measure mean value for sample x (Priness et al., 2007; Steuer et al., 2002; D'Haeseleer et al., 2000; Perou et al., 2000; Jarvis and Patrick, 1973; Page, 2000; Eisen et al., 1998).

13.3   CALCULATION RESULTS

Eight proximate [*moisture, ash, fat, protein,* CHO, *fiber, reducing sugar,* and *energy*] contents of seven bean varieties were taken from Adamu et al. The PCC matrix $\mathbf{R}$ was computed between beans; the upper triangle turns out to be

$$
\mathbf{R} = \begin{pmatrix}
1.000 & 1.000 & 0.998 & 1.000 & 0.998 & 0.999 & 0.999 \\
 & 1.000 & 0.999 & 0.999 & 0.998 & 0.998 & 0.999 \\
 & & 1.000 & 0.999 & 0.998 & 0.998 & 0.999 \\
 & & & 1.000 & 0.999 & 1.000 & 1.000 \\
 & & & & 1.000 & 0.999 & 0.999 \\
 & & & & & 1.000 & 1.000 \\
 & & & & & & 1.000
\end{pmatrix}
$$

Correlations are high between pairs of peas, pairs of beans, and between groups $R \approx 0.999$. All are illustrated in the partial correlation diagram (PCD), which could contain high ($r \geq 0.75$), medium ($0.50 \leq r < 0.75$), low ($0.25 \leq r < 0.50$), and *zero* ($r < 0.25$) partial correlations. All 21 pairs of beans present high partial correlations (cf. Fig. 13.8, grayscale). The corresponding interpretation is that all legumes present similar composition. The results are in agreement with previous outcomes (Figs. 13.6 and 13.7).

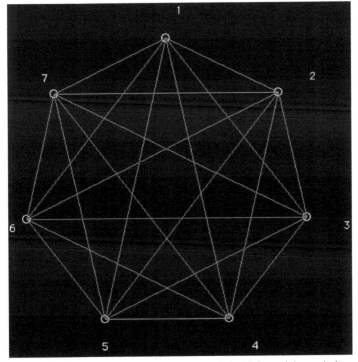

FIGURE 13.8 Partial correlation diagram showing all 21 high partial correlations.

The dendrogram of legumes according to proximate + mineral and proximate + mineral + phaseolin protein fraction contents (cf. Fig. 13.9) shows different behavior depending on *fat, fiber*, etc. Three classes are clearly recognized:

(1,4,6)(2,5,7)(3)

Mung bean in class 3 is separated from groupings 1 and 2: Bean, pigeon (red), and BEP are high in *moisture* and group into class 1; soya, AYB, and pigeon pea (gray variety) are high in *ash, protein*, CHO, *reducing sugar*, and *energy* and form cluster 2; mung bean in class 3 is rich in *fat* and *fiber*. Maximum similarity results between pigeon (red) and BEP. Beans in the same grouping appear highly correlated in PCD in qualitative agreement with previous results (Figs. 13.6–13.8).

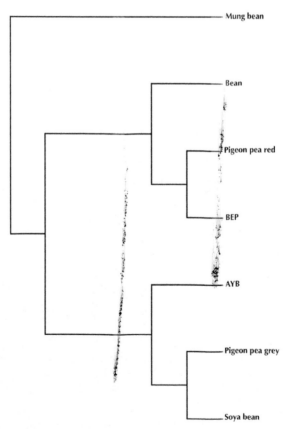

FIGURE 13.9 Dendrogram of some underutilized beans in according to proximates, minerals, and phaseolin.

The radial tree (cf. Fig. 13.10) shows different behavior of beans depending on *fat*, etc. The three classes above are clearly recognized in qualitative agreement with previous results, PCD, and dendrogram (Figs. 13.6–13.9). Again, maximum resemblance results between pigeon (red) and BEP.

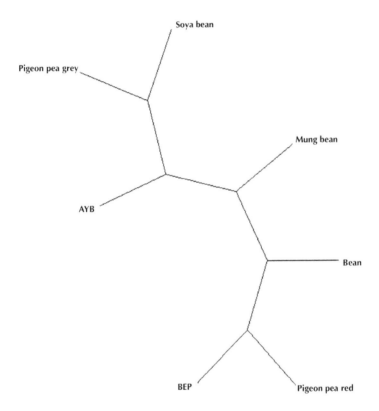

FIGURE 13.10 Radial tree of some underutilized beans according to proximates, minerals, and phaseolin.

Splits graph (cf. Fig. 13.11) for seven legumes in Table 13.1 reveals conflicting relations between classes because of interdependences (Huson, 1998). It indicates spurious relations between groupings resulting from base-composition effects. It illustrates different behavior of beans depending on *fat*, etc. in qualitative agreement with previous results, PCD, and binary/

radial trees (Figs. 13.6–13.10). One more time, maximum similarity results between pigeon (red) and BEP.

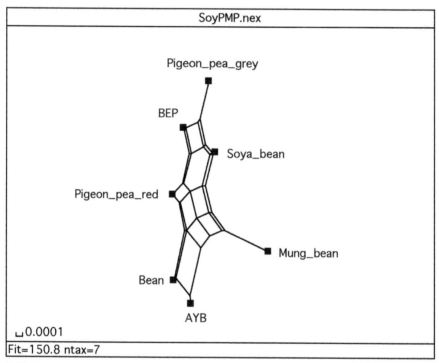

FIGURE 13.11 Splits graph of underutilized beans in according to proximates, minerals, and phaseolin.

PCA allows *summarizing* information contained in **X**-matrix. It decomposes **X**-matrix as product of matrices **P** and **T**. *Loading matrix* **P** with information about variables contains a few vectors: PCs, which are obtained as linear combinations of original *X*-variables. In *score matrix* **T** with information about objects, every object is described by projections on to PCs instead of original variables: $\mathbf{X} = \mathbf{TP'} + \mathbf{E}$, where ' denotes transpose matrix. Information not contained in matrices remains *unexplained* **X**-*variance* in *residual matrix* **E**. Every PC_i is a new coordinate expressed as linear combination of old x_j: $PC_i = S_j b_{ij} x_j$. New coordinates PC_i are *scores* (*factors*), while coefficients b_{ij} are *loadings*. Scores are ordered by information content versus total variance among objects. *Score–score plots* show compounds positions in new coordinate system, while *loading–loading plots* show location of features that represent compounds in new coordination. Properties of PCs follow: (1) they are extracted by decaying importance; (2) every PC is orthogonal to each other. A PCA was performed for beans. Importance of PCA factors F_{1-23} for nutrients (cf. Table 13.2) shows that F_1 explains 42% variance (58% error), $F_{1/2}$, 71% variance (29% error), F_{1-3}, 87% variance (13% error), etc.

TABLE 13.2 PCA Factors Importance for Proximates, Minerals, and Phaseolin Protein Fraction of Beans.

Factor	Eigenvalue	Percentage of variance	Cumulative percentage of variance
F_1	9.76200906	42.44	42.44
F_2	6.60938905	28.74	71.18
F_3	3.63377950	15.80	86.98
F_4	2.99482239	13.02	100.00
F_5	0.00000000	0.00	100.00
F_6	0.00000000	0.00	100.00
F_7	0.00000000	0.00	100.00
F_8	0.00000000	0.00	100.00
F_9	0.00000000	0.00	100.00
F_{10}	0.00000000	0.00	100.00
F_{11}	0.00000000	0.00	100.00
F_{12}	0.00000000	0.00	100.00
F_{13}	0.00000000	0.00	100.00
F_{14}	0.00000000	0.00	100.00
F_{15}	0.00000000	0.00	100.00
F_{16}	0.00000000	0.00	100.00
F_{17}	0.00000000	0.00	100.00
F_{18}	0.00000000	0.00	100.00
F_{19}	0.00000000	0.00	100.00
F_{20}	0.00000000	0.00	100.00
F_{21}	0.00000000	0.00	100.00
F_{22}	0.00000000	0.00	100.00
F_{23}	0.00000000	0.00	100.00

The PCA factors loadings are shown in Table 13.3.

TABLE 13.3 PCA Loadings for Proximates, Minerals, and Phaseolin Protein Fraction of Beans.

Property	PCA factor loadings			
	F_1	F_2	F_3	F_4
i_1	−0.22590359	−0.17352336	−0.28559738	0.04629024
i_2	0.22092641	0.15319768	0.31509330	0.05049684
i_3	0.16062148	−0.24178430	0.26814158	−0.18318493
i_4	0.23882151	0.15317234	−0.27890810	−0.04276636
i_5	−0.24165655	0.23081680	0.14533420	0.01864708
i_6	0.17945183	−0.31639153	0.03349598	0.08158517
i_7	0.10948926	0.24497211	−0.28657095	−0.25049562
i_8	0.05269382	0.24868623	0.31451615	−0.26142986
i_9	0.21183238	−0.28708594	−0.05934150	0.03841347
i_{10}	0.31134989	0.05879760	0.06383116	−0.07315650
i_{11}	0.28995128	−0.15189421	−0.01455354	0.09322975
i_{12}	−0.04553091	0.09579091	−0.03510978	0.55263538
i_{13}	0.24897246	0.17189271	−0.01958254	0.25725542
i_{14}	0.27649689	−0.15441892	−0.16110246	0.02435280
i_{15}	0.25531868	−0.14311471	0.22539716	−0.12073522
i_{16}	−0.00621197	−0.16374158	−0.47483811	0.03219519
i_{17}	0.10885766	0.35765327	−0.07947129	0.07292161
i_{18}	0.01163403	0.26909091	0.18927679	0.36082533
i_{19}	0.29468398	−0.13314347	0.08399695	−0.05625384
i_{20}	0.25360778	0.17813494	−0.20361108	0.06266525
i_{21}	0.28041176	0.18054394	−0.06362453	0.02744441
i_{22}	−0.10183307	0.14074013	−0.07622211	−0.49934292
i_{23}	0.16791831	0.26127559	−0.23239671	−0.16065997

TABLE 13.4 Profile of Principal Component Analysis Factors for Proximates, Minerals, and Phaseolin Protein Fraction of Some Underutilized Beans.

$\%i$	$\%i_1$	$\%i_2$	$\%i_3$	$\%i_4$	$\%i_5$	$\%i_6$	$\%i_7$	$\%i_8$	$\%i_9$	$\%i_{10}$	$\%i_{11}$	$\%i_{12}$	$\%i_{13}$	$\%i_{14}$	$\%i_{15}$	$\%i_{16}$	$\%i_{17}$	$\%i_{18}$	$\%i_{19}$	$\%i_{20}$	$\%i_{21}$	$\%i_{22}$	$\%i_{23}$
F_1	5.10	4.88	2.58	5.70	5.84	3.22	1.20	0.28	4.49	9.69	8.41	0.21	6.20	7.65	6.52	0.00	1.18	0.01	8.68	6.43	7.86	1.04	2.82
F_2	3.01	2.35	5.85	2.35	5.33	10.01	6.00	6.18	8.24	0.35	2.31	0.92	2.95	2.38	2.05	2.68	12.79	7.24	1.77	3.17	3.26	1.98	6.83
F_3	8.16	9.93	7.19	7.78	2.11	0.11	8.21	9.89	0.35	0.41	0.02	0.12	0.04	2.60	5.08	22.55	0.63	3.58	0.71	4.15	0.40	0.58	5.40
F_4	0.21	0.25	3.36	0.18	0.03	0.67	6.27	6.83	0.15	0.54	0.87	30.54	6.62	0.06	1.46	0.10	0.53	13.02	0.32	0.39	0.08	24.93	2.58

The PCA F_1–F_4 profile is listed in Table 13.4. For F_1, variable i_{10} shows the greatest weight in the profile; however, F_1 cannot be reduced to two variables $\{i_{10}, i_{11}\}$ without 82% error. For F_2, variable i_{17} presents greatest weight; notwithstanding, F_2 cannot be reduced to two variables $\{i_6, i_{17}\}$ without 77% error. For F_3, variable i_{16} assigns greatest weight; nevertheless, F_3 cannot be reduced to two variables $\{i_2, i_{16}\}$ without 68% error, etc.

The PCA scores plot F_2–F_1 for beans (cf. Fig. 13.12) shows that the three clusters above are clearly distinguished: class 1 with three legumes ($0 \approx F_1 > F_2$, *bottom*), grouping 2 with three vegetables ($F_1 < F_2 \approx 0$, *middle*) and cluster 3 with one bean ($F_1 \ll F_2$, *top*). Peas are closer between themselves than beans. Once more, maximum resemblance results between pigeon (red) and BEP.

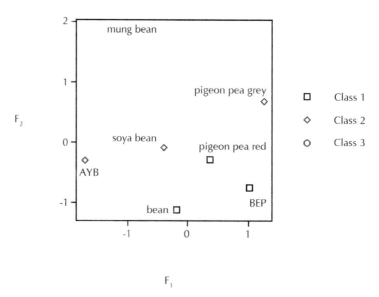

FIGURE 13.12   PCA scores plot of underutilized beans according to proximates, minerals, and phaseolin.

From PCA factors loading of beans, F_2–F_1 loadings plot (cf. Fig. 13.13) depicts 23 proximates, minerals, and phaseolin contents. Nutrients *ash*, *protein*, Fe, and *albumin* collapse, as well as macrominerals Mg, Zn, and *relative*. Seven clusters are clearly distinguished: class 1 with two nutrients ($\{1,16\}$, $0 > F_1 > F_2$, *bottom left*), grouping 2 with two proximates ($\{2,8\}$, $0 < F_1 < F_2$, *middle*), cluster 3 with five nutrients ($\{3,6,9,11,14\}$, $F_1 \gg F_2$, *bottom*), class 4 with four nutrients ($\{4,13,20,21\}$, $F_1 > F_2 > 0$, *top right*), grouping 5 with four nutrients ($\{5,12,18,22\}$, $0 \approx F_1 < F_2$, *top left*), cluster

6 with three nutrients ($\{7,17,23\}$, $0 < F_1 < F_2$, *top*) and class 7 with three macrominerals ($\{10,15,19\}$, $F_1 > F_2 \approx 0$, *right*). Constituents in class 4 are closer between themselves than groupings 1–3 and 5–7. Macronutrients *fat*, *protein*, and CHO result in separate classes 3–5: *protein* appears closer to *fat* than to CHO. Macromineral K is closer to Zn than to Na; Mg is closer to Zn than to Ca; Fe is closer to Ca than to microminerals Co and Mn. In addition, as a complement to scores diagram for loadings it is confirmed that beans in class 2, located in the middle, present a more pronounced contribution from nutrients in grouping 2 situated in the same position in Figure 13.12.

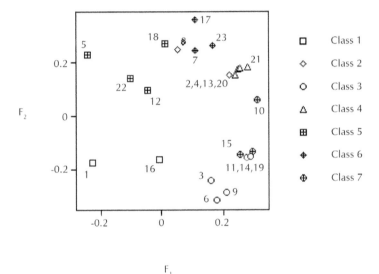

FIGURE 13.13 PCA loadings plot of underutilized beans in according to proximates, minerals, and phaseolin.

Instead of seven beans in the space $\mathfrak{R}^{23}$ of 23 nutrients, consider 23 nutrients in the space $\mathfrak{R}^7$ of seven legumes. Matrix **R** upper triangle results:

$$R = \begin{pmatrix}
1.000 & -0.983 & -0.381 & -0.419 & 0.120 & -0.056 & -0.280 & -0.764 & -0.071 & -0.830 & -0.437 & 0.104 & -0.690 & -0.262 & -0.650 & 0.699 & -0.558 & -0.481 & -0.592 & -0.544 & -0.756 & 0.073 & -0.451 \\
 & 1.000 & 0.381 & 0.344 & -0.118 & 0.117 & 0.118 & 0.686 & 0.104 & 0.793 & 0.469 & 0.042 & 0.727 & 0.259 & 0.646 & -0.718 & 0.517 & 0.569 & 0.588 & 0.504 & 0.719 & -0.240 & 0.336 \\
 & & 1.000 & -0.119 & -0.616 & 0.775 & -0.362 & 0.135 & 0.712 & 0.497 & 0.632 & -0.562 & -0.045 & 0.510 & 0.915 & -0.228 & -0.518 & -0.425 & 0.788 & -0.120 & 0.074 & -0.185 & -0.293 \\
 & & & 1.000 & -0.479 & 0.054 & 0.826 & 0.089 & 0.258 & 0.730 & 0.525 & -0.044 & 0.741 & 0.648 & 0.237 & 0.297 & 0.687 & 0.062 & 0.474 & 0.970 & 0.897 & 0.046 & 0.912 \\
 & & & & 1.000 & -0.417 & -0.453 & 0.974 & 0.412 & 0.847 & -0.149 & 0.137 & 0.794 & 0.744 & 0.282 & -0.549 & -0.431 & 0.791 & 0.062 & 0.113 & -0.604 & 0.387 & -0.129 \\
 & & & & & 1.000 & 0.328 & -0.205 & 0.416 & 0.009 & -0.272 & 0.372 & 0.195 & -0.103 & 0.199 & 0.723 & -0.020 & 0.054 & 0.725 & 0.638 & 0.573 & 0.965 \\
 & & & & & & 1.000 & -0.461 & 0.387 & -0.190 & -0.339 & 0.187 & -0.315 & 0.248 & -0.840 & 0.496 & 0.382 & 0.073 & 0.141 & 0.347 & 0.483 & 0.376 \\
 & & & & & & & 1.000 & 0.510 & 0.902 & -0.205 & 0.223 & 0.902 & 0.737 & 0.404 & -0.428 & -0.486 & 0.837 & 0.238 & 0.254 & -0.519 & -0.117 \\
 & & & & & & & & 1.000 & 0.798 & -0.230 & 0.763 & 0.738 & 0.799 & -0.200 & 0.435 & 0.105 & 0.876 & 0.779 & 0.902 & -0.163 & 0.593 \\
 & & & & & & & & & 1.000 & -0.069 & 0.605 & 0.955 & 0.821 & 0.181 & -0.026 & -0.146 & 0.948 & 0.567 & 0.623 & -0.565 & 0.180 \\
 & & & & & & & & & & 1.000 & 0.426 & -0.160 & -0.433 & 0.013 & 0.309 & 0.738 & -0.319 & 0.130 & 0.043 & -0.682 & -0.145 \\
 & & & & & & & & & & & 1.000 & 0.527 & 0.349 & -0.143 & 0.733 & 0.599 & 0.516 & 0.882 & 0.912 & -0.467 & 0.598 \\
 & & & & & & & & & & & & 1.000 & 0.694 & 0.431 & -0.019 & -0.328 & 0.878 & 0.626 & 0.612 & -0.410 & 0.311 \\
 & & & & & & & & & & & & & 1.000 & -0.261 & -0.158 & -0.201 & 0.950 & 0.274 & 0.466 & -0.269 & 0.039 \\
 & & & & & & & & & & & & & & 1.000 & -0.250 & -0.584 & -0.024 & 0.149 & -0.100 & -0.063 & 0.093 \\
 & & & & & & & & & & & & & & & 1.000 & 0.673 & -0.038 & 0.763 & 0.749 & 0.137 & 0.828 \\
 & & & & & & & & & & & & & & & & 1.000 & -0.206 & 0.273 & 0.339 & -0.353 & 0.150 \\
 & & & & & & & & & & & & & & & & & 1.000 & 0.500 & 0.624 & -0.356 & 0.209 \\
 & & & & & & & & & & & & & & & & & & 1.000 & 0.959 & -0.124 & 0.865 \\
 & & & & & & & & & & & & & & & & & & & 1.000 & -0.134 & 0.812 \\
 & & & & & & & & & & & & & & & & & & & & 1.000 & 0.381 \\
 & & & & & & & & & & & & & & & & & & & & & 1.000
\end{pmatrix}$$

Correlation is greater between phaseolins $R \approx 0.460$ than minerals $R \approx$ 0.281 and proximates $R \approx -0.071$. Some correlations between groups are high (e.g., *fiber*-K $R_{6,9} = 0.974$. Most correlations of *moisture* $R_{1,i}$ are negative, especially with *ash* $R_{1,2} = -0.983$), as well as *fat* $R_{3,i}$, especially with CHO $R_{3,5} = -0.616$, CHO $R_{5,i}$, especially with Zn $R_{5,14} = -0.972$, Na $R_{12,i}$, especially with *prolamin* $R_{12,22} = -0.682$, Co $R_{1,16}$, especially with *energy* $R_{8,16} = -0.840$, and *prolamin* $R_{1,22}$, especially with Na $R_{12,22} = -0.682$. The dendrogram for 23 proximates, minerals, and phaseolins of beans (cf. Fig. 13.14) separates the seven clusters above in agreement with PCA loadings plot (Fig. 13.13). Again, macronutrients appear in separate classes 3–5; *protein* is closer to *fat* than to CHO; K is closer to Zn than to Na; Mg is closer to Zn than to Ca; and Fe is closer to Ca than to Co.

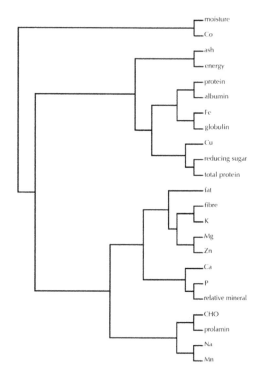

FIGURE 13.14   Dendrogram of proximates, minerals, and phaseolin for some underutilized beans.

Radial tree for 23 proximates, minerals, and phaseolins of beans (cf. Fig. 13.15) separates the seven clusters above in agreement with PCA loadings plot and dendrogram (Figs. 13.13 and 13.14). Once more, macronutrients result in separate classes 3–5; *protein* is closer to *fat* than to CHO; K is closer to Zn than to Na; Mg is closer to Zn than to Ca; Fe is closer to Ca than to Co.

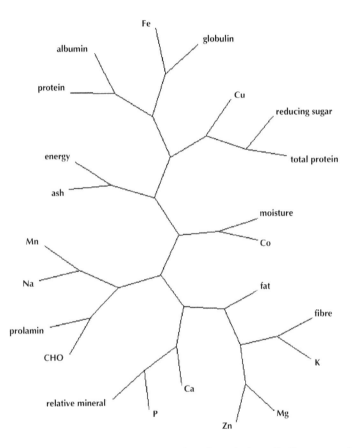

FIGURE 13.15   Radial tree of proximates, minerals, and phaseolin for some underutilized beans.

Splits graph for 23 proximates, minerals, and phaseolins of beans (cf. Fig. 13.16) shows that nutrients 4, 10, 11, 13, 14, 17, and 19–21 collapse, as well as 6 and 9. It reveals conflicting relationships between classes. It separates the seven clusters above in agreement with loadings plot and binary/radial trees (Figs. 13.13–13.15). Once more, macronutrients split into classes 3–5; *protein* is closer to *fat* than to CHO; K is closer to Zn than to Na; Mg is closer to Zn than to Ca; and Fe is closer to Ca than to Co.

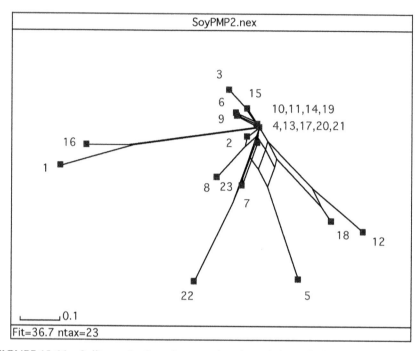

FIGURE 13.16 Splits graph of proximates, minerals, and phaseolin for some underutilized beans.

A PCA was performed for nutrients. Factor F_1 explains 99.9% variance (0.1% error), $F_{1/2}$, 99.96% variance (0.04% error), F_{1-3}, 99.99% variance (0.01% error), etc. Scores plot of PCA F_2–F_1 showed that nutrients 1–4, 6, 7, 12–14, 16–18, and 20–22 collapse. It distinguished the seven clusters above: class 1 ($F_1 < F_2 < 0$, cf. Fig. 13.17, *left*), grouping 2 ($0 \approx F_1 > F_2$, *bottom*), cluster 3 ($0 \approx F_1 < F_2$, *top left*), class 4 ($F_1 < F_2 < 0$, *left*), grouping 5 ($0 \approx F_1 > F_2$, *bottom left*), cluster 6 ($0 \approx F_1 > F_2$, *middle left*), and class 7 ($0 \approx F_1 < F_2$, *top*). Now constituents in class 1 result closer between themselves than groupings 2–7. Again, macronutrients appear in separate classes 3–5; *protein* is closer to *fat* than to CHO; and K is closer to Zn than to Na. The results are in qualitative agreement with loadings plot, binary/radial trees, and splits graph (Figs. 13.13–13.16).

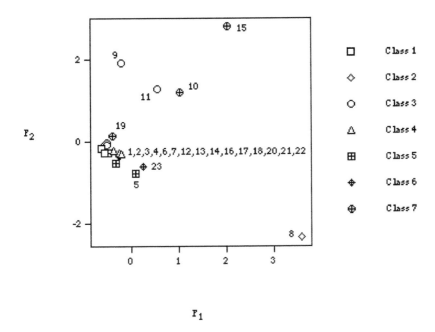

FIGURE 13.17 PCA scores plot of proximates, minerals, and phaseolin for some underutilized beans.

Soybean oil is in disadvantage versus Mediterranean diet: although contents in oleic or linoleic acid of sunflower (*Helianthus annuus*) oil be similar to the soybean one, sunflower oil lacks linolenic acid, which is easily oxidized and the component that makes inadvisable to make fried foods with soybean oil. Disadvantages versus virgin olive *Olea europaea* L. (VOO) or high-oleic sunflower oil are greater. Soybean oil is not a source of isoflavones because it does not contain them. The VOO classes with high-oleic sunflower oil, and sunflower, with soybean oil (cf. Fig. 13.18).

Main VOO hydrophilic phenols are secoiridoids [oleuropein (cf. Fig. 13.19e), ligstroside, elenolic acid], flavones [luteolin (Fig. 13.19a), apigenin (Fig. 13.19b)], and phenolic alcohols and their derivatives (hydroxytyrosol, its acetate). Structure of lipophilic methylated phenol Vit E α-tocopherol (α-TCP, Fig. 13.19c) of VOO and sunflower oils is close to γ-TCP (Fig. 13.19d) of soybean and maize oils. The health-benefiting mechanisms of VOO phenolic compounds were revised (Parkinson and Cicerale, 2016). The potential of liquid chromatography coupled to fluorescence detection (LC-FD) in food metabolomics was applied to the determination of phenolic compounds in VOO (Monasterio et al., 2016). The state of the art on functional VOOs enriched with bioactive compounds was reviewed (Reboredo-Rodríguez et al., 2017). Analytical evaluation and antioxidant properties of secondary metabolites were informed in Northern Italian mono and multi-varietal VOOs from early and late harvested olives (Trombetta et al., 2017). Erythrodiol increased the half-life of ABCA1 and enhanced CHOL efflux from THP-1-derived macrophages (Wang et al., 2017).

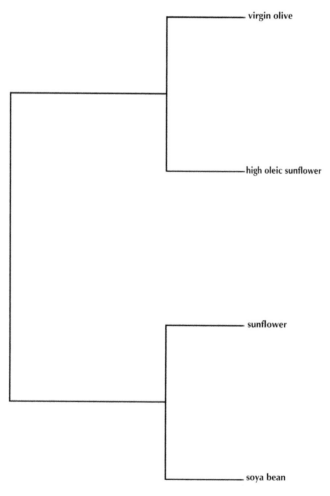

FIGURE 13.18 Dendrogram of Mediterranean diet and soybean oils according to unsaturated oily acids.

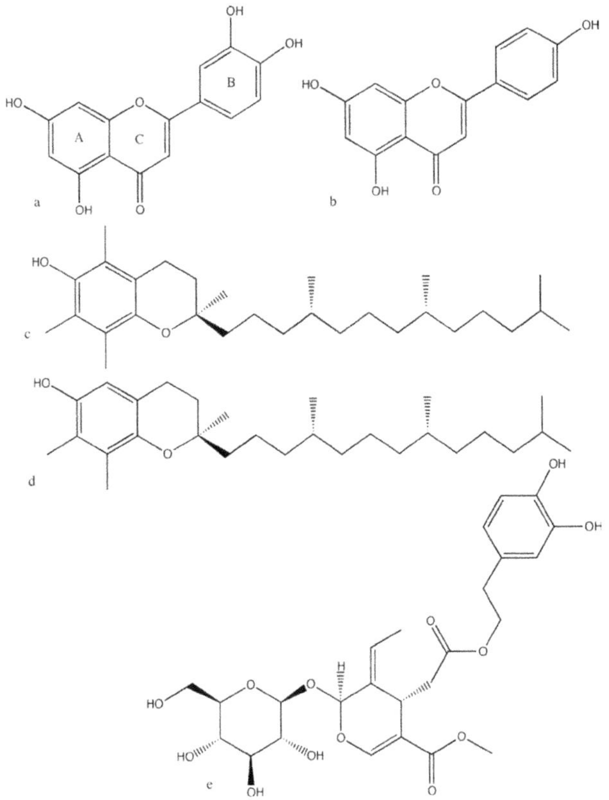

FIGURE 13.19 VOO flavones: (a) luteolin and (b) apigenin, (c) α-tocopherol, (d) γ-tocopherol, and (e) oleuropein.

Fatty acid (FA), phospholipid, and sterol compositions of breadfruit and wonderful kola seeds were informed (Aremu et al., 2017).

13.4 DISCUSSION

Culture enhancement is promoted extending the genetic basis and knowledge of available resources. The knowledge of the genetic diversity occurring in cultivars will contribute to expand the basis for improvement programs and may contribute to the efforts performed to increase bean per capita consumption while decreasing fast food consumption. According to the definition of functional food, the common bean is considered as such because it contains a number of bioactive compounds that present beneficial impacts on consumer's health. The analyzed evidence confirms that the grain is a source of bioactive compounds that impact health in a number of ways. The content of the compounds is correlated to the seed coat color, and dark-colored or black beans are distinguished by their elevated content of polyphenols, anthocyanins, condensed tannins, and flavonoids, for example, quercetin, which confer them with a higher antioxidant capacity. The bean is a food frequently consumed in its cooked form, which thermal treatment impacts on content of bioactive compounds. Contrastingly, the process is necessary to decrease the levels of antinutrients, for example, lectins, phytic acid, tannins, saponins, and protease inhibitors that, according to evidence, are potential neutraceutical compounds that may be beneficial to treat cancer, cardiovascular diseases, human immunodeficiency virus, diabetes, obesity, degenerative diseases, etc.

High correlation was reported between lipid peroxide radical and tumor-promoter effect (Maeda et al., 1992). Preparation of function-enhanced vegetable oils was informed (Maeda et al., 2016). Flavan-3-ol epicatechin (EC, Fig. 13.7f) is able to cross blood–brain barrier better than more hydrophilic stilbenoid resveratrol (RES, Fig. 13.7g). The antioxidants [e.g., stilbenoids DES, RES, 4-nitrostilbene (4-NO_2TS, Fig. 13.7i)] show OES activity. The possibility of breast cancer prevention was reviewed via use of soy isoflavones and fermented soy beverage produced via probiotics (Takagi et al., 2015). Soy and breast cancer were revised with focus on angiogenesis (Varinska et al., 2015). Genomic history of the origin and domestication of common bean unveiled its closest sister species (Rendón-Anaya et al., 2017). Short-term local adaptation of historical common bean varieties and implications for in situ management of bean diversity was reported (Klaedtke et al., 2017).

Beans are important legumes in human nutrition. They contain essential nutrients and minerals required for human growth and body functioning. Their high *protein* qualifies as alternatives and affordable *protein* source in comparison to the expensive price of animal-meat *protein*. Some are used as *protein*-rich condiments to enhance traditional-soups flavor in Nigeria. *Ash* content is an indication of the mineral nutrients present in a food material. Mineral content does not translate to absolute bioavailability when beans are consumed: the presence of antinutritional factors following bean processing is part of factors to be taken into consideration in determining absolute nutritive value of bean seeds. Beans are a source of complex CHOs (e.g., dietary *fiber*, resistant starch). They are low in *fat* and CHOL. Beans compared to common bean *P. vulgaris* in proximates but minerals differed between bean types. Ca, Mg, P, and K (especially in mung bean) were in high amounts.

Phaseolin protein fractions were dissimilar among bean kinds. Underutilized beans compare in nutritional composition to habitual beans in Nigeria. The AYB and mung beans are alternatives from a *protein*-rich standpoint. Fats are nutrients necessary for the growth and maintenance of human body. CHOL should be reduced, as well as both unnatural *trans* FAs and saturated FAs. The immediate solution to it is to reduce fast food.

13.5 FINAL REMARKS

From the present results and discussion, the following final remarks can be drawn:

(1) Some criteria reduced the analysis to a manageable quantity from the enormous set of beans compositions: they refer to proximates, minerals, and phaseolin. The meta-analysis was useful to rise the number of samples and variety of analyzed data. Different behavior of beans depends on *fat, fiber*, etc. With regard to components, macronutrients result unconnected in three groupings and *protein* was closer to *fat* than to CHO. Beans compared to common bean *P. vulgaris* in proximates but minerals and phaseolin protein fractions differed among bean types. Soybean presents adequate proximate, minerals, and phaseolin (especially *prolamin*) contents, good anti-oxidant capacity and may be used as a functional food. It represents a legume useful as a natural source for nutraceutical formulations.

(2) PCAs of compositions and beans CAs allowed classifying them and agreed. Phytochemistry, cytochemistry, and understanding of computational methods are essential for tackling associated *data-mining* tasks.

(3) The analyzed evidence shows that more studies are needed to expand the knowledge of the nutraceutical quality of bean geno-types, grown or wild type, and their impact on health, to be used in genetic improvement programs or as a strategy to encourage their consumption, which is based on the potential it presents for health preservation and disease prevention. Even though beans are traditionally included in the diet and are a potential functional food contributing to health preservation and disease prevention, a consid-erable amount of research needs to be conducted to characterize the diverse genotypes regarding their content of bioactive compounds and their effect on consumer's health status. This study points out the

necessity to study wild-type genotypes to know their nutraceutical qualities to be used as valuable resource in enhancement programs and strategies aimed to encourage its consumption.

ACKNOWLEDGMENTS

The authors thank support from Generalitat Valenciana (Project No. PROMETEO/2016/094) and Universidad Catolica de Valencia San Vicente Martir (Project No. UCV.PRO.17-18.AIV.03).

KEYWORDS

- beans
- protein
- cholesterol
- beverage
- cake

REFERENCES

Achinewhu, S. Composition and Food Potential of African Oil Bean (*Pentaclethra macrophylla*) and Velvet Bean (*Mucuna uriens*). *J. Food Sci.* **1982,** *47,* 1736–1737.

Adamu, G. O. L.; Ezeokoli, O. T.; Dawodu, A. O.; Adebayo-Oyetoro, A. O.; Ofodili, L. N. Macronutrients and Micronutrients Profile of Some Underutilized Beans in South Western Nigeria. *Int. J. Biochem. Res. Rev.* **2015,** *7,* 80–89.

Aremu, M. O. Haruna, A.; Oko, O. J.; Ortutu, S. C. Fatty Acid, Phospholipid and Sterol Compositions of Breadfruit (*Artocarpus altilis*) and Wonderful Kola (*Buchholzia aoriacea*) Seeds. *Int. J. Sci.* **2017,** *6* (4), 116–123.

Broughton, W.; Hernández, G.; Blair, M.; Beebe, S.; Gepts, P.; Vanderleyden, J. Beans (*Phaseolus* spp.)—Model Food Legumes. *Plant Soil* **2003,** *252,* 55–128.

Castellano, G.; Torrens, F. Information Entropy-Based Classification of Triterpenoids and Steroids from *Ganoderma*. *Phytochemistry* **2015a,** *116,* 305–313.

Castellano, G.; Torrens, F. Quantitative Structure–Antioxidant Activity Models of Isoflavonoids: A Theoretical Study. *Int. J. Mol. Sci.* **2015b,** *16,* 12891–12906.

Castellano, G.; Tena, J.; Torrens, F. Classification of Polyphenolic Compounds by Chemical Structural Indicators and Its Relation to Antioxidant Properties of *Posidonia oceanica* (L.) Delile. *MATCH Commun. Math. Comput. Chem.* **2012,** *67,* 231–250.

Castellano, G.; González-Santander, J. L.; Lara, A.; Torrens, F. Classification of Flavonoid Compounds by Using Entropy of Information Theory. *Phytochemistry* **2013,** *93,* 182–191.

Castellano, G.; Lara, A.; Torrens, F. Classification of Stilbenoid Compounds by Entropy of Artificial Intelligence. *Phytochemistry* **2014**, *97*, 62–69.

Chang, Y. L.; Liu, T. C.; Tsai, M. L. Selective Isolation of Trypsin Inhibitor and Lectin from Soybean Whey by Chitosan/Tripolyphosphate/Genipin Co-Crosslinked Beads. *Int. J. Mol. Sci.* **2014**, *15*, 9979–9990.

Chávez-Mendoza, C.; Sánchez, E. Bioactive Compounds from Mexican Varieties of the Common Bean (*Phaseolus vulgaris*): Implications for Health. *Molecules* **2017**, *22*, 1360-1–1360-32.

D'Haeseleer, P.; Liang, S.; Somogyi, R. Genetic Network Inference: From Co-Expression Clustering to Reverse Engineering. *Bioinformatics* **2000**, *16*, 707–726.

Edem, D.; Amugo, C.; Eka, O. Chemical Composition of Yam Beans (*Sphenostylis stenocarpa*). *Trop. Sci.* **1990**, *30*, 59–63.

Eisen, M. B.; Spellman, P. T.; Brown, P. O.; Botstein, D. Cluster Analysis and Display of Genome-Wide Expression Patterns. *Proc. Natl. Acad. Sci. U.S.A.* **1998**, *95*, 14863–14868.

Hotelling, H. Analysis of a Complex of Statistical Variables into Principal Components. *J. Educ. Psychol.* **1933**, *24*, 417–441.

Huson, D. H. SplitsTree: Analyzing and Visualizing Evolutionary Data. *Bioinformatics* **1998**, *14*, 68–73.

IMSL. *Integrated Mathematical Statistical Library (IMSL)*; IMSL: Houston, 1989.

Jarvis, R. A.; Patrick, E. A. Clustering Using a Similarity Measure Based on Shared Nearest Neighbors. *IEEE Trans. Comput.* **1973**, *C22*, 1025–1034.

Jolliffe, I. T. *Principal Component Analysis*; Springer: New York, 2002.

Kaufman, P. B.; Duke, J. A.; Brielmann, H. Boik, J.; Hoyt, J. E. A Comparative Survey of Leguminous Plants as Sources of the Isoflavones, Genistein and Daidzein: Implications for Human Nutrition and Health. *J. Altern. Complement. Med.* **1997**, *3*, 7–12.

Khalil, I. A.; Khan, S. Protein Quality of Asian Beans and Their Wild Progenitor, *Vigna sublobata* (Roxb). *Food Chem.* **1995**, *52*, 327–330.

Klaedtke, S. M.; Caproni, L.; Klauck, J.; de la Grandville, P.; Dutartre, M.; Stassart, P. M.; Chable, V.; Negri, V.; Raggi, L. Short-Term Local Adaptation of Historical Common Bean (*Phaseolus vulgaris* L.) Varieties and Implications for *in situ* Management of Bean Diversity. *Int. J. Mol. Sci.* **2017**, *18*, 493-1–493-19.

Kramer, R. *Chemometric Techniques for Quantitative Analysis*; Marcel Dekker: New York, 1998.

López-Pedrouso, M.; Alonso, J.; Santalla Ferradás, M.; Pedreira, R.; Álvarez, G.; Zapata, C. In-Depth Characterization of the Phaseolin Protein Diversity of Common Bean (*Phaseolus vulgaris* L.) Based on Two-Dimensional Electrophoresis and Mass Spectrometry. *Food Technol. Biotechnol.* **2012**, *50*, 315–325.

Maeda, H.; Katsuki, T.; Akaike, T.; Yasutake, R. High Correlation between Lipid Peroxide radical and Tumor–Promoter Effect: Suppression of Tumor Promotion in the Epstein–Barr Virus/B-Lymphocyte System and Scavenging of Alkyl Peroxide Radicals by Various Vegetable Extracts. *Jpn. J. Cancer Res.* **1992**, *83*, 923–928.

Maeda, H.; Satoh, T.; Islam, W. M. Preparation of Function-Enhanced Vegetable Oils. *Funct. Foods Health Dis.* **2016**, *6*, 33–41.

Maria John, K. M.; Jung, E. S.; Lee, S.; Kim, J. S.; Lee, C. H. Primary and Secondary Metabolites Variation of Soybean Contaminated with *Aspergillus sojae*. *Food Res. Int.* **2013**, *54*, 487–494.

Maria John, K. M.; Enkhtaivan, G.; Lee, J.; Thiruvengadam, M.; Keum, Y. S.; Kim, D. H. Spectroscopic Determination of Metabolic and Mineral Changes of Soya-Chunk Mediated by *Aspergillus sojae*. *Food Chem.* **2015**, *170*, 1–9.

Mazur, W. M.; Duke, J. A.; Wähälä, K.; Rasku, S.; Adlercreutz, H. Isoflavonoids and Lignans in Legumes: Nutritional and Health Aspects in Humans. *J. Nutr. Biochem.* **1998**, *9*, 193–200.

Monasterio, R. P.; Olmo-García, L.; Bajoub, A.; Fernández-Gutiérrez, A.; Carrasco-Pancorbo, A. Potential of LC Coupled to Fluorescence Detection in Food Metabolomics: Determination of Phenolic Compounds in Virgin Olive Oil. *Int. J. Mol. Sci.* **2016**, *17*, 1627-1–1627-17.

Page, R. D. M. *Program TreeView*; Universiy of Glasgow: Glasgow, UK, 2000.

Parkinson, L.; Cicerale, S. The Health Benefiting Mechanisms of Virgin Olive Oil Phenolic Compounds. *Molecules* **2016**, *21*, 1734-1–1734-12.

Patra, S. K.; Mandal, A. K.; Pal, M. K. J. State of Aggregation of Bilirubin in Aqueous Solution: Principal Component Analysis Approach. *Photochem. Photobiol. Sect. A* **1999**, *122*, 23–31.

Perou, C. M.; Sørlie, T.; van Eisen, M. B.; de Rijn, M.; Jeffrey, S. S.; Rees, C. A.; Pollack, J. R.; Ross, D. T.; Johnsen, H.; Akslen, L. A.; Fluge, O.; Pergamenschikov, A.; Williams, C.; Zhu, S. X. Lønning, P. E.; Børresen-Dale, A. L.; Brown, P. O.; Botstein, D. Molecular Portraits of Human Breast Tumours. *Nature (London)* **2000**, *406*, 747–752.

Priness, I.; Maimon, O.; Ben-Gal, I. Evaluation of Gene-Expression Clustering via Mutual Information Distance Measure. *BMC Bioinf.* **2007**, *8*, 111-1–111-12.

Reboredo-Rodríguez, P.; Figueiredo-González, M.; González-Barreiro, C.; Simal-Gándara, J.; Salvador, M. D.; Cancho-Grande, B.; Fregapane, G. State of the Art on Functional Virgin Olive Oils Enriched with Bioactive Compounds and Their Properties. *Int. J. Mol. Sci.* **2017**, *18*, 668-1–668-28.

Rendón-Anaya, M.; Montero-Vargas, J. M.; Saburido-Álvarez, S.; Vlasova, A.; Capella-Gutiérrez, S.; Ordaz-Ortiz, J. J.; Aguilar, O. M.; Vianello-Brondani, R. P.; Santalla, M.; Delaye, L.; Gabaldón, T.; Gepts, P.; Winkler, R.; Guigó, R.; Delgado-Salinas, A.; Herrera-Estrella, A. Genomic History of the Origin and Domestication of Common Bean Unveils Its Closest Sister Species. *Genome Biol.* **2017**, *18*, 60-1–60-17.

Shaw, P. J. A. *Multivariate Statistics for the Environmental Sciences*; Hodder-Arnold: New York, 2003.

Steuer, R.; Kurths, J.; Daub, C. O.; Weise, J.; Selbig, J. The Mutual Information: Detecting and Evaluating Dependencies between Variables. *Bioinformatics* **2002**, *18* (Suppl. 2), S231–S240.

Takagi, A.; Kano, M.; Kaga, C. Possibility of Breast Cancer Prevention: Use of Soy Isoflavones and Fermented Soy Beverage Produced Using Probiotics. *Int. J. Mol. Sci.* **2015**, *16*, 10907–10920.

Torrens, F.; Castellano, G. Bisphenol, Diethylstilbestrol, Polycarbonate and the Thermomechanical Properties of Epoxy–Silica Nanostructured Composites. *J. Res. Upd. Polym. Sci.* **2013**, *2*, 183–193.

Torrens, F.; Castellano, G. A Tool for Interrogation of Macromolecular Structure. *J. Mater. Sci. Eng. B* **2014a**, *4* (2), 55–63.

Torrens, F.; Castellano, G. From Asia to Mediterranean: Soya Bean, Spanish Legumes and Commercial *Soya Bean* Principal Component, Cluster and Meta-Analyses. *J. Nutr. Food Sci.* **2014b**, *4* (5), 98–98.

Torrens, F.; Castellano, G. Mucoadhesive Polymer Hyaluronan as Biodegradable Cationic/ Zwitterionic-Drug Delivery Vehicle. *ADMET DMPK* **2014c**, *2*, 235–247.

Torrens, F.; Castellano, G. Computational Study of Nanosized Drug Delivery from *Cyclo*dextrins, Crown Ethers and Hyaluronan in Pharmaceutical Formulations. *Curr. Top. Med. Chem.* **2015**, *15*, 1901–1913.

Torrens, F.; Castellano, G. Principal Component, Cluster and Meta-analyses of Soya Bean, Spanish Legumes and Commercial Soya Bean. In *High-Performance Materials and Engineered Chemistry*; Haghi, A. K., Pogliani, L., Ribeiro, A. C. F., Torrens, F., Balköse D., Thomas, S., Eds.; Apple Academic–CRC: Waretown, NJ, 2018; pp 267–294.

Torrens, F.; Redondo, L.; Castellano, G. Artemisinin: Tentative Mechanism of Action and Resistance. *Pharmaceuticals* **2017**, *10*, 20-4–20-4.

Torrens, F.; Redondo, L.; Castellano, G. Reflections on Artemisinin, Proposed Molecular Mechanism of Bioactivity and Resistance. In *Applied Physical Chemistry with Multidisciplinary Approaches*; Haghi, A. K., Balköse, D., Thomas, S., Eds.; Apple Academic–CRC: Waretown, NJ, 2018; pp 189–215.

Torrens-Zaragozá, F. Polymer Bisphenol-A, the Incorporation of Silica Nanospheres into Epoxy–Amine Materials and Polymer Nanocomposites. *Nereis* **2011**, *2011* (3), 17–23.

Torrens-Zaragozá, F. Molecular Categorization of Yams by Principal Component and Cluster Analyses. *Nereis* **2013**, *2013* (5), 41–51.

Torrens-Zaragozá, F. Classification of Lactic Acid Bacteria against Cytokine Immune Modulation. *Nereis* **2014**, *2014* (6), 27–37.

Torrens Zaragozá, F. Classification of Fruits Proximate and Mineral Content: Principal Component, Cluster, Meta-Analyses. *Nereis* **2015**, *2015* (7), 39–50.

Torrens-Zaragozá, F. Classification of Food Spices by Proximate Content: Principal Component, Cluster, Meta-Analyses. *Nereis* **2016**, *2016* (8), 23–33.

Trombetta, D.; Smeriglio, A.; Marcoccia, D.; Giofrè, S. V.; Toscano, G.; Mazzotti, F.; Giovanazzi, A.; Lorenzetti, S. Analytical Evaluation and Antioxidant Properties of Some Secondary Metabolites in Northern Italian Mono- and Multi-Varietal Extra Virgin Olive Oils (EVOOs) from Early and Late Harvested Olives. *Int. J. Mol. Sci.* **2017**, *18*, 797-1–797-14.

Tryon, R. C. J. A Multivariate Analysis of the Risk of Coronary Heart Disease in Framingham. *Chronic Dis.* **1939**, *20*, 511–524.

Varinska, L.; Gal, P.; Mojzisova, G.; Mirossay, L.; Mojzis, J. Soy and Breast Cancer: Focus on Angiogenesis. *Int. J. Mol. Sci.* **2015**, *16*, 11728–11749.

Wang, L.; Wesemann, S.; Krenn, L.; Ladurner, A.; Heiss, E. H.; Dirsch, V. M.; Atanasov, A. G. Erythrodiol, an Olive Oil Constituent, Increases the Half-Life of ABCA1 and Enhances Cholesterol Efflux from THP-1-Derived Macrophages. *Front. Pharmacol.* **2017**, *8*, 375-1–375-8.

Xu, J.; Hagler, A. Chemoinformatics and Drug Discovery. *Molecules* **2002**, *7*, 566–600.

Yang, C. Y.; Hsu, C. H.; Tsai, M. L. Effect of Crosslinked Condition on Characteristics of Chitosan/Tripolyphosphate/Genipin Beads and Their Application in the Selective Adsorption of Phytic Acid from Soybean Whey. *Carbohydr. Polym.* **2011**, *86*, 659–665.

CHAPTER 14

MOLECULAR CLASSIFICATION OF ALGAE BY PHENOLICS AND ANTIOXIDANTS

FRANCISCO TORRENS[1*] and GLORIA CASTELLANO[2]

[1]*Institut Universitari de Ciència Molecular, Universitat de València, Edifici d'Instituts de Paterna, PO Box 22085, E-46071 Valưncia, Spain*

[2]*Departamento de Ciencias Experimentales y Matemáticas, Facultad de Veterinaria y Ciencias Experimentales, Universidad Catylica de Valencia San Vicente Mártir, Guillem de Castro-94, E-46001 València, Spain*

Corresponding author. E-mail: torrens@uv.es

ABSTRACT

One of the challenges which the society faces in 21st century is that of being able to feed a rising world population, and algae are a possibility to relieve the problem. Their fast growth and easiness of adapting to the environment can allow the production, at great scale, of some important compounds from the nutritional viewpoint, and substances with bioactivity that help with the prevention of certain diseases. It is described that some unique characteristics that algae present differentiate them from other living organisms and could be the basis of a revolution in feeding in future. A useful classification of algae with regard to food is provided. Phenolic constituents of nine edible algal products are clustered by principal component analyses (PCAs) of content and algae cluster analyses (CAs), which agree, Samples group into four classes. Compositional PCA and algae CA allow classifying them and concur. The first PCA axis explains 31%, the first two, 56%, the first three, 74% variance, etc. Different algae behavior depends on *4-hydroxybenzoic acid*, etc. *Undaria pinnatifida* wakame (W)-instant is closer to *Chlorella pyrenoidosa* than to *U. pinnatifida* W. *Porphyra tenera* is closer to *Laminaria japonica* than to *Palmaria palmata*. Algal products are functional foods rich in polyphenols, antioxidants, and bioactives. A hypothesis stated that antagonisms and synergisms exist between phenolics expecting more antagonisms than synergies.

14.1 INTRODUCTION

Algae contain high-quality proteins with essential amino acids, dietary fiber, essential fatty acids (FAs), minerals, vitamins (Vits), and phenolics (Ambrozova et al., 2014). Oxidative stress is associated with diseases [e.g., cardiovascular ones, cancer, atherosclerosis, hypertension, ischemia, diabetes, hyperoxaluria, neurodegenerative ones (Alzheimer's, Parkinson's), rheumatoid arthritis, ageing], but it is not a primary cause (Rop et al., 2012). Polyphenols are split into classes [e.g., phenolic acids (hydroxybenzoic and hydroxycinnamic acids), flavonoids (flavones, flavonols, flavanones, flavanonols, flavanols, and anthocyanins), isoflavonoids (isoflavones, coumestans), stilbenes, lignans, and phenolic polymers (proanthocyanidins: condensed and hydrolyzable tannins)] (Manach et al., 2004). Their roles in plants are protection versus ultraviolet (UV) radiation/pathogens, pigmentation, reproduction, and growth (Zern and Fernandez, 2005). They are integral structural components of algal cell walls being studied because of ecological (e.g., protection from UV radiation, reproductive role, protective mechanism vs. biotic factors) and therapeutic properties [anticancer, antioxidative, antibacterial, anti-allergic, antidiabetes, anti-ageing, anti-inflammatory, antihuman immunodeficiency virus activities] (Li et al., 2011; Thomas and Kim, 2011). Photochemiluminescent quantification of Vit C in plasma is based on the antioxidant capacity of water solubles (ACW) to scavenge O_2 (Popov and Lewin, 1999; Craft et al., 2012). Studies confirmed phenolic features and ACW in algae (Yuan and Walsh, 2006; Jiménez-Escrig et al., 2001). Machu et al. (2015) reported phenolic content and ACW in algal products (cf. Table 14.1). Highest *total phenolic content* (TPC) and ACW were in *Eisenia bicyclis*. Highest *catechin* (C) and sum of selected phenolics were in *Porphyra tenera*.

Earlier publications in *Nereis* classified yams (Torrens-Zaragozá, 2013), lactic acid bacteria (Torrens-Zaragozá, 2014), fruits (Torrens-Zaragozá, 2015), food spices (Torrens-Zaragozá, 2016), and legumes (Torrens and Castellano, 2014b, 2018, in press) by principal component (PCA), cluster (CA), and meta-analyses. The molecular classifications of 33 phenolic compounds derived from the cinnamic and benzoic acids from *Posidonia oceanica* (Castellano et al., 2012), 74 flavonoids (Castellano et al., 2013), 66 stilbenoids (Castellano et al., 2014), 71 triterpenoids and steroids from *Ganoderma* (Castellano and Torrens, 2015a) and 17 isoflavonoids from *Dalbergia parviflora* (Castellano and Torrens, 2015b) and artemisinin derivatives (Torrens et al., 2017, 2018) were reported. A tool for interrogation of macromolecular structure was published (Torrens and Castellano, 2014a). Mucoadhesive polymer hyaluronan favors transdermal penetration absorption of caffeine (Torrens and Castellano, 2014c, 2015). Endocrine disruptor diethylstilbestrol (DES), bisphenol-A (similar to DES), polycarbonate, and epoxy-silica nanocomposites were reported (Torrens-Zaragozá, 2011; Torrens and Castellano, 2013).

TABLE 14.1 Total Phenolic Content, Phenolic Compounds, and Antioxidant Capacity of the Water-soluble Compounds in Edible Algal Products.

Algae	TPC	GA	HBA[a]	C	EC	Cg	ECg[b]	EGC	EGCg	PC	Sum[c]	ACW
Brown												
1. *Eisenia bicyclis*	192.6	2.8	0.0	0.0	3.2	2.9	0.0	0.0	0.0	0.0	8.9	7.53
2. *Hizikia fusiformis*	34.5	14.1	0.0	0.0	8.2	0.0	0.0	0.0	0.0	0.0	22.3	1.70
3. *Laminaria japonica*	8.7	0.0	0.0	0.0	3.1	0.0	0.0	4.0	0.0	0.0	7.1	0.21
4. *Undaria pinnatifida* Wakame (W)	8.6	0.0	1.9	0.0	0.0	0.0	0.0	4.8	0.0	0.0	6.7	0.31
5. *Undaria pinnatifida* W-instant (Wi)	8.0	0.0	8.1	0.0	6.3	2.0	0.0	21.4	7.5	0.0	45.3	0.22
Red												
6. *Palmaria palmata*	31.8	0.0	5.8	0.0	0.0	0.0	0.0	0.0	0.0	0.0	5.8	0.56
7. *Porphyra tenera*	18.2	3.5	1.6	128.8	16.4	0.0	0.0	16.0	4.0	0.0	170.3	0.13
Green												
8. *Chlorella pyrenoidosa*	18.0	5.0	20.5	0.0	0.0	0.0	0.0	20.2	0.0	0.0	45.7	0.21
Cyanobacterium												
9. *Spirulina platensis*	43.2	0.0	0.0	22.7	27.5	0.0	0.0	0.0	0.0	28.9	79.1	0.76

[a]Total phenolic content (TPC) [(‰ gallic acid (GA) equivalent (GAE)]: i_1: TPC; phenolic compounds (μg/g): hydroxybenzoic acids [i_2: GA; i_3: 4-hydroxybenzoic acid (HBA)]; flavanols [i_4: catechin hydrate (C); i_5: epicatechin (EC); i_6: catechin gallate (Cg); i_7: epigallocatechin (EGC); i_8: epigallocatechin gallate (EGCg); i_9: pyrocatechol (PC)]; i_{10}: antioxidant capacity of the water solubles (ACW, μmol vitamin C/g).

[b]ECg, epicatechin gallate.

[c]Sum, sum of selected phenolic compounds.

A useful classification of algae with regard to food is provided. The main aim this chapter is to develop code-learning potentialities and, since algae phenolics are more naturally described via varying size-structured representation, find general approaches to information processing. In view of algae medicinal benefits, the objective was to cluster them with PCA/CA, which differentiated contents.

14.2 CLASSIFICATION OF ALGAE WITH REGARD TO FOOD

Taking into account the most relevant algae with regard to food, a distinction can be made between microalgae and macroalgae (cf. Table 14.2) (Ibáñez and Herrero, 2017).

TABLE 14.2 Proposed Classification for Algae with Regard to Food.

Group	Class	Characteristics
Microalgae	Diatomophyceae	Unicellular diatoms with one or two flagella
		Resistant cell wall made of silica (SiO_2)
	Dinoflagellata	Flagellates
		Part of marine plankton
	Cyanobacteria	Blue-green algae
		Unicellular or colonial
Macroalgae	Rhodophyta	Red algae
		Pluricellular in specialized tissues
		Almost all are marine
	Phaeophyceae	Brown algae
		Almost all are marine
		Some of the greatest-sized algae
	Chlorophyceae	Green algae
		Unicellular or pluricellular
		The majority in fresh water

Source: Ibáñez and Herrero (2017).

In the microalgae group, there are diatoms, dinoflagellates, and cyanobacteria. Many of the organisms are considered as part of marine plankton and are mainly isolated unicellular organisms. Diatoms and dinoflagellates can present flagella, which are structures that allow that they move in the aquatic medium. Some species show loud characteristics, for example, some dinoflagellates are phosphorescent and are responsible for some of the night images of tropical seas so spectacular that people see in the news. As regards to cyanobacteria, generically known as blue-green algae, they are mainly unicellular organisms, although they can form colonies or associations of independent cells in a number of forms, for example, filaments. As their own name indicates, they show characteristics similar to bacteria, although, unlike these, cyanobacteria can make their own food via solar radiation thanks to the chlorophyll present in them, while bacteria are heterotrophic organisms, which imply that they should feed on other organic matters. Cyanobacteria are found in all imaginable ecosystems, and are, in fact, one of the groups of organisms most widely distributed on the Earth. Some of them were used as food by ancient civilizations.

As for macroalgae, three main groups exist, popularly known as red (*Rhodophyta*), brown (*Phaeophyta*), and green (*Chlorophyta*) algae. These three groups were found in the 20th century versus the color of the stipe, which is one of the parts in which the structure of macroscopic algae divides: holdfast, stipe, and blades. The holdfast is the part of the alga that is in contact with the substratum where it is fixed, like a root, although, unlike this, it has no function of absorption and transport of nutrients, but it is limited to keep the alga anchored to a certain surface. In some species, the holdfast is substituted with a basal disc of fixation. In the form of either holdfast or fixation disc, behind this structure, the stipe grows, which can present a variable length and would be alike the blade of higher plants. Finally, the blades are the most visible parts of the alga and can have varied morphology, from filamentous to tubular, thin, ribbon, laminar, or even calcified in some cases. Together with the stipe, they conform the frond (similar to what the leaves of higher plants would be). In the group of green algae, not only pluricellular but also unicellular algae can be found. It is a wide group that contains more than 7000 different species, most of them of fresh water, although species typical of marine environments exist. Green algae present chlorophyll-*a* and -*b*, which are the main pigments to carry out photosynthesis in the chloroplasts, although they contain some accessory pigments called xanthophylls (a type of carotenoids), which contributes them their green color. On their part, red algae are usually multicellular organisms, although, as the rest of algae, they do not form specialized tissues like higher plants. They are typical of mild environments, for example, tropical climates, and are found mainly in marine ecosystems. It was calculated that 4000 species belonging to this group exist. Generally, as well as chlorophyll, they contain in their chloroplasts phycoerythrin, which is a pigment associated with a certain type of proteins, phycoerythrobilins, which confer them their characteristic red color. Red algae are the most important industrial sources of carrageenans, that is, linear sulfated polysaccharides that are used as ingredients of foods. The third group of macroalgae is formed by brown algae. In this case, it is considered that 1500 different species exist; between them, some of the most widespread and known, and those that present a greater size. Brown algae have an accessory pigment, which is found in chloroplasts together with chlorophyll, called fucoxanthin, to thank for their color. The compound is an important carotenoid because of its shown anticancer effect, etc. Generally, independently of the group that be considered, a great morphological variety exists, with varied types of growth, grades of specialization, types of reproduction, and chemical composition, which presents a great importance from the viewpoint of food.

14.3   COMPUTATIONAL METHOD

PCA is a dimension-reduction technique (Hotelling, 1933; Kramer, 1998 Patra et al., 1999; Jolliffe, 2002; Xu and Hagler, 2002; Shaw, 2003). From starting variables X_j, PCA builds orthogonal variables $\tilde{P}_j$, linear combinations of mean-centered ones $\tilde{X}_j = X_j - \overline{X}_j$, corresponding to eigenvectors of sample covariance matrix $S = 1/(n-1)\sum_{i=1}^{n}(x_i - \overline{x})(x_i - \overline{x})'$. For every loading vector $\tilde{l}_j$, matching eigenvalue $\tilde{l}_j$ of S tells how much data variability is explained: $\tilde{l}_j = \mathrm{Var}(\tilde{P}_j)$. Loading vectors are sorted in decaying eigenvalues. First k PCs explain most variability. After selecting k, one projects p-dimensional data on to subspace spanned by k loading vectors and computes coordinates versus $\tilde{P}_j$, yielding scores:

$$\tilde{t}_i = \tilde{P}'(x_i - \overline{x})$$

(14.1)

for every $i = 1, \ldots, n$ having trivially zero mean. With regard to starting coordinate system, projected data point is computed fitting:

$$\hat{x}_i = \overline{x} + \tilde{P}\tilde{t}_i$$

(14.2)

Loading matrix $\tilde{P}$ ($p \times k$) contains loadings column-wise and diagonal one $\tilde{L} = (\tilde{l}_j)_j$ ($k \times k$), eigenvalues. Loadings k explain variation:

$$\frac{\left(\sum_{j=1}^{k} \tilde{l}_j\right)}{\left(\sum_{j=1}^{p} \tilde{l}_j\right)^3} \geq 80 \tag{14.3}$$

CA starting point is $n \times p$ data matrix $\mathbf{X}$ containing p components measured in n samples (IMSL, 1989; Tryon, 1939). One assumes that data were preprocessed to remove artifacts, and missing values, imputed. The CA organizes samples into a small number of clusters such that examples within a bunch are similar. Distances l_q between samples $x, x' \in \mathfrak{R}^p$ result:

$$\|x - x'\|_q = \left(\sum_{i=1}^{p} |x_i - x_i'|^q\right)^{1/q} \tag{14.4}$$

(e.g., Euclidean l_2, Manhattan l_1 distances). *Pearson's correlation coefficient* (PCC) relates samples:

$$r(x - x') = \frac{\sum_{i=1}^{p}(x_i - \bar{x})(x_i' - \bar{x}')}{\left[\sum_{i=1}^{p}(x_i - \bar{x})^2 \sum_{i=1}^{p}(x_i' - \bar{x}')^2\right]^{1/2}} \tag{14.5}$$

where $\bar{x} = \left(\sum_{i=1}^{p} x_i\right)/p$ is measure mean value for sample x (Priness et al., 2007; Steuer et al., 2002; D'Haeseleer et al., 2000; Perou et al., 2000; Jarvis and Patrick, 1973; Page, 2000; Eisen et al., 1998).

14.4 CALCULATION RESULTS

Nine commercial algal food products from brown (entries 1–5) and red seaweeds (entries 6–7), freshwater green alga (entry 8), and cyanobacterium (entry 9 in Table 14.1) from Machu et al. were evaluated for phenolics content and ACW; matrix $\mathbf{R}$ was computed between algae; upper triangle is

$$\mathbf{R} = \begin{pmatrix} 1.000 & 0.909 & 0.858 & 0.842 & 0.124 & 0.981 & -0.018 & 0.427 & 0.650 \\ & 1.000 & 0.799 & 0.705 & 0.022 & 0.873 & -0.074 & 0.343 & 0.602 \\ & & 1.000 & 0.909 & 0.511 & 0.826 & -0.028 & 0.560 & 0.612 \\ & & & 1.000 & 0.591 & 0.859 & -0.056 & 0.788 & 0.415 \\ & & & & 1.000 & 0.162 & -0.159 & 0.721 & -0.141 \\ & & & & & 1.000 & -0.037 & 0.538 & 0.608 \\ & & & & & & 1.000 & -0.154 & 0.314 \\ & & & & & & & 1.000 & -0.006 \\ & & & & & & & & 1.000 \end{pmatrix}$$

Correlations between pairs of brown algae $R \sim 0.627$ are higher than between red ones $|R_{6,7}| = 0.037$. Correlations are high between most brown algae and red alga *Palmaria palmata* $R_{1-4,6} \approx 0.885$. Some correlations between groups are high, for example, $R_{1,6} = 0.981$ and $R_{5,8} = 0.721$. They are illustrated in the partial correlation diagram (PCD) that contains high ($r \geq 0.75$), medium ($0.50 \leq r < 0.75$), low ($0.25 \leq r < 0.50$), and *zero* ($r < 0.25$) partial correlations. It includes 10 high (cf. Fig. 14.1, grayscale), 10 medium, four low, and 12 *zero* partial correlations. Pairs of algae with high partial correlation show similar contents.

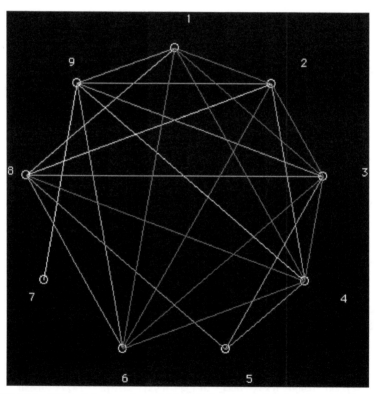

FIGURE 14.1   PCD of algal products: high (1-2, 1-3, 1-4, 1-6, 2-3, 2-6, 3-4, 3-6, 4-6, 4-8), medium (1-9, 2-4, 2-9, 3-5, 3-8, 3-9, 4-5, 5-8, 6-8, 6-9) and low (1-8, 2-8, 4-9, 7-9) partial correlations.

The dendrogram of algae according to phenolics content and phenolics content + ACW (cf. Fig. 14.2) shows different behavior depending on *4-hydroxybenzoic acid* (HBA), *catechin hydrate* (C), *epigallocatechin* (EGC), and *epigallocatechin gallate* (EGCg). Four classes are recognized:
    (1,2,6)(3,4)(5,7,8)(9)

Class 3 and, then, grouping 4 are separated from branchings 1 and 2: *E. bicyclis*, *Hizikia fusiformis*, and *P. palmata* present high TPC, *gallic acid* (GA), *catechin gallate* (Cg), and ACW and are grouped into class 1; *Laminaria japonica* and *Undaria pinnatifida* wakame (W) with low TPC are grouped into cluster 2; *U. pinnatifida* W-instant (Wi), *P. tenera*, and *Chlorella pyrenoidosa* with high HBA, C, EGC, EGCg, and sum of selected phenolics are grouped into class 3; *Spirulina platensis* with high *epicatechin* (EC) and *pyrocatechol* (PC) are grouped into cluster 4. Algae are classed according to TPC, GA, HBA, C, EC, Cg, EGC, EGCg, and PC; brown algae split into 3 classes and red algae, into 2 clusters; brown *U. pinnatifida* Wi is closer to green *C. pyrenoidosa* than to *U. pinnatifida* W; red *P. tenera* is closer to brown *L. japonica* than to *P. palmata*. Algae in the same class appear highly correlated in PCD (Fig. 14.1).

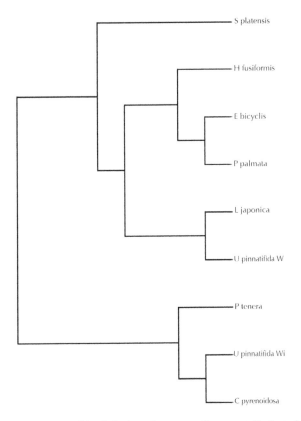

FIGURE 14.2 Dendrogram of food algal products according to total/selected phenolics and water solubles.

The radial tree (cf. Fig. 14.3) shows different behavior of algae depending on HBA, etc. The same classes above are clearly recognized in agreement with PCD and dendrogram (Figs. 14.1 and 14.2). Again, algae with high TPC, etc. are grouped into class 1, etc.; *U. pinnatifida* Wi is closer to *C. pyrenoidosa* than to *U. pinnatifida* W; *P. tenera* is closer to *L. japonica* than to *P. palmata*.

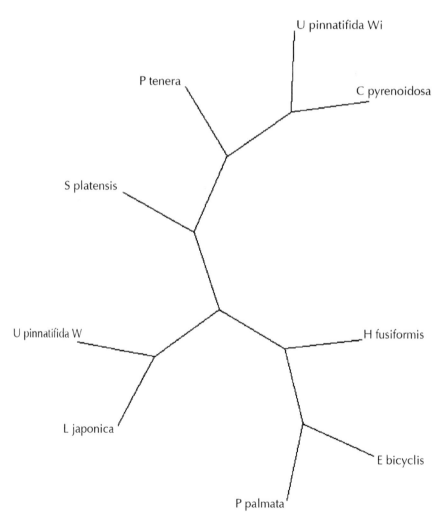

FIGURE 14.3   Radial tree of food algal products according to total/selected phenolics and water solubles.

The splits graph for nine algae in Table 14.1 (cf. Fig. 14.4) shows that *E. bicyclis* and *P. palmata* collapse. It reveals conflicting relations between classes 1, 2, and 3 because of interdependences (Huson, 1998). It indicates spurious relations between groupings resulting from base-composition effects. It illustrates different behavior of algae depending on HBA, etc. One more time, algae with high TPC, etc. are grouped into class 1, etc.; *U. pinnatifida* Wi is closer to *C. pyrenoidosa* than to *U. pinnatifida* W; *P. tenera* is closer to *L. japonica* than to *P. palmata*. Split graph is in qualitative agreement with PCD and binary/radial trees (Figs. 14.1–14.3).

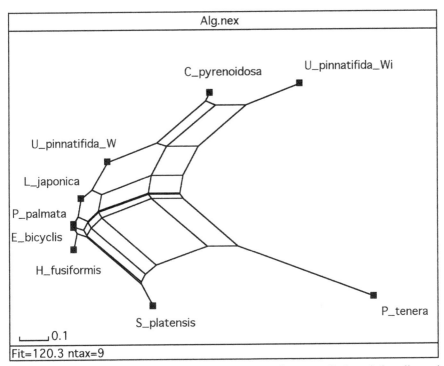

FIGURE 14.4 Splits graph of food algal products according to total/selected phenolics and water solubles.

Usually in quantitative structure–property relations (QSPRs), data contain <100 objects and >1000 *X*-variables. Nobody discovers by *inspection* patterns. PCA allows *summarizing* information contained in **X**-matrix. It decomposes **X**-matrix as product of matrices **P** and **T**. *Loading matrix* **P** with information about variables contains a few vectors: PCs that are obtained as linear combinations of original *X*-variables. In *score matrix*

T with information about objects, every object is described in terms of projections on to PCs instead of the original variables: $\mathbf{X} = \mathbf{TP'} + \mathbf{E}$, where $'$ denotes transpose matrix. Information not contained in matrices remains *unexplained* X-*variance* in *residual matrix* **E**. Every PC_i is a new coordinate expressed as linear combination of the old x_j: $PC_i = \Sigma_j b_{ij} x_j$. New coordinates PC_i are *scores (factors)*, while coefficients b_{ij} are *loadings*. Scores are ordered according to information content versus total variance among objects. *Score–score plots* show positions of compounds in new coordinate system, while *loading–loading plots* show location of features that represent compounds in new coordination. The PCs present properties: (1) they are extracted by decaying importance, and (2) every PC is orthogonal to each other. A PCA was performed for algae. Importance of PCA factors F_{1-9} for phenolics (cf. Table 14.3) shows that first factor F_1 explains 31% variance (69% error), first two factors $F_{1/2}$, 56% variance (44% error), first three factors F_{1-3}, 74% variance (26% error), etc. For F_1 variable, i_{10} shows greater weight; however, F_1 cannot be reduced to two variables $\{i_1, i_{10}\}$ without 46% error. For F_2, variable i_5 presents greater weight; notwithstanding, F_2 cannot be reduced to two variables $\{i_5, i_9\}$ without 45% error. For F_3 variable, i_8 assigns greater weight; nevertheless, F_3 cannot be reduced to two variables $\{i_6, i_8\}$ without 50% error, etc.

TABLE 14.3 Importance of PCA Factors for TPC, Selected Phenolics, and ACW in Algal Products.

Factor	Eigenvalue	Percentage of variance	Cumulative percentage of variance
F_1	3.13670432	31.37	31.37
F_2	2.43106794	24.31	55.68
F_3	1.85591469	18.56	74.24
F_4	1.15980235	11.59	85.83
F_5	0.82396808	8.24	94.07
F_6	0.54381688	5.44	99.51
F_7	0.04601029	0.46	99.97
F_8	0.00271544	0.03	100.00
F_9	0.00000000	0.00	100.00
F_{10}	0.00000000	0.00	100.00

Scores plot of PCA F_2–F_1 for algae (cf. Fig. 14.5) shows that *U. pinnatifida* Wi and *C. pyrenoidosa* collapse. It illustrates different behavior depending on HBA, etc. The four clusters above are distinguished: class 1 with three algae ($F_1 > F_2 \approx 0$, *right*), grouping 2 with two algae ($F_1 < F_2 \approx 0$, *middle*), cluster 3 with three algae ($F_1 << F_2$, *left*), and class 4 with one alga ($F_1 >> F_2$, *bottom*). Again, algae with high TPC, etc. are grouped into class 1, etc.; *U. pinnatifida* Wi is closer to *C. pyrenoidosa* than to *U. pinnatifida* W; *P. tenera* is closer to *L. japonica* than to *P. palmata*. The diagram is in qualitative agreement with PCD, binary/radial trees, and splits graph (Figs. 14.1–14.4).

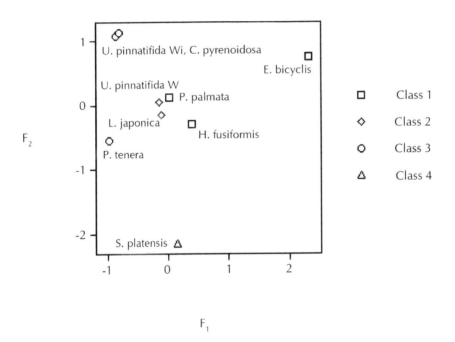

FIGURE 14.5 PCA scores plot of algal products according to total/selected phenolics and water solubles.

From PCA factors loading of algae, F_2–F_1 *loadings plot* (cf. Fig. 14.6) depicts 10 contents. Four clusters are clearly distinguished: class 1 with three components {1,6,10} ($F_1 > F_2 > 0$, *right*), grouping 2 with two constituents {2,3} ($0 \approx F_1 < F_2$, *middle*), cluster 3 with three contents {4,7,8} ($F_1 < F_2 \approx 0$, *left*), and class 4 with two phenolics {5,9} ($0 \approx F_1 > F_2$, *bottom*). HBAs group together into class 2 while flavanols split into three clusters. The Cg is closer to ACW than to EGCg. In addition, as a complement to score diagram for loadings, it is confirmed that algae in class 1, located in the right side, present a more pronounced contribution from phenolics in grouping 1 situated in the same side of Figure 14.5. Algae in cluster 3 in the left side show a contribution from compositions in class 3 in the same side. Algae in grouping 4 at the bottom indicate a contribution from content in cluster 4 in the same position. The diagram is in qualitative agreement with PCD, binary/radial trees, and splits graph (Figs. 14.1–14.5).

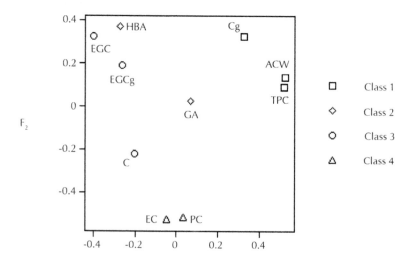

FIGURE 14.6 PCA loadings plot of algal products according to total/selected phenolics and water solubles.

Instead of nine algae in space $\mathfrak{R}^{10}$ of 10 contents, consider 10 components in space $\mathfrak{R}^{9}$ of nine algae. Matrix **R** upper triangle results:

$$
\mathbf{R} = \begin{pmatrix}
1.000 & 0.071 & -0.274 & -0.140 & -0.035 & 0.730 & -0.416 & -0.264 & 0.018 & 0.986 \\
 & 1.000 & 0.005 & 0.015 & -0.002 & -0.139 & -0.089 & -0.185 & -0.228 & 0.155 \\
 & & 1.000 & -0.187 & -0.389 & -0.074 & 0.691 & 0.129 & -0.233 & -0.300 \\
 & & & 1.000 & 0.521 & -0.219 & 0.302 & 0.351 & 0.052 & -0.198 \\
 & & & & 1.000 & -0.163 & -0.100 & 0.152 & 0.822 & -0.116 \\
 & & & & & 1.000 & 0.082 & 0.369 & -0.185 & 0.756 \\
 & & & & & & 1.000 & 0.711 & -0.302 & -0.414 \\
 & & & & & & & 1.000 & -0.179 & -0.247 \\
 & & & & & & & & 1.000 & -0.084 \\
 & & & & & & & & & 1.000
\end{pmatrix}
$$

Correlations between pairs of HBAs $R_{2,3} = 0.005$ are smaller than between flavanols $R \sim 0.148$. Some correlations between flavanols are high, for example, C–EC $R_{4,5} = 0.521$, EC–PC $R_{5,9} = 0.822$ and EGC–EGCg $R_{7,8} = 0.711$. Some correlations between groups are high, for example, TPC–Cg $R_{1,6} = 0.730$, TPC–ACW $R_{1,10} = 0.986$, and HBA–EGC $R_{3,7} = 0.691$. Most correlations are negative, for example, TPC–EGC $R_{1,7} = -0.416$. The TPC correlates positively with ACW indicating a close relationship between TPC and ACW. The Cg correlates positively with TPC and ACW. The trend lines of selected phenolics and their sum versus TPC and ACW show either negative or positive slopes. The corresponding interpretation allows the hypothesis that antagonisms and synergisms exist between selected phenolics. More antagonisms than synergies are expected but some synergisms can be high, for example, those involving CG. Correlations are illustrated in PCD that contains three high (cf. Fig. 14.7, grayscale), four medium, three low, and 35 zero partial correlations. Contents with high partial correlation are present in similar algae.

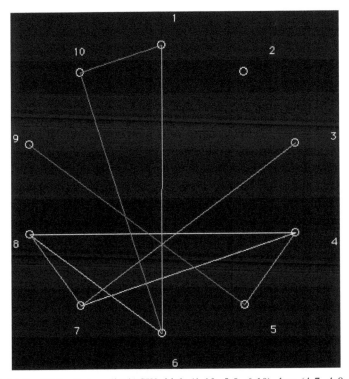

FIGURE 14.7 PCD of phenolics/ACW: high (1-10, 5-9, 6-10), low (4-7, 4-8, 6-8) and medium (the remaining) partial correlations.

The dendrogram for 10 contents of algae (cf. Fig. 14.8) separates the same four classes above depending on *E. bicyclis*:

(1,6,10)(2,3)(4,7,8)(5,9)

The diagram agrees with PCA loadings plot and PCD (Figs. 14.6 and 14.7). Again, HBAs group together while flavanols split into three classes; Cg is closer to ACW than to EGCg.

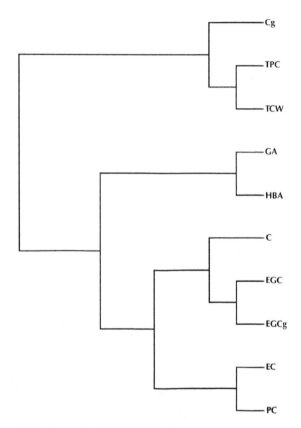

FIGURE 14.8 Dendrogram of total/selected phenolic and water-soluble compounds for food algal products.

The radial tree for 10 contents of algae (cf. Fig. 14.9) separates the same four classes above in agreement with PCA loadings plot, PCD, and dendrogram (Figs. 14.6–14.8). Again, HBAs group together, while flavanols split into three classes; Cg is closer to ACW than to EGCg.

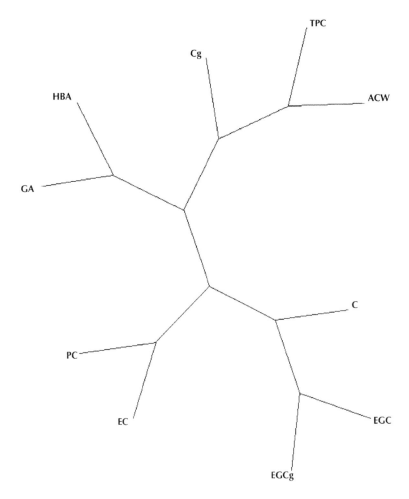

FIGURE 14.9   Radial tree of total/selected phenolic and water-soluble compounds for food algal products.

The splits graph for 10 contents of algae (cf. Fig. 14.10) reveals conflicting relations between all classes. It separates the same four classes above in agreement with PCA loadings plot, PCD and binary/radial trees (Figs. 14.6–14.9). One more time, HBAs group together while flavanols split into three classes; Cg is closer to ACW than to EGCg.

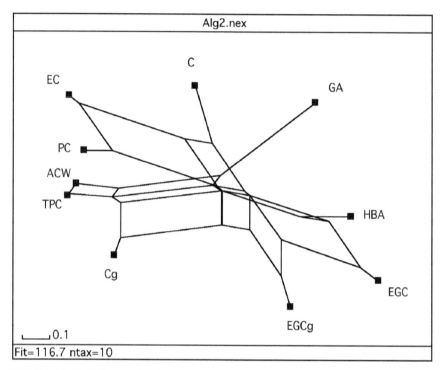

FIGURE 14.10 Splits graph of total/selected phenolics and water solubles for food algal products.

A PCA was performed for contents. Factor F_1 explains 59% variance (41% error), $F_{1/2}$, 80% variance (20% error), F_{1-3}, 91% variance (9% error), etc. Scores plot of PCA F_2–F_1 for phenolics (cf. Fig. 14.11) shows the four clusters above: class 1 with two constituents ($0 \approx F_1 > F_2$, *right*), grouping 2 with two components ($F_1 << F_2$, *middle*), cluster 3 with three contents ($F_1 < F_2 \approx 0$, *left*), and class 4 with two phenolics ($0 > F_1 > F_2$, *bottom*). Once more, HBAs group together while flavanols split into three classes; Cg is closer to ACW than to EGCg. The diagram is in qualitative agreement with PCA loadings plot, PCD, binary/radial trees, and splits graph (Figs. 14.6–14.10).

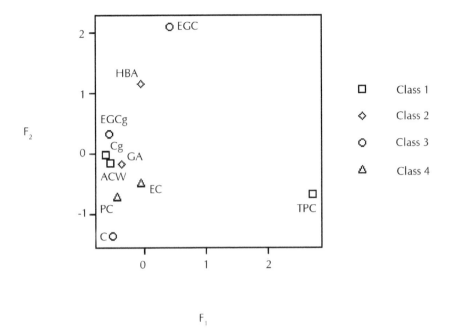

FIGURE 14.11   PCA scores plot of total/selected phenolics and water solubles for food algal products.

14.5   DISCUSSION

One of the challenges which the society faces in 21st century is that of being able to feed a rising world population, and algae (an abundant and least exploited marine resources) are a possibility to relieve the problem. Their fast growth and easiness of adapting to the environment can allow the production, at great scale, of some important compounds from the nutritional viewpoint, and substances with bioactivity that help with the prevention of certain diseases. It is described that some unique characteristics that algae present differentiate them from other living organisms and could be the basis of a revolution in feeding in future. It is interesting: (1) the development of processes respectful to the environment for obtaining compounds of high value added from natural sources; (2) the study and characterization of new functional ingredients and its relationship with health, for example, the use of clean extraction technologies and processes, and advanced multidimensional analytical techniques. A good classification should be practical. For edible algae, a useful classification with regard to food is provided.

Uses of algae are as follows: (1) Agar, a gelatin derived from red algae, is a good medium to grow bacteria and fungi. (2) Alginate from brown algae is used in food-gelling agents and medical dressings. (3) As energy source, algae-based biofuels produce the greatest biomass per unit area. (4) Algae are used as fertilizers, soil conditioners, and livestock feed. Algaculture is an aquaculture. (5) In nutrition, algae provide Vits [e.g., A, thiamine (B_1), B_2, B_6, niacin (B_3), C] being rich in I, K, Fe, Mg, and Ca. Algae oils contain high levels of unsaturated FAs being the source of essential ω-3 FAs in fish oil. (6) In pollution control, sewage is treated with algae-reducing toxic chemicals. Algae are used to capture fertilizers in runoff from farms. Aquariums are filtered via algae in an algae turf scrubber. (7) In bioremediation, algae colonize silicone resins. (8) Algal natural pigments (carotenoids, chlorophylls) are used as an alternative to chemical dyes. (9) Some algae are used as a stabilizer in milk products. Algae and the waste of some foods (e.g., peels, seeds, and stems) contain high level of antioxidants, which provides an opportunity for industry to obtain ingredients for functional food or medicine. The method showed present limitations. (1) No standard method exists in measuring antioxidant capacity and results measured with different assays are incomparable. (2) It is not addressed to what extent natural antioxidants are absorbed by target organs/tissues and their antioxidant capacity retains in vivo. (3) It is unknown how much natural antioxidants contribute to the total beneficial effects of algae in health promotion/disease prevention and exert beneficial effects via reducing directly oxidative damage rather than other biofunctions, since most clinical trials in the treatment of various diseases produced negative results. Brown *E. bicyclis* is the alga with highest TPC and ACW. Almost all TPC-rich extracts are made from parts of plants that are not usually considered in raw form, that is, bark and wood, but cocoa, red wine, and green tea. Algal-product TPCs compare to black tea, but they do not match up to fresh algae because of different extraction conditions and samples characteristics distinguishing fresh algal samples from processed products. The same applies to selected phenolics, their sum, and ACW. Most abundant phenolic in algal products is EC. Red *P. tenera* is the alga with highest C and sum of selected phenolics. Algal food products contain many phenolics. They are functional foods rich in polyphenols. They contain high levels of antioxidants and other bioactive compounds.

The TPC compares with ACW. Algal products present higher ACW than fruits rich in Vit. C, for example, bilberry, strawberry, kiwi, etc. The TPC correlates positively with ACW indicating a close relationship between TPC and ACW. The Cg correlates positively with TPC and ACW. The trend lines of selected phenolics and their sum versus TPC and ACW show either negative or positive slopes. The corresponding interpretation allows the hypothesis that antagonisms and synergisms exist between selected phenolics. More antagonisms than synergies are expected between selected phenolics. However, some synergisms can be high, especially those involving CG.

Algal product arame from brown *E. bicyclis* is a functional food and phenolic source. It is harvested for food, alginate, fertilizer, and I. As food, it is high in Ca, I, Fe, Mg, and Vit A. Algal product wakame from brown *U. pinnatifida* is harvested for food, health, and aquaculture. Its compound fucoxanthin helps burn fatty tissue, and wakame is used in topical beauty treatments. It is rich in ω-3 FA eicosapentaenoic acid, Na, Ca, I, and Vits (B_1, B_3). Marine green microalga *Chlorella* sp. sustainably photoproduces H_2 under P-deprived conditions. In presence of CO_2, the system accumulates starch during the initial, photosynthetic stage of P-deprivation and produces more H_2. The P-separation/recovery from wastewater via zeolite adsorption and its use as P-source for the cultivation of *C. vulgaris*/cyanobacterium *S. platensis* are promising for biomass production, avoiding the direct use of wastewaters as cultivation medium.

14.6 FINAL REMARKS

From the present results and discussion, the following final remarks can be drawn:

(1) One of the challenges which the society faces in 21st century is that of being able to feed a rising world population, and algae are a possibility to relieve the problem. Their fast growth and adaptation to the environment allow the production, at great scale, of some important compounds from the nutritional viewpoint, and substances with bioactivity that help with the prevention of certain diseases. It is described that some unique characteristics that algae present differentiate them from other living organisms and could be the basis of a revolution in feeding in future.

(2) A good classification should be practical. For edible algae, a useful classification with regard to food is provided.

(3) Phytoremediation is a green technology and bioremediation by algae can be a better option to remove toxicants from the polluted environment because algae present the ability to detoxify poisonous elements and to grow in degraded ecosystems.

(4) The final objective of studies of algae is the sustainable production of foods, health, and energy.

(5) The comparison between qualitative and quantitative classifications shows good correlation for most studied brown algae and bad association for examined red ones.

(6) A hypothesis stated that antagonisms and synergisms exist between phenolics. More antagonisms than synergies are expected but some synergisms can be high.

(7) Algal food products deserve detailed studies.

(8) Algal product arame is a phenolic source deserving detailed study.

(9) It is essential to create awareness in society about the reliability of medicinal properties of certain algae via scientific research with analysis of constituents, toxicology, and physiological effects. The provided information is more detailed than is available anywhere else and can be used to improve health and give patients with chronic diseases adjunctive therapies.

ACKNOWLEDGMENTS

The authors thank support from Generalitat Valenciana (Project No. PROMETEO/2016/094) and Universidad Catolica de Valencia San Vicente Martir (Project No. UCV.PRO.17-18.AIV.03).

KEYWORDS

- algae
- dietary fiber
- essential fatty acids
- minerals
- phenolics

REFERENCES

Ambrozova, J. V.; Mißurcová, L.; Vicha, R.; Machu, L.; Samek, D.; Baron, M.; Mlcek, J.; Sochor, J.; Jurikova, T. Influence of Extractive Solvents on Lipid and Fatty Acids Content of Edible Freshwater Algal and Seeweed Products, the Green Microalga *Chlorella kessleri* and the Cyanobacterium *Spirulina platensis*. *Molecules* **2014**, *19*, 2344–2360.

Castellano, G.; González-Santander, J. L.; Lara, A.; Torrens, F. Classification of Flavonoid Compounds by Using Entropy of Information Theory. *Phytochemistry* **2013**, *93*, 182–191.

Castellano, G.; Lara, A.; Torrens, F. Classification of Stilbenoid Compounds by Entropy of Artificial Intelligence. *Phytochemistry* **2014**, *97*, 62–69.

Castellano, G.; Tena, J.; Torrens, F. Classification of Polyphenolic Compounds by Chemical Structural Indicators and Its Relation to Antioxidant Properties of *Posidonia oceanica* (L.) Delile. *MATCH Commun. Math. Comput. Chem.* **2012**, *67*, 231–250.

Castellano, G.; Torrens, F. Information Entropy-Based Classification of Triterpenoids and Steroids from *Ganoderma*. *Phytochemistry* **2015a**, *116*, 305–313.

Castellano, G.; Torrens, F. Quantitative Structure–Antioxidant Activity Models of Isoflavonoids: A Theoretical Study. *Int. J. Mol. Sci.* **2015b**, *16*, 12891–12906.

Craft, B. D.; Kerrihard, A. L.; Amarowicz, R.; Pegg, R. B.; Phenol-Based Antioxidants and the In Vitro Methods Used for their Assessment. *Compr. Rev. Food Sci. Food Saf.* **2012**, *11*, 148–173.

D'Haeseleer, P.; Liang, S.; Somogyi, R. Genetic Network Inference: From Co-Expression Clustering to Reverse Engineering. *Bioinformatics* **2000**, *16*, 707–726.

Eisen, M. B.; Spellman, P. T.; Brown, P. O.; Botstein, D. Cluster Analysis and Display of Genome-Wide Expression Patterns. *Proc. Natl. Acad. Sci. U.S.A.* **1998**, *95*, 14863–14868.

Hotelling, H. Analysis of a Complex of Statistical Variables into Principal Components. *J. Educ. Psychol.* **1933**, *24*, 417–441.

Huson, D. H. SplitsTree: Analyzing and Visualizing Evolutionary Data. *Bioinformatics* **1998**, *14*, 68–73.

Ibáñez, E.; Herrero, M. *Las Algas que Comemos*; Qúe Sabemos de? No. 81; CSIC–La Catarata: Madrid, 2017.

IMSL. *Integrated Mathematical Statistical Library (IMSL)*; IMSL: Houston, 1989.

Jarvis, R. A.; Patrick, E. A. Clustering Using a Similarity Measure Based on Shared Nearest Neighbors. *IEEE Trans. Comput.* **1973**, *C22*, 1025–1034.

Jiménez-Escrig, A.; Jiménez-Jiménez, I.; Pulido, R.; Saura-Calixto, F. Antioxidant Activity of Fresh and Processed Edible Seaweeds. *J. Sci. Food Agric.* **2001**, *81*, 530–534.

Jolliffe, I. T. *Principal Component Analysis*; Springer: New York, 2002.

Kramer, R. *Chemometric Techniques for Quantitative Analysis*; Marcel Dekker: New York, 1998.

Li, Y. X.; Wijesekara, I.; Li, Y.; Kim, S. K. Phlorotannins as Bioactive Agents from Brown Algae. *Process Biochem.* **2011**, *46*, 2219–2224.

Machu, L.; Mißurcová, L.; Ambrozova, J. V.; Orsavova, J.; Mlcek, J.; Sochor, J.; Jurikova, T. Phenolic Content and Antioxidant Capacity in Algal Food Products. *Molecules* **2015**, *20*, 1118–1133.

Manach, C.; Scalbert, A.; Morand, C.; Rémésy, C.; Jiménez, L. Polyphenols: Food Sources and Bioavailability. *Am. J. Clin. Nutr.* **2004**, *79*, 727–747.

Page, R. D. M. *Program TreeView*; Universiy of Glasgow: Glasgow, UK, 2000.

Patra, S. K.; Mandal, A. K.; Pal, M. K. J. State of Aggregation of Bilirubin in Aqueous Solution: Principal Component Analysis Approach. *Photochem. Photobiol., Sect. A* **1999**, *122*, 23–31.

Perou, C. M.; Sørlie, T.; van Eisen, M. B.; de Rijn, M.; Jeffrey, S. S.; Rees, C. A.; Pollack, J. R.; Ross, D. T.; Johnsen, H.; Akslen, L. A.; Fluge, O.; Pergamenschikov, A.; Williams,

C.; Zhu, S. X. Lønning, P. E.; Børresen-Dale, A. L.; Brown, P. O.; Botstein, D. Molecular Portraits of Human Breast Tumours. *Nature (London)* **2000**, *406*, 747–752.

Popov, I.; Lewin, G. Antioxidative Homeostasis: Characterization by Means of Chemiluminescent Technique. *Methods Enzymol.* **1999**, *300*, 437–456.

Priness, I.; Maimon, O.; Ben-Gal, I. Evaluation of Gene-Expression Clustering via Mutual Information Distance Measure. *BMC Bioinf.* **2007**, *8*, 111-1–111-12.

Rop, O.; Mlcek, J.; Jurikova, T.; Neugebauerova, J.; Vabkova, J. Edible Flowers—A New Promising Source of Mineral Elements in Human Nutrition. *Molecules* **2012**, *17*, 6672–6683.

Shaw, P. J. A. *Multivariate Statistics for the Environmental Sciences*; Hodder-Arnold: New York, 2003.

Steuer, R.; Kurths, J.; Daub, C. O.; Weise, J.; Selbig, J. The Mutual Information: Detecting and Evaluating Dependencies between Variables. *Bioinformatics* **2002**, *18* (Suppl. 2), S231–S240.

Thomas, N. V.; Kim, S. K. Potential Pharmacological Applications of Polyphenolic Derivates from Marine Brown Algae. *Environ. Toxicol. Pharmacol.* **2011**, *32*, 325–335.

Torrens, F.; Castellano, G. Bisphenol, Diethylstilbestrol, Polycarbonate and the Thermomechanical Properties of Epoxy–Silica Nanostructured Composites. *J. Res. Upd. Polym. Sci.* **2013**, *2*, 183–193.

Torrens, F.; Castellano, G. A Tool for Interrogation of Macromolecular Structure. *J. Mater. Sci. Eng. B* **2014a**, *4* (2), 55–63.

Torrens, F.; Castellano, G. From Asia to Mediterranean: Soya Bean, Spanish Legumes and Commercial *Soya Bean* Principal Component, Cluster and Meta-Analyses. *J. Nutr. Food Sci.* **2014b**, *4* (5), 98–98.

Torrens, F.; Castellano, G. Mucoadhesive Polymer Hyaluronan as Biodegradable Cationic/ Zwitterionic–Drug Delivery Vehicle. *ADMET DMPK* **2014c**, *2*, 235–247.

Torrens, F.; Castellano, G. Computational Study of Nanosized Drug Delivery from *Cyclo*dextrins, Crown Ethers and Hyaluronan in Pharmaceutical Formulations. *Curr. Top. Med. Chem.* **2015**, *15*, 1901–1913.

Torrens, F.; Castellano, G. Principal Component, Cluster and Meta-Analyses of Soya Bean, Spanish Legumes and Commercial Soya Bean. In *High-Performance Materials and Engineered Chemistry*; Torrens, F., Balköse, D., Thomas, S., Eds.; Apple Academic–CRC: Waretown, NJ, 2018; pp 267–294.

Torrens, F.; Castellano, G. Meta-Analysis of Underutilized Beans with Nutrients Profile. In *Applied Food Science and Engineering with Industrial Applications*; Aguilar, C. N., Carvajal-Millan, C., Eds.; Apple Academic–CRC: Waretown, NJ, in press.

Torrens, F.; Redondo, L.; Castellano, G. Artemisinin: Tentative Mechanism of Action and Resistance. *Pharmaceuticals* **2017**, *10*, 20–24.

Torrens, F.; Redondo, L.; Castellano, G. Reflections on Artemisinin, Proposed Molecular Mechanism of Bioactivity and Resistance. In *Applied Physical Chemistry with Multidisciplinary Approaches*; Haghi, A. K., Balköse, D., Thomas, S., Eds.; Apple Academic–CRC: Waretown, NJ, 2018; pp 189–215.

Torrens-Zaragozá, F. Polymer Bisphenol-A, the Incorporation of Silica Nanospheres into Epoxy-Amine Materials and Polymer Nanocomposites. *Nereis* **2011**, *2011* (3), 17–23.

Torrens-Zaragozá, F. Molecular Categorization of Yams by Principal Component and Cluster Analyses. *Nereis* **2013**, *2013* (5), 41–51.

Torrens-Zaragozá, F. Classification of Lactic Acid Bacteria against Cytokine Immune Modulation. *Nereis* **2014,** *2014* (6), 27–37.

Torrens-Zaragozá, F. Classification of Fruits Proximate and Mineral Content: Principal Component, Cluster, Meta-Analyses. *Nereis* **2015,** *2015* (7), 39–50.

Torrens-Zaragozá, F. Classification of Food Spices by Proximate Content: Principal Component, Cluster, Meta-Analyses. *Nereis* **2016,** *2016* (8), 23–33.

Tryon, R. C. J. A Multivariate Analysis of the Risk of Coronary Heart Disease in Framingham. *Chronic Dis.* **1939,** *20*, 511–524.

Xu, J.; Hagler, A. Chemoinformatics and Drug Discovery. *Molecules* **2002,** *7*, 566–600.

Yuan, Y. V.; Walsh, N. A. Antioxidant and Antiproliferative Activities of Extracts from a Variety of Edible Seaweeds. *Food Chem. Toxicol.* **2006,** *44*, 1144–1150.

Zern, T. L.; Fernandez, M. L. Cardioprotective Effects of Dietary Polyphenols. *J. Nutr.* **2005,** *135*, 2291–2294.

CHAPTER 15

EXTRACTION OF NATURAL PRODUCTS FOUND IN VEGETAL SPECIES: CLOVE/CITRUS

FRANCISCO TORRENS[1*] and GLORIA CASTELLANO[2]

[1]Institut Universitari de Ciència Molecular, Universitat de València, Edifici d'Instituts de Paterna, PO Box 22085, E-46071 València, Spain

[2]Departamento de Ciencias Experimentales y Matemáticas, Facultad de Veterinaria y Ciencias Experimentales, Universidad Catylica de Valencia San Vicente Mórtir, Guillem de Castro-94, E-46001 València, Spain

*Corresponding author. E-mail: torrens@uv.es

ABSTRACT

An essential oil is a class of volatile oils that give plants their characteristic odor. Essential oils are also known as ethereal oil, volatile oil, aetherolea or, simply, as the oil of the plant from which they are extracted, for example, oil of clove. The oil is *essential* in the sense that it contains the *essence of* vegetal specie fragrance, the characteristic fragrance of the plant from which it is derived. These can be extracted from different parts of plants, for example, stems, leaves, and flowers. Plants provide the inspiration and foundation for modern medicines. Natural product isolation is a key component of the process of drug discovery from plants. The relationship between organic chemistry and biochemistry is introduced.

15.1 INTRODUCTION

Clove *Syzygium aromaticum* L. (family *Myrtaceae*) essential oil (EO) presents bioactivities, for example, antibacterial, antifungal, insecticidal, and antioxidant properties and, traditionally, is used as flavoring agent and antimicrobial material in food (Alma et al., 2007; Sebaaly et al., 2015; Deng et al., 2015). Phenylpropanoid eugenol (Eug, $C_{10}H_{12}O_2$, 4-allyl-2-methoxyphenol) is a major component in clove EO, which contains smaller amounts of monoterpene ester eugenyl acetate (Eug-Ac) and sesquiterpene β-caryophyllene (Crph, cf. Fig. 15.1) (Aguilar-González and López-Malo, 2013). When mixed with ZnO, Eug forms a cement used in dentistry, which results in the characteristic odor of a dentist's office. It is volatile, soluble in organic solvents but slightly soluble in water. Its 1-octanol/water partition coefficient is $\log P = 2.27$.

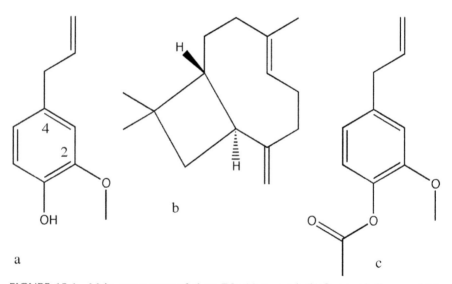

FIGURE 15.1 Main components of clove EO: (a) eugenol, (b) β-caryophyllene, and (c) eugenyl acetate.

Eugenol belongs to the class of volatile phenylpropanoids (e.g., Eug methyl ether, chavicol, estragole, methylcinnamate), which are metabolites derived from L-phenylalanine (cf. Fig. 15.2) and widely distributed via the plant kingdom (Anand et al., 2016). A number of Eug therapeutic effects were shown, for example, antivirus, antibacterial, antipyresis, analgesia, anti-inflammatory, anticoagulation, anti-oxidation, antihypoxia, and anti-ulcer (Mahapatra et al., 2009).

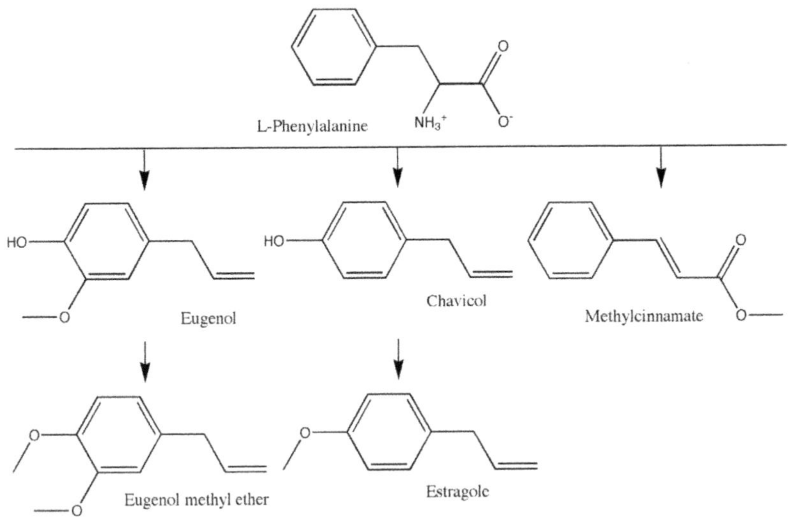

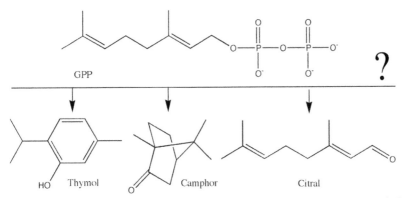

FIGURE 15.2   Phenylpropanoid biosynthetic pathway derived from L-phenylalanine.

Extraction was applied to undergraduate isolation experiments, for example, natural product (NP) isolation from spices (O'Shea et al., 2012), anise (LeFevre, 2000; Garin, 1980), cloves (Ntamila and Hassanali, 1976; Guntero et al., 2017), citrus fruits (Greenberg, 1968; Glidewell, 1991), thyme, sagebrush, lemongrass (cf. Fig. 15.3) (McLain et al., 2015), etc. (Runquist, 1969; Griffin, 1974; Garin, 1976; Craveiro et al., 1976). The NP isolation was introduced to organic chemistry students via steam distillation and liquid-phase extraction (Garin, 1976). The extraction of NPs in vegetal species was introduced to organic chemistry students (Craveiro et al., 1976).

FIGURE 15.3   Geranyl pyrophosphate (GPP) proposed biosynthetic precursor of citral, thymol, and camphor.

Eugenyl acetate can be prepared from Eug from clove (cf. Fig. 15.4).

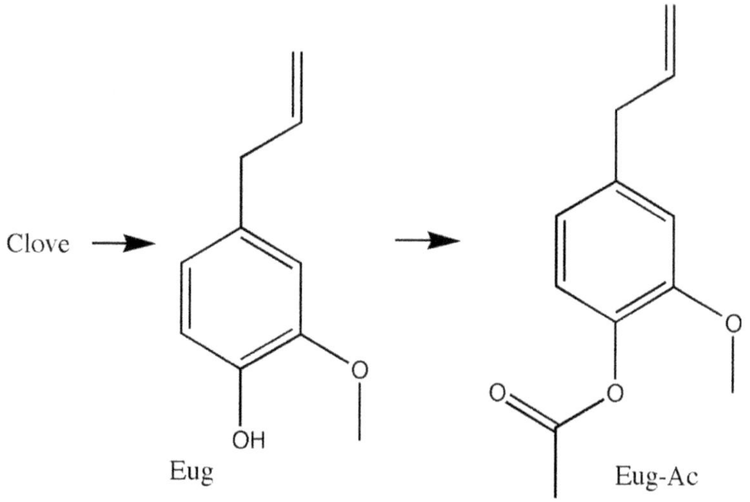

FIGURE 15.4 Eugenyl acetate prepared from Eug from clove.

Earlier publications in *Nereis*, etc. classified yams (Torrens-Zaragozá, 2013), lactic acid bacteria (Torrens-Zaragozá, 2014), fruits (Torrens-Zaragozá, 2015), algae (Torrens and Castellano, in press(a)), food spices (Torrens-Zaragozá, 2016), oil legumes (Torrens and Castellano, 2014, 2018a, in press(b)), and *Citrus* spp. (Torrens and Castellano, 2018b) by principal component, cluster, and meta-analyses. The molecular classifications of 33 phenolic compounds derived from the cinnamic and benzoic acids from *Posidonia oceanica* (Castellano et al., 2012), 74 flavonoids (Castellano et al., 2013), 66 stilbenoids (Castellano et al., 2014), 71 triterpenoids and steroids from *Ganoderma* (Castellano and Torrens, 2015a), 17 isoflavonoids from *Dalbergia parviflora* (Castellano and Torrens, 2015b), and artemisinin derivatives (Torrens et al., 2017; Torrens et al., 2018) were informed. A tool for interrogation of macromolecular structure was reported (Torrens and Castellano, 2014a). Mucoadhesive polymer hyaluronan favors transdermal penetration absorption of caffeine (Torrens and Castellano, 2014b, 2015). Endocrine disruptor diethylstilbestrol (DES), bisphenol-A (similar to DES), polycarbonate, and epoxy-silica nanocomposites were discussed (Torrens-Zaragozá, 2011; Torrens and Castellano, 2013). Polyphenolic phytochemicals in cancer prevention and therapy, bioavailability, and bioefficacy were informed (Estrela et al., 2017).

15.2 BIOLOGICAL EFFECTS OF ESSENTIAL OILS

The biological effects of EOs were reviewed (Bakkali et al., 2008). Their major components include two groups of distinct biosynthetical origin. The main group is composed of terpenes and terpenoids, and the other of aromatic and aliphatic constituents, all characterized by low molecular weight (cf. Fig. 15.5).

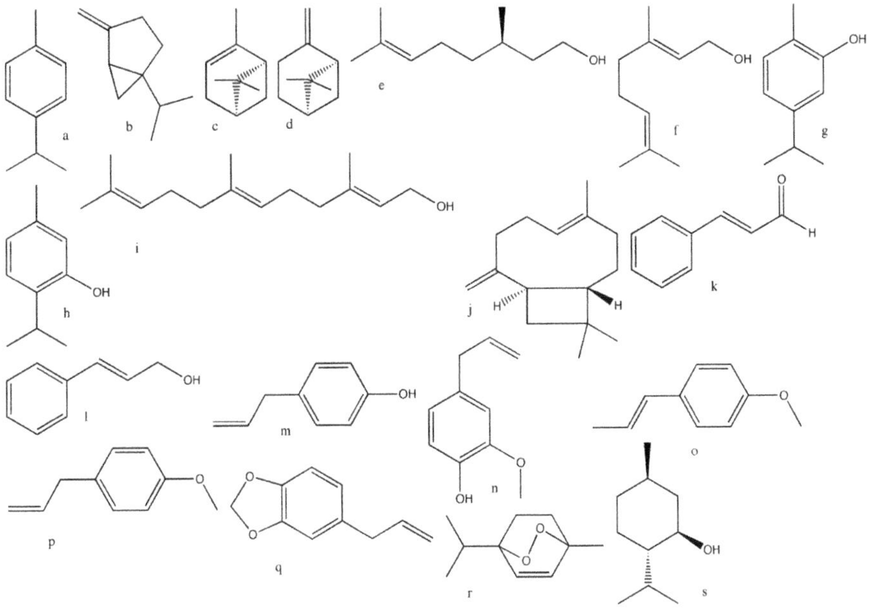

FIGURE 15.5 Chemical structures of selected EOs components: (1) terpenes—monoterpenes. Carbure monocyclic: (a) cymene-*y* or *p*-cymene and (b) sabinene; carbure bicyclic: (c) α-pinene, (d) β-pinene; alcohol acyclic: (e) citronellol, (f) geraniol; phenol: (g) carvacrol, (h) thymol—sesquiterpenes; carbure: (i) farnesol; alcohol: (j) caryophyllene. (2) Aromatic compounds: aldehyde (k) cinnamaldehyde; alcohol: (l) cinnamyl alcohol; phenol (m) chavicol; phenol: (n) Eug; methoxy derivative: (o) anethole, (p) estragole; methylene dioxy compound: (q) safrole. (3) Terpenoids (isoprenoids): (r) ascaridole and (s) menthol.

15.3 *CITRUS*

Citrus fruits, the generic collective name for oranges, mandarins, limes, lemons, grapefruits, and citrons, belong to genus *Citrus* (family Rutaceae). Extraction was applied to undergraduate isolation experiments, for example, natural product isolation from *Citrus* fruits (Greenberg, 1968; Glidewell, 1991). Organic acids, sugars, phenolic compounds and antioxidant capacities of orange juice and wine were determined (Kelebek et al., 2009). Experimental adjustment on drug interactions via intestinal cytochrome P450 (CYP)3A activity in rat was reported (Kinoshita et al., 2011). Neuroprotective effects of citrus flavonoids were informed (Hwang et al., 2012). EOs used in aromatherapy were reviewed (Ali et al., 2015). EOs were revised as antimicrobials in food systems (Calo et al., 2015). The chemistry and pharmacology of citrus limonoids were reviewed (Gualdani et al., 2016). Volatile and nonvolatile constituents and antioxidant capacity of oleoresins in three Taiwan citrus varieties were determined by supercritical fluid extraction (Chen and Huang, 2016). Innovative *green* and novel strategies were revised for the extraction of bioactive added-value compounds from *Citrus* wastes (Putnik et al., 2017). The chemical structures of the most abundant flavonoids in *Citrus* spp. are shown in Figure 15.6.

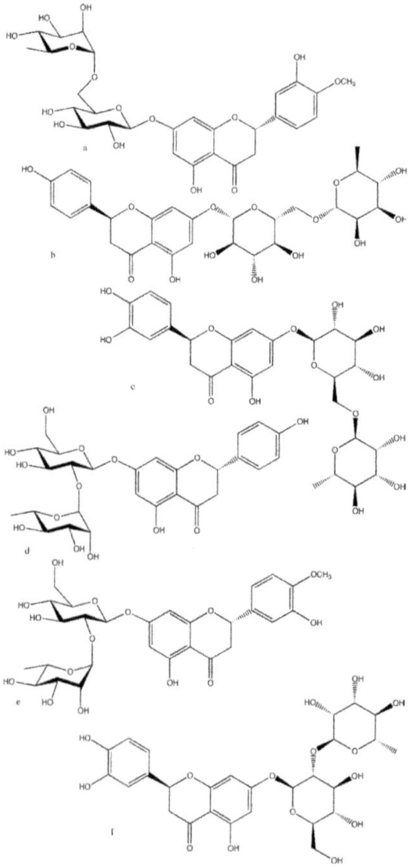

FIGURE 15.6   Flavonoids: (a) hesperidin, (b) narirutin, (c) eriocitrin, (d) naringin, (e) neohesperidin, and (f) neoeriocitrin.

The most abundant carotenoids in *Citrus* spp. peels are displayed in Figure 15.7.

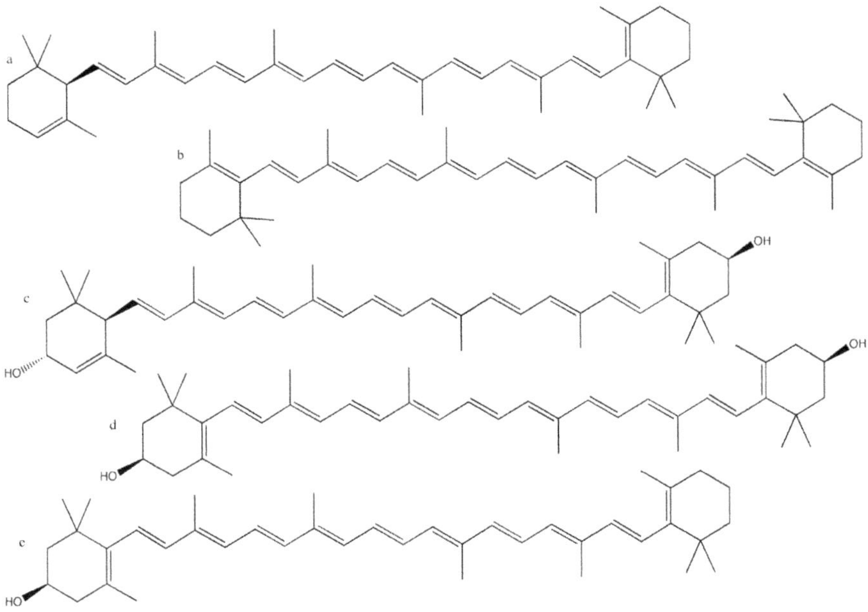

FIGURE 15.7 Carotenoids: (a) α-carotene, (b) β-carotene, (c) lutein, (d) zeaxanthin, and (e) β-cryptoxanthin.

Bioactivities of EOs were reviewed from plant chemoecology to traditional healing systems (Sharifi-Rad et al., 2017). Immature orange (*Aurantii fructus immaturus*) and citrus unshiu peel (*Citri unshiu pericarpium*) induced *p*-glycoprotein and CYP3A4 expression (Okada et al., 2017). Preclinical and clinical studies of anticancer potential of *Citrus* juices and extracts were revised (Cirmi et al., 2017). Effect of *Citrus* EOs and constituents was informed on growth of *Xanthomonas citri citri* (Mirzaei-Najafgholi et al., 2017). Bergamot EO attenuated anxiety-like behavior in rats (Rombolà et al., 2017). Bioavailable *Citrus sinensis* extract, polyphenolic composition, and biological activity were reported (Pepe et al., 2017). Volatile composition was published in two pummelo cultivars from different regions in China (Zhang et al., 2017). Five new limonoids were informed from peels of satsuma orange *C. reticulata* (Kikuchi et al., 2017). Anti-inflammatory and neuroprotective constituents were reported from *Citrus grandis* peels (Kuo et al., 2017). Antioxidant capacity, anticancer ability, and flavonoids composition of 35 *Citrus* were published (Wang et al., 2017). The anti-inflammatory properties of *Citrus wilsonii* extract in LPS-induced RAW 264.7 and primary mouse bone marrow-derived dendritic cells were informed (Cheng et al., 2017). Characterization and purification of bergamottin from *C. grandis*, and its antiproliferative activity and effect on glucose consumption in HepG2 cells were reported (Liu et al., 2017). Bergamottin (5-geranoxypsolaren) is the major component of bergamot (*Citrus bergamia*) oil. It is a natural furanocoumarin compound with weak polarity (cf. Fig. 15.8). It increases drug bioavailability via interaction with isoforms of the cytochrome P450 (CYP) enzymes.

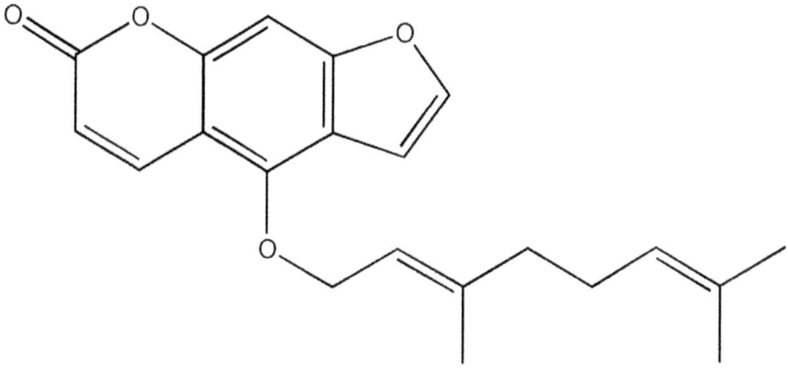

FIGURE 15.8   Structure of bergamottin.

Antioxidant and anticancer activities of EO from Gannan navel orange peel were informed (Yang et al., 2017). It is reported fruits constituents of *Citrus medica*, and the effect of 6,7-dimethoxy-coumarin on superoxide anion formation and elastase release (Chan et al., 2017). Cardiovascular activity of the chemical constituents of EOs was reviewed (De Andrade et al., 2017).

15.4   DISCUSSION ON *Citrus*

Health benefits derived from *C. sinensis* juice, which were mainly based on the antioxidant protection provided in vitro, revealed positive influence on digestion processes.

Citrus fruits, which are rich in abundant sources of hesperetin and other flavonoids, are stable and promising for the development of general food-type neuroprotection and brain foods.

Citrus waste is a good source of bioactive compounds usually discarded but can be used as food additives and/or nutraceuticals. The use of biologically active compounds, for example, common antioxidant compounds, from *Citrus* waste for the development of new functional products or nutraceuticals lies on the border between pharmacy and health and presents a growing interest.

15.5 GENERAL DISCUSSION

Plant EOs and their other products from secondary metabolism presented a great usage in folk medicine, food flavoring, fragrance, and pharmaceutical industries. The cytotoxic capacity of EOs based on a pro-oxidant activity can make them excellent antiseptic and antimicrobial agents for personal use, that is, purifying air, personal hygiene, or even internal use via oral consumption, and insecticidal use for the preservation of crops or food stocks. The antimicrobial activity of phenolic clove EO is attributed to its phenolic compounds, which can denaturalize proteins and react with cell membrane phospholipids, changing its permeability and causing microbial death. Clove EO and extracts present compounds with a vast and effective antimicrobial activity versus a great variety of organisms (e.g., bacteria, molds, yeasts), which can be exploited in the different fields of food industry as natural additives alternative to synthetic natural antimicrobials and extend the lifetime of processed foods. EO can be a natural alternative as disinfectant or preservative in hortifruitcultural industry during postharvest.

Aromatherapy is a natural and noninvasive gift of nature for humans. It is not only the disease symptoms that are eradicated but the whole body is also rejuvenated by the use of aroma. Aromatherapy regulates the physiological, spiritual, and psychological upliftment for the new phase of life. The therapy is not only preventive but also can be used in the acute and chronic stages of disease. Pharmaceutical industries try for environmental friendly, alternative, and natural medicine for disease associated with pathogens and metabolism. There may be a possibility of enhancing the rate of reaction and bioavailability of drugs from the use of EOs. If properly studied, EOs may have the synergistic effect with the drugs used in the treatment of central nervous system disorder. The time at which the plant contains the maximum amount of EO with various chemical constituents also is a matter of discussion. EOs can be a useful nonmedical option or can be combined with conventional care for some health conditions, provided safety, and quality issues be considered. The tilt of the scientific community toward complementary and alternative medicine gave the hope to reduce the unwanted effects of modern medicine by EOs and, if properly explored to their full potential, the therapy can be a boon not only to patients but also to a common man.

Many EOs exhibit activity versus foodborne pathogens and spoilage organisms in vitro, and, to a smaller degree, in foods. Gram-positive organisms seem to be much more susceptible to EOs than Gram-negative ones. Future research should focus on the effectiveness of different EOs in a number of food matrices. Synergism between EOs and other compounds, or with other processing techniques, will need to be investigated before they could be applied commercially. This work was able to emphasize that a continuing dialogue exists between the *laboratory of evolution* and the *laboratory of the scientist*, a dialogue that uses chemicals as its language. In pharmacology, an understanding of this relationship and the role played by plants in drug discovery is important.

15.6 FINAL REMARKS

From the preceding discussions, the following final remarks can be drawn:

(1) Phenolics are the main contributor to the antioxidant capacity of *Citrus* fruit. The same happens with the peach fruit, Chinese bayberry, vegetables, and grains, indicating that this is a common phenomenon in nature.

(2) Further, in vivo studies are necessary to explore the bioavailability of the *Citrus* flavonoids along with their antioxidant and anti-inflammatory effects, but usually in vitro models were well correlated to animal and human studies.

(3) Work is in progress on proximate classification of spices, seeds, and oleaginous seeds, soybean, Spanish legumes, commercial *soybean* and underutilized beans, and oils components, antifungals, antimicrobials, and anti-inflammatories of *Eucalyptus camaldulensis* and *Mentha pulegium*.

ACKNOWLEDGMENTS

The authors thank support from Generalitat Valenciana (Project No. PROMETEO/2016/094) and Universidad Catolica de Valencia San Vicente Martir (Project No. UCV.PRO.17-18.AIV.03).

KEYWORDS

- clove
- antibacterial
- antifungal
- insecticidal
- antioxidant

REFERENCES

Aguilar-González, A. E.; López-Malo, A. A. Extractos y Aceite Esencial del Clavo de Olor (*Syzygium aromaticum*) y su Potencial Aplicación Como Agentes Antimicrobianos en Alimentos. *Temas Sel. Ing. Aliment.* **2013**, *7* (2), 35–41.

Ali, B.; Al-Wabel, N. A.; Shams, S.; Ahamad, A.; Khan, S. A.; Anwar, F. Essential Oils Used in Aromatherapy: A Systematic Review. *Asian Pac. J. Trop. Biomed.* **2015**, *5*, 601–611.

Alma, M. H.; Ertas, M.; Nitz, S.; Kollmannsberger, H. Chemical Composition and Content of Essential Oil from the Bud of Cultivated Turkish Clove (*Syzygium aromaticum* L.). *BioResources* **2007**, *2* (2), 265–269.

Anand, A.; Jayaramaiah, R. H.; Beedkar, S. D.; Singh, P. A.; Joshi, R. S.; Mulani, F. A.; Dholakia, B. B.; Punekar, S. A.; Gade, W. N.; Thulasiram, H. V.; Giri, A. P. Comparative Functional Characterization of Eugenol Synthase from Four Different *Ocimum* Species: Implications on Eugenol Accumulation. *Biochim. Biophys. Acta* **2016**, *1864*, 1539–1547.

Bakkali, F.; Averbeck, S.; Averbeck, D.; Idaomar, M. Biological Effects of Essential Oils—A Review. *Food Chem. Toxicol.* **2008**, *46*, 446–475.

Calo, J. R.; Crandall, P. G.; O'Bryan, C. A.; Ricke, S. C. Essential Oils as Antimicrobials in Food Systems—A Review. *Food Control* **2015**, *54*, 111–119.

Castellano, G.; González-Santander, J. L.; Lara, A.; Torrens, F. Classification of Flavonoid Compounds by Using Entropy of Information Theory. *Phytochemistry* **2013**, *93*, 182–191.

Castellano, G.; Lara, A.; Torrens, F. Classification of Stilbenoid Compounds by Entropy of Artificial Intelligence. *Phytochemistry* **2014**, *97*, 62–69.

Castellano, G.; Tena, J.; Torrens, F. Classification of Polyphenolic Compounds by Chemical Structural Indicators and Its Relation to Antioxidant Properties of *Posidonia oceanica* (L.) Delile. *MATCH Commun. Math. Comput. Chem.* **2012**, *67*, 231–250.

Castellano, G.; Torrens, F. Information Entropy-Based Classification of Triterpenoids and Steroids from *Ganoderma*. *Phytochemistry* **2015a**, *116*, 305–313.

Castellano, G.; Torrens, F. Quantitative Structure–Antioxidant Activity Models of Isoflavonoids: A Theoretical Study. *Int. J. Mol. Sci.* **2015b**, *16*, 12891–12906.

Chan, Y. Y.; Hwang, T. L.; Kuo, P. C.; Hung, H. Y.; Wu, T. S. Constituents of the Fruits of *Citrus medica* L. var. *sarcodactylis* and the Effect of 6,7-Dimethoxy-Coumarin on Superoxide Anion Formation and Elastase Release. *Molecules* **2017**, *22*, 1454-1–1454-9.

Chen, M. H.; Huang, T. C. Volatile and Nonvolatile Constituents and Antioxidant Capacity of Oleoresins in Three Taiwan Citrus Varieties as Determined by Supercritical Fluid Extraction. *Molecules* **2016**, *21*, 1735-1–1735-12.

Cheng, L.; Ren, Y.; Lin, D.; Peng, S.; Zhong, B.; Ma, Z. The Anti-Inflammatory Properties of *Citrus wilsonii* Tanaka Extract in LPS-Induced RAW 264.7 and Primary Mouse Bone Marrow-Derived Dendritic Cells. *Molecules* **2017**, *22*, 1213-1–1213-14.

Cirmi, S.; Maugeri, A.; Ferlazzo, N.; Gangemi, S.; Calapai, G.; Schumacher, U.; Navarra, M. Anticancer Potential of *Citrus* Juices and their Extracts: A Systematic Review of Both Preclinical and Clinical Studies. *Front. Pharmacol.* **2017**, *8*, 420-1–420-11.

Craveiro, A. A.; Matos, F. J. A.; de Alencar, J. W. A Simple and Inexpensive Steam Generator for Essential Oils Extraction. *J. Chem. Educ.* **1976**, *53*, 652–652.

De Andrade, T. U.; Brasil, G. A.; Endringer, D. C.; da Nóbrega, F. R.; de Sousa, D. P. Cardiovascular Activity of the Chemical Constituents of Essential Oils. *Molecules* **2017**, *22*, 1539-1–1539-18.

Deng, J.; Yang, B.; Chen, C.; Liang, J. Renewable Eugenol-Based Polymeric Oil-Absorbent Microspheres: Preparation and Oil Absorption Ability. *ACS Sustain. Chem. Eng.* **2015**, *3*, 599–605.

Estrela, J. M.; Mena, S.; Obrador, E.; Benlloch, M.; Castellano, G.; Salvador, R.; Dellinger, R. W. Polyphenolic Phytochemicals in Cancer Prevention and Therapy: Bioavailability *versus* Bioefficacy. *J. Med. Chem.* **2017**, *60*, 9413–9436.

Garin, D. L. Steam Distillation of Essential Oils Anethole from Anise Followed by Permanganate Oxidation to Anisic Acid. *J. Chem. Educ.* **1980**, *57*, 138–138.

Garin, D. L. Steam Distillation of Essential Oils Carvone from Caraway. *J. Chem. Educ.* **1976**, *53*, 105–105.

Glidewell, C. Monoterpenes: An Easily Accessible but Neglected Class of Natural Products. *J. Chem. Educ.* **1991**, *68*, 267–269.

Greenberg, F. H. Natural Products Isolation Orange Oil: An Undergraduate Organic Experiment. *J. Chem. Educ.* **1968**, *45*, 537–538.

Griffin, R. W. Natural Products. An Independent Study Project. *J. Chem. Educ.* **1974**, *51*, 601–602.

Gualdani, R.; Cavalluzzi, M. M.; Lentini, G.; Habtemariam, S. The Chemistry and Pharmacology of Citrus Limonoids. *Molecules* **2016**, *21*, 1530-1–1530-39.

Guntero, V. A.; Mancini, P. M.; Kneeteman, M. N. Introducing Organic Chemistry Students to the Extraction of Natural Products Found in Vegetal Species. *World J. Chem. Educ.* **2017**, *5*, 142–147.

Hwang, S. L.; Shih, P. H.; Yen, G. C. Neuroprotective Effects of Citrus Flavonoids. *J. Agric. Food Chem.* **2012**, *60*, 877–885.

Kelebek, H.; Selli, S.; Canbas, A.; Cabaroglu, T. HPLC Determination of Organic Acids, Sugars, Phenolic Compounds and Antioxidant Capacity of Orange Juice and Orange Wine Made from a Turkish cv. Kozan. *Microchem. J.* **2009**, *91*, 187–192.

Kikuchi, T.; Ueno, Y.; Hamada, Y.; Furukawa, C.; Fujimoto, T.; Yamada, T.; Tanaka, R. Five New Limonoids from Peels of Satsuma Orange (*Citrus reticulata*). *Molecules* **2017**, *22*, 907-1–907-10.

Kinoshita, N.; Yamaguchi, Y.; Hou, X. L.; Takahashi, K.; Takahashi, K. Experimental Adjustment on Drug Interactions through Intestinal CYP3A Activity in Rat: Impacts of Kampo Medicines Repeat Administered. *Evid.-Based Complement. Alternat. Med.* **2011**, *2011*, 827435-1–827435-10.

Kuo, P. C.; Liao, Y. R.; Hung, H. Y.; Chuang, C. W.; Hwang, T. L.; Huang, S. C.; Shiao, Y. J.; Kuo, D. H.; Wu, T. S. Anti-Inflammatory and Neuroprotective Constituents from the Peels of *Citrus grandis*. *Molecules* **2017**, *22*, 967-1–967-11.

LeFevre, J. W. Isolating *Trans*-Anethole from Anise Seeds and Elucidating Its Structure: A Project Utilizing One- and Two-Dimensional NMR Spectrometry. *J. Chem. Educ.* **2000**, *77*, 361–363.

Liu, Y.; Ren, C.; Cao, Y.; Wang, Y.; Duan, W.; Xie, L.; Sun, C.; Li, X. Characterization and Purification of Bergamottin from *C. grandis* (L.) Osbeck cv. Yongjiazaoxiangyou and Its Antiproliferative Activity and Effect on Glucose Consumption in HepG2 Cells. *Molecules* **2017**, *22*, 1227-1–1227-13.

Mahapatra, S. K.; Chakraborty, S. P.; Majumdar, S.; Bag, B. G.; Roy, S. Eugenol Protects Nicotine-Induced Superoxide Mediated Oxidative Damage in Murine Peritoneal Macrophages In Vitro. *Eur. J. Pharmacol.* **2009**, *623*, 132–140.

McLain, K. A.; Miller, K. A.; Collins, W. R. Introducing Organic Chemistry Students to Natural Product Isolation Using Steam Distillation and Liquid Phase Extraction of Thymol, Camphor, and Citral, Monoterpenes Sharing a Unified Biosynthetic Precursor. *J. Chem. Educ.* **2015**, *92*, 1226–1228.

Mirzaei-Najafgholi, H.; Tarighi, S.; Golmohammadi, M.; Taheri, P. The Effect of Citrus Essential Oils and their Constituents on Growth of *Xanthomonas citri* subsp. *citri*. *Molecules* **2017**, *22*, 591-1–591-14.

Ntamila, M. S.; Hassanali, A. Isolation of Oil of Clove and Separation of Eugenol and Acetyl Eugenol. An Instructive Experiment for Beginning Chemistry Undergraduates. *J. Chem. Educ.* **1976**, *53*, 263–263.

O'Shea, S. K.; Von Riesen, D. D.; Rossi, L. L. Isolation and Analysis of Essential Oils from Spices. *J. Chem. Educ.* **2012**, *89*, 665–668.

Okada, N.; Murakami, A.; Urushizaki, S.; Matsuda, M.; Kawazoe, K.; Ishizawa, K. Extracts of Immature Orange (*Aurantii fructus immaturus*) and Citrus Unshiu Peel (*Citri unshiu pericarpium*) Induce *P*-Glycoprotein and Cytochrome P450 3A4 Expression via Upregulation of Pregnane X Receptor. *Front. Pharmacol.* **2017**, *8*, 84-1–84-9.

Pepe, G.; Pagano, F.; Adesso, S.; Sammonella, E.; Ostacolo, C.; Manfra, M.; Chieppa, M.; Sala, M.; Russo, M.; Marzocco, S.; Campiglia, P. Bioavailable *Citrus sinensis* Extract: Polyphenolic Composition and Biological Activity. *Molecules* **2017**, *22*, 623-1–623-15.

Putnik, P.; Kovaçevi, D. B.; Jambrak, A. R.; Barba, F. J.; Cravotto, G.; Binello, A.; Lorenzo, J. M.; Shpigelman, A. Innovative *Green* and Novel Strategies for the Extraction of Bioactive Added Value Compounds from Citrus Wastes—A Review. *Molecules* **2017**, *22*, 680-1–680-24.

Rombolà, L.; Tridico, L.; Scuteri, D.; Sakurada, T.; Sakurada, S.; Mizoguchi, H.; Avato, P.; Corasaniti, M. T.; Bagetta, G.; Morrone, L. A. Bergamot Essential Oil Attenuates Anxiety-Like Behaviour in Rats. *Molecules* **2017**, *22*, 614-1–614-11.

Runquist, O. The Essential Oils: A Series of Laboratory Experiments. *J. Chem. Educ.* **1969**, *46*, 846–847.

Sebaaly, C.; Jraij, A.; Fessi, H.; Charcosset, C.; Greige-Gerges, H. Preparation and Characterization of Clove Essential Oil-Loaded Liposomes. *Food Chem.* **2015**, *178*, 52–62.

Sharifi-Rad, J.; Sureda, A.; Tenore, G. C.; Daglia, M.; Sharifi-Rad, M.; Valussi, M.; Tundis, R.; Sharifi-Rad, M.; Loizzo, M. R.; Ademiluyi, A. O.; Sharifi-Rad, R.; Ayatollahi, S. A.; Iriti, M. Biological Activities of Essential Oils: From Plant Chemoecology to Traditional Healing Systems. *Molecules* **2017**, *22*, 70-1–70-55.

Torrens, F.; Castellano, G. Bisphenol, Diethylstilbestrol, Polycarbonate and the Thermomechanical Properties of Epoxy–Silica Nanostructured Composites. *J. Res. Updates Polym. Sci.* **2013**, *2*, 183–193.

Torrens, F.; Castellano, G. From Asia to Mediterranean: Soya Bean, Spanish Legumes and Commercial *Soya Bean* Principal Component, Cluster and Meta-Analyses. *J. Nutr. Food Sci.* **2014**, *4* (5), 98.

Torrens, F.; Castellano, G. A Tool for Interrogation of Macromolecular Structure. *J. Mater. Sci. Eng. B* **2014a**, *4* (2), 55–63.

Torrens, F.; Castellano, G. Mucoadhesive Polymer Hyaluronan as Biodegradable Cationic/ Zwitterionic-Drug Delivery Vehicle. *ADMET DMPK* **2014b**, *2*, 235–247.

Torrens, F.; Castellano, G. Computational Study of Nanosized Drug Delivery from Cyclodextrins, Crown Ethers and Hyaluronan in Pharmaceutical Formulations. *Curr. Top. Med. Chem.* **2015**, *15*, 1901–1913.

Torrens, F.; Castellano, G. Principal Component, Cluster and Meta-Analyses of Soya Bean, Spanish Legumes and Commercial Soya Bean. In *High-Performance Materials and Engineered Chemistry*; Torrens, F., Balköse, D., Thomas, S., Eds.; Apple Academic–CRC: Waretown, NJ, 2018a; pp 267–294.

Torrens, F.; Castellano, G. Classification of Citrus: Principal Components, Cluster, and Meta-Analyses. In *Applied Physical Chemistry with Multidisciplinary Approaches*; Haghi, A. K., Balköse, D., Thomas, S., Eds.; Apple Academic–CRC: Waretown, NJ, 2018b; pp 217–234.

Torrens, F.; Castellano, G. Molecular Classification of Algae by Phenolics and Antioxidants. In *Industrial Chemistry and Chemical Engineering*; Haghi, A. K., Ed.; Apple Academic–CRC: Waretown, NJ, in press(a).

Torrens, F.; Castellano, G. *Meta-Analysis of Underutilized Beans with Nutrients Profile.* In *Industrial Chemistry and Chemical Engineering*; Haghi, A. K., Ed.; Apple Academic–CRC: Waretown, NJ, in press(b).

Torrens, F.; Redondo, L.; Castellano, G. Artemisinin: Tentative Mechanism of Action and Resistance. *Pharmaceuticals* **2017**, *10*, 20–24.

Torrens, F.; Redondo, L.; Castellano, G. Reflections on Artemisinin, Proposed Molecular Mechanism of Bioactivity and Resistance. In *Applied Physical Chemistry with Multidisciplinary Approaches*; Haghi, A. K., Balköse, D., Thomas, S., Eds.; Apple Academic–CRC: Waretown, NJ, 2018; pp 189–215.

Torrens-Zaragozá, F. Classification of Food Spices by Proximate Content: Principal Component, Cluster, Meta-Analyses. *Nereis* **2016**, *2016* (8), 23–33.

Torrens-Zaragozá, F. Classification of Fruits Proximate and Mineral Content: Principal Component, Cluster, Meta-Analyses. *Nereis* **2015**, *2015* (7), 39–50.

Torrens-Zaragozá, F. Classification of Lactic Acid Bacteria against Cytokine Immune Modulation. *Nereis* **2014**, *2014* (6), 27–37.

Torrens-Zaragozá, F. Molecular Categorization of Yams by Principal Component and Cluster Analyses. *Nereis* **2013**, *2013* (5), 41–51.

Torrens-Zaragozá, F. Polymer Bisphenol-A, the Incorporation of Silica Nanospheres into Epoxy–Amine Materials and Polymer Nanocomposites. *Nereis* **2011**, *2011* (3), 17–23.

Wang, Y.; Qian, J.; Cao, J.; Wang, D.; Liu, C.; Yang, R.; Li, X.; Sun, C. Antioxidant Capacity, Anticancer Ability and Flavonoids Composition of 35 Citrus (*Citrus reticulata* Blanco) Varieties. *Molecules* **2017**, *22*, 1114-1–1114-20.

Yang, C.; Chen, H.; Chen, H.; Zhong, B.; Luo, X.; Chun, J. Antioxidant and Anticancer Activities of Essential Oil from Gannan Navel Orange Peel. *Molecules* **2017**, *22*, 1391-1–1391-10.

Zhang, M.; Li, L.; Wu, Z.; Wang, Y.; Zang, Y.; Liu, G. Volatile Composition in Two Pummelo Cultivars (*Citrus grandis* L. Osbeck) from Different Cultivation Regions in China. *Molecules* **2017**, *22*, 716-1–716-17.

CHAPTER 16

FRUIT WINES: OPPORTUNITIES FOR MEXICAN MANGO WINE

JOSÉ JUAN BUENROSTRO-FIGUEROA[1,2], CRISTIAN TORRES-LEÓN[1], RAÚL RODRÌGUEZ[1], HELIODORO DE LA GARZA-TOLEDO[1], and CRISTÓBAL NOÉ AGUILAR[1*]

[1]*Group of Bioprocesses and Bioproducts, Food Research Department, School of Chemistry, Universidad Autonoma de Coahuila, 25280 Saltillo, Coahuila, Mexico*

[2]*Research Center in Food and Development AC, 33089 Cd. Delicias, Chihuahua, Mexico*

Corresponding author. E-mail: cristobal.aguilar@uadec.edu.mx

ABSTRACT

The preservation of foods by the alcoholic fermentation is an activity closely linked to most of the cultures of the world. Nowadays, a large quantity of alcoholic beverages is produced, mainly beer and wine, by an alcoholic fermentation process of a sugary substrate or must, with the yeast *Saccharomyces cerevisiae* under anaerobic conditions. In the spontaneous fermentation, the yeast confers particular characteristics to each wine making, to produce alcoholic beverages from any fruit containing sufficient fermentable carbohydrates. During fermentation, in addition to producing ethanol, secondary compounds are produced enriching and characterizing the wines, preserving the flavor of the fruit origin, and giving special characteristics. In Mexico, mangoes fruits are produced in large quantities but not fully exploited due to lack of markets, value-added options, and postharvest losses of up to 30% annually, which demands new processes and conservation methods to extend the mango shelf life. Thus, mango wine becomes an alternative product to face the issues to use surplus mango crops, attending the new trends and demands of alcoholic fermented beverages.

16.1 INTRODUCTION

The production of alcoholic beverages has been an activity linked to most cultures for millennia. Since antiquity, human developed techniques to ferment musts containing carbohydrates to convert them into beer and wine, and even learned to distill alcohol to increase its concentration in beverages, but without knowing the role or existence of microorganisms. This sounds inexplicable that a sugary liquid be transformed spontaneously into a different liquid, obtaining a drink with pleasant sensory properties that promote euphoria. Probably, the first alcoholic fermented beverages were made from sugary substrates, such as fruit juices, since it is only necessary to contact the juice with the wild yeasts present on the surface of fruit itself.

Later, we achieved the elaboration of beverages from starch substrates (like cereals) that require the amylolytic activity of enzymes generated during the germination; so, it was necessary that we learned first the art of the malting before being able to manufacture the beer (Desrosier, 1997).

Currently, the primary focus of production for biotechnology companies is the alcoholic beverage industry. The main alcoholic beverage, according to its production volume, is the beer, followed by table wines. However, there are other traditional Mexican beverages that are obtained by fermentation, such as tequila, pulque, sotol, colonche, tepache, tesgüino, chicha, chilocle, coyote, and pozol (Buenrostro et al., 2010).

Recent efforts have been made to create alcoholic beverages from fruit. Apart from grapes, there are many other fruits available that can be used as substrates for winemaking (Jagtap and Bapat, 2015), and there are already reports of fruit wines, such as apple (Joshi and Bhutani, 1991), palm sap (Joshi and Sharma, 1995), and apricot (Joshi et al., 1990), among others. Akubor et al. (2003) reported the production of banana wine containing 5% v/v alcohol content in 10 days of fermentation, with acceptable sensory characteristics compared to commercial wine from Germany. Likewise, the blackberry juice fermentation, with different strains of yeast reached up to 8.5% v/v (Calderón-Santoyo et al., 2008); Ezeronye (2004) used an isolated strain of palm wine for the fermentation of cashew, mango, papaya, and pineapple, reporting alcohol values in a range of 10.6–12.6% w/v. In the production of mango wine, Reddy and Reddy (2005) reported the production of a wine with an alcohol content between 7% and 8.5% w/v, considered of good quality, at 30°C and a pH of 5, promoting the formation of esters and aromatic groups. Although Mexico is one of the main producers of mango in the world (FAOSTAT, 2017), the productive potential has not been fully exploited, as there are tropical fruits that occur in large quantities in the country, such as the banana (*Musa sapientum*) and mango (*Mangífera indica* L.). The purpose of this chapter is to review relevant information about the fruit wines, emphasizing on the opportunities for Mexican mango wine.

16.2 ALCOHOLIC FERMENTATION

Alcoholic fermentation is a biological process in the absence of oxygen, produced by the activity of some microorganisms that process sugars (glucose, fructose, sucrose, starch, etc.) to obtain as final products: alcohol in the form of ethanol ($CH_3–CH_2–OH$), carbon dioxide (CO_2) in the form of gas, and adenosine triphosphate molecules that consume microorganisms themselves in their cellular anaerobic energy metabolism (Howard and Piggott, 2003).

In traditional systems, the alcoholic fermentation is produced by the action of the natural microorganisms present, either on the surface of the fruit or in the environment (Calderón-Santoyo et al., 2008). In this, biological process usually predominates as a producer microorganism, the yeast *Saccharomyces cerevisiae*, which grows and is rapidly reproduced in media containing glucose and fructose (Táboas, 2002).

In spontaneous fermentations, to produce wine by means of autochthonous or indigenous microorganisms, *Saccharomyces* yeasts play a fundamental role. Other yeasts have also been found in the first 2 or 4 days of fermentation and contribute to give particular characteristics to each wine. However, as the concentration of alcohol increases, they die (Fleet and Heard, 1993).

S. cerevisiae is a unicellular fungus that is divided by budding and can have asexual reproduction. In its haploid form, or in a sexual way when form a zygote, it forms an ascus containing four haploid ascospores. Figure 16.1 shows how this yeast ferments the sugars if they are in anaerobic conditions in ethanol and CO_2 (Scragg, 2004).

Alcoholic fermentation depends on several factors: biological diversity (yeast strain, concentration), physical (temperature), and chemical (pH, activator nutrients) (Torrija et al., 2002). Oxygen is the initial trigger of the fermentation, as the yeasts will need it in its growth phase. However, at the end of the fermentation, it is advisable that the presence of oxygen is low to avoid the loss of ethanol and the appearance instead of acetic acid or acetyl.

The simplified or fermentation process is

Sugars + yeasts = $CH_3–CH_2–OH + CO_2$ + heat + other substances

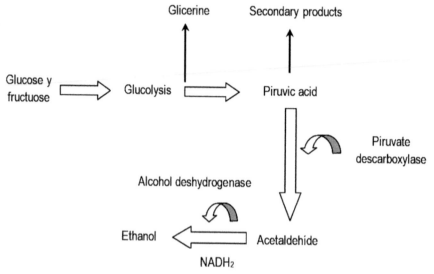

FIGURE 16.1   Transformation from glucose to ethanol.

Ethanol is the largest volatile component of alcoholic beverages, confers body, apparently reduces acidity, increases sweetness, and has a global effect of softness on other flavor characteristics, but the flavor and aroma of these beverages are influenced, in addition to ethanol, by the presence of a wide variety of organic compounds present in much lower quantities, but they are responsible for the characteristics that make it different from other beverages. These compounds are mainly methanol, superior alcohols, carbonyls, organic acids, esters, and sulfur compounds, which together are called congenerics (De la Garza-Toledo, 2008). One of these compounds, glycerol, determines the body in wines and probably confers a slight sensation of sweetness (Reed, 1991).

16.3 THE WINE

The wine is defined exclusively like the liquid product obtained by alcoholic fermentation, total or partial, of the grape juice, without the addition of any substance. In other words, although wines of several fruits can be produced, they must have the corresponding allowance for the raw material from which they were produced (Desrosier, 1997). Alcoholic fermentation is the basics of winemaking, since it is closely related to the activity of the wine and its enzymatic load in the must as well as the sugar content.

The strain of yeast used influences the production of compounds and consequently the quality of the wine; so, one should have the idea in the future that the oenological industry will be inclined by the vine that provides certain characteristic in a wine, avoid the formation of compounds that are unfavorable on the quality of the final product as well as the strain being as close to 100% purity (Táboas, 2002). Without the yeast role, a "wine" would never be wine, because alcoholic graduation must be at least 9% of volume.

A wine is obtained by the fermentation of the sugars contained in the fruit must, which are transformed into alcohol, mainly. However, its importance is not only in obtaining ethanol from grape sugars; in addition, during the fermentation process, many secondary products that influence the aroma and quality of the wine will be formed (Mallouchos, 2003), depending on the presence and concentration of sugar fruit (Mauricio et al., 1997).

Starting from this principle, it is possible to make fruit wine from sweet fruits mainly with strong and pleasant aromas as well flavors to obtain different types of fruit wine with organoleptic characteristics according to the formed compounds.

16.4 FERMENTED ALCOHOLIC PRODUCTS IN MEXICO

In Mexico, there are important wine regions: Coahuila, Queretaro, Aguascalientes-Zacatecas, and Baja California, the latter is responsible for about 90% of the national production. There are several wine-producing companies that employ different grape varieties (Table 16.1).

TABLE 16.1 Wine Producers in Mexico.

Company	Place	Products
Adobe Guadalupe	Valle de Guadalupe, B.C.	Varietal wine
Cachola	Zacatecas	Red and white wine
Casa Bibayoff	Valle de Guadalupe, B.C.	Red and white wine
Casa de Piedra	Ensenada, B.C.	Author wine
Casa Ferriño	Cuatrociénegas, Coahuila	Generous wine
Casa Madero	Parras, Coahuila	Varietal, red, and white wine
Cavas Atonelli	Ezequiel Montes, Qro.	Varietal table wine
Cavas Valmar	Ensenada, B.C.	Red wine
Cuna de Tierra	Dolores Hidalgo, Gto.	Red and white wine

TABLE 16.1 *(Continued)*

Company	Place	Products
Chateau Camou	Ensenada, B.C.	Red and white wine
Dinastía	Aguascalientes	Wine, brandy
Dos búhos	San Miguel de Allende, Gto.	Red wine
Freixenet de México	Ezequiel Montes, Qro.	Red, sparkling, autor, and young wine
Hacienda de Letras	Aguascalientes	Red wine
La Bordaleza	Aguascalientes	Wine, brandy
LA Cetto	Valle de Guadalupe, B.C.	Red and white wine
La Madrileña	San Juan del Río, Qro.	Table wine
La Redonda	Ezequiel Montes, Qro.	Young wine, sparkling wine
Mogor-Badan	Ensenada, B.C.	Red and white wine
Monte-Xanic	Valle de Guadalupe, B.C.	Red and white wine
Pedro Domecq	Valle de Calafia, B.C.	Varietal wine
Pedro Domecq	Ramos Arizpe, Coahuila	Brandy, table wine
Santo Tomás	Valle de Santo Tomás, B.C.	Red and white wine
Vinícola del Vergel	Gomes Palacio, Durango.	Young wine
Viña de Liceaga	San Antonio de las Minas, B.C.	Red and white wine

Some of fermented alcoholic products based on fruits made in Mexico are as follows:

Colonche: it is a drink made in Zacatecas obtained by fermentation of Cardona red tuna juice.

Mosco (traditional of Toluca city, State of Mexico): strong liquor made from Capulín, Apple, orange, or nanche by process of fermentation; climax is reached when the presence of mosquitoes in the ferment is detected.

Cider: produced from the crushed and fermented apples and consumed in Christmas festivities; traditionally it is made in Puebla State.

Tepache: fermented shell beverage and pineapple juice (fruit) sweetened with piloncillo; it is traditional beverage of the Jalisco State and is also made of apple, orange, and guava and consumed in other regions of the country.

Tuba: a palm wine obtained by fermentation of the sap of the stem of several species, mainly coconut palm, from Colima state.

Wine: product resulting from the fermentation of fresh grapes or grape juice. The whites and pinks come from the fermentation of the grape juice, and the reds from the whole grape bean; cultivated and cared for in the traditional European way; produced in regions of Coahuila, Queretaro, Aguascalientes-Zacatecas, and Baja California.

Fruit wines: homemade with various fruits such as Capulín, BlackBerry, apple, plum, peach, orange, and acachul, typical of the state of Mexico, Hidalgo, and Puebla.

Others are the quince and walnut wine produced in the state of Zacatecas.

16.5 MANGO FRUIT AND ITS POTENTIAL FOR WINE PRODUCTION

Mango (*M. indica* L.) belongs to the family Anacardiaceae and is cultivated in tropical and subtropical regions. Mango is one of the most favored fruits due to its delicacy, delicious flavor, attractive fragrance, beautiful color, and nutritional value. Mango has a high content of bioactive compounds such as ascorbic acid, lupeol, β-carotene, and polyphenolic compounds (Robles-Sánchez et al., 2013). Based on its chemical composition and organoleptic characteristics, it is regarded as "the king of fruits" (Tharanathan et al., 2006), a distinction that makes it the second most traded tropical fruit in the world and fifth in total production (FAOSTAT, 2017).

The world production of mango is estimated at 45,225,211 millions of tons per year; India is the largest producer of mango with 12,691,950 t/year, followed by China (3330,313 t/year), Thailand (2047,424 t/year), and Mexico (1593,605 t/year). Furthermore, Mexico is the largest exporter (FAOSTAT, 2017). The mainly varieties cultivated in Mexico are Haden, Kent, Tommy-Atkins, Ataulfo, and Keit. Additionally, the variety Ataulfo (*Mangifera caesia* Jack ex Wall) has a denomination of origin since it is endemic to Mexico (NOM-188-SCFI, 2012). Mango producers suffer large annual losses that come to exceed 30% of annual production by "lack of marketing options."

The consumption of mango-based products such as mango-flavored beverages, single or combined with other fruits, is rapidly increasing in Europe. In India, Mango is the most popular and selected fruit; there are about 25 cultivars.

The chemical composition of mango pulp varies with the location of cultivation, variety, and stage of maturity. The chemical composition of the mango pulp is shown on Table 16.2. Mango has a carbohydrate content of 16–17%, which is considered high (Reddy and Reddy, 2007); this value is very important to produce wine by alcoholic fermentation. The total sugar content of mangoes varied between 11.5% and 25% (fresh weight) and up to 15% of the fresh pulp of green, mature fruits was starch (Tharanathan et al., 2006). The soluble sugars of the fruit pulp consisted mainly of glucose, fructose, and sucrose. The content of fermentable sugars in mango must is 182 g/L. Besides, the mango pulp has an important composition of phytochemical compounds. The secondary metabolites (phenolic extracts obtained from mango samples) in showed in Table 16.3. The content of sugars and polyphenols are very important in the formulation of wines as they contribute to the characteristics of smell and taste (Li et al., 2012). Polyphenols also increase the antioxidant activity of wines. The mango pulp is a material with functional characteristics for the formulation of wines with antioxidant properties due the compounds of high biological power (Table 16.3).

TABLE 16.2 Proximal Analysis of Ripe Mango Pulp (Value Per 100 g).

Calories	62.1–63.7 Cal
Moisture	78.9–82.8 g
Protein	0.36–0.40 g
Fat	0.30–0.53 g
Carbohydrates	16.20–17.18 g
Fiber	0.85–1.06 g
Ash	0.34–0.52 g
Calcium	6.1–12.8 mg
Phosphorus	5.5–17.9 mg
Iron	0.20–0.63 mg
Vitamin A (carotene)	0.135–1.872 mg
Thiamine	0.020–0.073 mg
Riboflavin	0.025–0.068 mg
Niacin	0.025–0.707 mg
Ascorbic acid	7.8–172.0 mg
Tryptophan	3–6 mg
Methionine	4 mg
Lysine	32–37 mg

Source: Some parts taken from Gopalan et al. (1977).

TABLE 16.3 Chemical Species from ESI (−) FT-ICR MS Data for the Phenolic Fraction of Mango Pulp.

	$m/z_{theoretical}$	Molecular formula [M–H]⁻	Compound
1	183.02991	$C_8H_7O_5$	Methylgallate
2	300.99909	$C_{14}H_5O_8$	Ellagic acid
3	331.06738	$C_{13}H_{15}O_{10}$	Glucogallin
4	335.04117	$C_{15}H_{11}O_9$	Methyldigallate ester
5	421.07807	$C_{19}H_{17}O_{11}$	Mangiferin
6	483.07861	$C_{20}H_{19}O_{14}$	Digalloyl glucose
7	635.08996	$C_{27}H_{23}O_{18}$	Trigalloyl glucose
8	787.1014	$C_{34}H_{27}O_{22}$	Tetragalloyl glucose

Source: Some parts taken from Oliveira et al. (2016).

Despite the nutritional and functional potential of the mango, it is necessary to implement conservation methods that allow extending the useful life of these fruits because after harvest and under native climatic conditions, mango fruit ripen within a week, but reach an overripe stage and spoil within 2 weeks (Tafolla-Arellano et al., 2017). An alternative is the fermentation, using the surplus of these fruits for the production of fruit wines, thus solving the bottleneck that arises between May 15 and July 15, when production is saturated and the problems of marketing and competition begin between some states of Mexico because they send the product at the same time to the markets of supply and/or processing industries, which causes the low price received and sometimes up to the delay in payment, damaging mainly to small producers and those who are not members of any grouping.

Several authors carried out studies for the production of mango wine (Li et al., 2012; Pino and Queris, 2011; Reddy and Reddy, 2010). Coelho et al. (2015) reported an ethanol yield in mango wine production of 101 ± 1.78 g/L; this yield was higher than that calculated for other fruits such as orange (72.3 ± 2.08 g/L), cherry (66.1 ± 4.02 g/L), and banana (98.2 ± 7.88 g/L). The authors justify the good performance of the mango fruit by the chemical composition (Table 16.2). Mango wine had good acceptance from trained panelists, demonstrating its suitability as food-grade product. Reddy and Reddy (2005) selected six varieties (Banginapalli, Alphono, Bangalora, Allampur Baneshan, Neelam, and Raspuri) available in abundance in southern India, to optimize the conditions of juice extraction and wine production, obtaining a good acceptance with those obtained with the first three varieties. Subsequently, Reddy and Reddy (2010) report a complete optimization study using yeast strain (*S. cerevisiae* 101) of the variables that affect the obtaining of good quality mango wine and good health benefits. Furthermore, Pino and Queris (2011) reported a complete evaluation of the mango wine aroma, identified and quantified a total of 102 compounds. The wine had 8.97 mg/L volatile compounds, which included esters (40), alcohols (15), terpenes (12), acids (8), aldehydes and ketones (6), lactones (4), phenols (2), furans (2), and miscellaneous compounds (13). Aroma compounds are especially important in fruit wine as they contribute to the quality of the final product.

16.6   FINAL COMMENTS

The alcoholic fermentation is a viable alternative to use the surpluses of all those fruits that by action of environment of the tropical countries require some method of conservation to prolong their shelf life, offering new products to taste such as a wide range of fruit wines, all of them flavored and unique qualities of each fruit. The elaboration of these alcoholic beverages from tropical fruits has a great impact in helping the economy of the different countries of the tropics.

Based on the information analyzed, it is proposed to use the surplus of mango varieties that are most produced in Mexico, for the elaboration of Mango wine, using both commercial and isolated yeast strains from other sources, including the isolation of a mango strain for subsequent application and/or comparison with other alcohol-producing yeasts, which would strongly impact the economy of the producing areas of this fruit, resulting in an integral use and therefore a growing demand for cultivation. It is recommended to continue scientific research to evaluate factors as the influence of mango variety on the quality of formulated wines.

KEYWORDS

- alcoholic fermentation
- wine
- sugar substrate
- *Saccharomyces cerevisiae*
- *Mangifera indica*

REFERENCES

Akubor, P. I.; Obio, S. O.; Nwadimere, K. A.; Obioman, E. Production and Quality Evaluation of Banana Wine. *Plants Food Hum. Nutr.* **2003**, *58*, 1–6.

Buenrostro, J.; Rodríguez, R.; de la Garza-Toledo, H.; Aguilera-Carbó, A.; Nevárez-Moorillón, G. V.; Prado-Baragán, L. A.; Aguilar, C. N. Mexican Fermented Foods and Beverages. In *Advances in Food Science and Technology*; Haghi, A. K., Ed.; Nova Science Publishers: Hauppauge, NY, 2010; pp 245–270. ISBN: 978-1-61942-120-2.

Calderón-Santoyo, M.; Godina-Galindo, G.; Bautista-Rosales, P. U.; Gutierrez-Martínez, P.; Ragazzo-Sánchez, J. A. Blackberry (*Rubus* spp.) Juice Alcoholic Fermentation: Kinetics and Aromatic Aspects. In *Memories of 3rd International Congress Food Science and Food Biotechnology in Developing Countries*, Querétaro, Qro, 2008; pp 616–621.

Coelho, E.; Vilanova, M.; Genisheva, Z.; Oliveira, J.; Teixeira, J.; Domingues, L. Systematic Approach for the Development of Fruit Wines from Industrially Processed Fruit Concentrates, Including Optimization of Fermentation Parameters, Chemical Characterization and Sensory Evaluation. *LWT—Food Sci. Technol.* **2015**, *62*, 1043–1052. DOI:10.1016/j.lwt.2015.02.020.

De la garza-Toledo, H. *Aspectos fundamentales y tecnológicos para la producción de sotol*; Universidad Autónoma de Coahuila: Saltillo, Coahuila, 2008; p 199.

Desrosier, N. W. *Elementos de biotecnología alimentaria*; CECSA: Mexico, DF, 1997; pp 263–305.

Ezeronye, O. U. Nutrient Utilization Profile of *Sacharomyces cerevisiae* from Palm Wine in Tropical Fruit Fermentation. *Anton. van Leeuwnh.* **2004**, *86*, 235–240.

FAOSTAT. F. and A. Organization of the United Nations. *FAO [WWW Document]. Statistical Database—Agriculture*, 2017. http://www.fao.org/faostat/es/#data/QC/visualize.

Fleet, G. H.; Heard, G. M. Yeasts Growth during Fermentation. In *Wine Microbiology and Biotechnology*; Fleet, G. H., Ed.; Harwood Academic Publishers: Camberwell, VIC, 1993; pp 27–54.

Gopalan, C.; Rama Shastri, B.; Balasubramanian, S. *Nutritive Value of Indian Foods*, National Institute of Nutrition: Hyderabad, India, 1977.

Howard, L. A. G.; Piggott, J. R. *Fermented Beverage Production*; Kluwer Academic/Plenum Publishers: New York, 2003; p 423.

Jagtap, U. B.; Bapat, V. A. Wines from Fruits Other Than Grapes: Current Status and Future Prospectus. *Food Biosci.* **2015**, *9*, 80–96. DOI:10.1016/j.fbio.2014.12.002.

Joshi, V. K.; Bhutani, V. P.; Sharma, R. C. Effect of Dilution and Addition of Nitrogen Source in Chemical, Mineral and Sensory Qualities of Wild Apricot Wine. *Am. J. Enol. Vitic.* **1990**, *41*, 229–231.

Joshi, V. K.; Bhutani, V. P. The Influence of Enzymatic Clarification in Fermentation Behavior and Qualities of Apple Wine. *Sci. Alim.* **1991**, *11*, 491–498.

Joshi, V. K.; Sharma, S. K. Comparative Fermentation Behavior and Physico-Chemical Sensory Characteristics of Plum Wine as Effected by the Type of Preservatives. *Chem. Microb. Technol. (Lebensm.)* **1995**, *17*, 65.

Li, X.; Jie, L.; Yu, B.; Curran, P.; Liu, S. Fermentation of Three Varieties of Mango Juices with a Mixture of *Saccharomyces cerevisiae* and *Williopsis saturnus* var. *mrakii*. *Int. J. Food Microbiol.* **2012**, *158*, 28–35. DOI:10.1016/j.ijfoodmicro.2012.06.015.

Mallouchos, A.; Komaitis, M.; Koutinas, A.; Kanellaki, M. Wine Fermentations by Immobilized and Free Cells at Different Temperatures. Effect of Immobilization and Temperature on Volatile By-Products. *Food Microbiol.* **2003**, *80*, 109–113.

Mauricio, J. C.; Moreno, J.; Zea, L.; Ortega, J. M.; Medina, M. The Effects of Grape Must Fermentation Conditions on Volatile Alcohols and Esters Formed by *Saccharomyces cerevisiae*. *J. Sci. Food Agric.* **1997**, *75*, 155–160.

NOM-188-SCFI. Mango Ataulfo del Soconusco, Chiapas (*Mangifera caesia* Jack ex Wall) Especificaciones y métodos de prueba, 2012.

Oliveira, B. G.; Costa, H. B.; Ventura, J. A.; Kondratyuk, T. P.; Barroso, M. E. S.; Correia, R. M.; Pimentel, E. F.; Pinto, F. E.; Endringer, D. C.; Romão, W. Chemical Profile of Mango (*Mangifera indica* L.) Using Electrospray Ionisation Mass Spectrometry (ESI-MS). *Food Chem.* **2016**, *204*, 37–45. DOI:10.1016/j.foodchem.2016.02.117.

Pino, J. A.; Queris, O. Analysis of Volatile Compounds of Mango Wine. *Food Chem.* **2011**, *125*, 1141–1146. DOI:10.1016/j.foodchem.2010.09.056.

Reddy, L. V. A.; Reddy, O. V. S. Production and Characterization of Wine from Mango Fruit (*Mangifera indica* L.). *World J. Microbiol. Biotechnol.* **2005**, *21*, 1345–1350.

Reddy, L. V. A.; Reddy, O. V. S. Production of Ethanol from Mango (*Mangifera indica* L.) Fruit Juice Fermentation. *Res. J. Microbiol.* **2007**, *2* (10), 763–769.

Reddy, L.; Reddy, O. Effect of Fermentation Conditions on Yeast Growth and Volatile Composition of Wine Produced from Mango (*Mangifera indica* L.) fruit juice. *Food Bioprod. Process.* **2010**, *89*, 487–491. DOI:10.1016/j.fbp.2010.11.007.

Reed, G.; Nagodawithana, T. W. *Yeast Technology*, 2nd ed.; Van Nostrand Reinhold: New York; 1991, 186.

Robles-Sánchez, R. M.; Rojas-Graü, M. A.; Odriozola-Serrano, I.; González-Aguilar, G.; Martin-Belloso, O. Influence of Alginate-Based Edible Coating as Carrier of Antibrowning Agents on Bioactive Compounds and Antioxidant Activity in Fresh-Cut Kent Mangoes. *LWT—Food Sci. Technol.* **2013**, *50*, 240–246. DOI:10.1016/j.lwt.2012.05.021.

Scragg, A. *Biotecnología para Ingenieros*; Limusa: Noriega, Mexico, DF, 2004; pp 126–136.

Táboas, R. Study of the Alcoholic Fermentation in Wine, Influence of the Yeast *Saccharomyces cerevisiae* and SO₂. *Electr. J. Environ. Agric. Food Chem.* **2002**, *1* (2), 126–136.

Tafolla-Arellano, J. C.; et al. Transcriptome Analysis of Mango (*Mangifera indica* L.) Fruit Epidermal Peel to Identify Putative Cuticle-Associated Genes. *Sci. Rep.* **2017**, *7*, 46163. DOI:10.1038/srep46163.

Tharanathan, R.; Yashoda, H.; Prabha, T. Mango (*Mangifera indica* L.), "The King of Fruits"— An Overview. *Food Rev. Int.* **2006**, *22*, 95–123. DOI:10.1080/87559120600574493.

Torrija, M. J.; Beltrán, G.; Novo, M.; Poblet, M.; Guillamón, J. M.; Más, A.; Rozés, N. Effects of Fermentation Temperature and *Saccharomyces* Species on the Cell Fatty Acid Composition and Presence of Volatile Compounds in Wine. *Int. J. Food Microbiol.* **2002**, *2658*, 1–10.

INDEX

Milton Keynes UK
Ingram Content Group UK Ltd.
UKHW031142141024
449569UK00024B/1142